NON-DESTRUCTIVE TESTING

Edited by **Fausto Pedro Garcia Marquez, Mayorkinos Papaelias** and **Noor Zaman**

Non-Destructive Testing

http://dx.doi.org/10.5772/61596
Edited by Fausto Pedro Garcia Marquez, Mayorkinos Papaelias and Noor Zaman

Contributors

Dongsheng Li, Mengdao Jin, Yanbin Wang, Deli Gao, Jun Fang, Slim Soua, Ángela Angulo, Tat Hean Gan, Beatriz Menendez, Yanhua Sun, Liu Shiwei, Guillermo Roa Rodriguez, Willian Aperador, Tomasz Boczar, Daria Wotzka, Paweł Frącz, Michał Kozioł, Michał Kunicki, Andrzej Cichoń, Paulo Cezar Monteiro Lamim Filho, Lane Baccarini, Robson Pederiva, Jie Xu, Qinghua Han, Ying Xu, Cédric Le Bastard, Khaled Chahine, Yide Wang, Vincent Baltazart, Nicolas Pinel, Christophe Bourlier, Xavier Derobert

CBS, Edition **2017**

Published by InTech

Janeza Trdine 9, 51000 Rijeka, Croatia

Publishing Process Manager Dajana Pemac

Technical Editor SPi Global

Cover Designer InTech Design Team

Additional hard copies can be obtained from orders@intechopen.com

Non-Destructive Testing, Edited by Fausto Pedro Garcia Marquez, Mayorkinos Papaelias and Noor Zaman
p. cm.
Print ISBN 978-953-51-2501-3
Online ISBN 978-953-51-2502-0

INTECH
open science | open minds

Contents

Preface

Non-destructive testing (NDT) is based on inspection methodologies that do not require the change or destruction of the component or system under evaluation. Numerous NDT techniques are increasingly used, thanks to the recent advances in sensing technologies, data acquisition, data storage and signal processing. Inspection information is widely employed in order to make effective maintenance decisions based on the defects identified, their location and severity. This book presents the main advances recently made on different NDT techniques, together with the principal approaches employed to process the signals obtained during inspection.

The first chapter introduces the development of a new remote monitoring technique for the evaluation of corrosion in reinforced concrete structures employing embeddable variations of known ASTM standards and telecommunication technologies. This is a new way for estimating the rate of mass loss in steel rebars. Corrosion of steel rebars is one of the most common causes of structural degradation in civil infrastructure. The technique is based on the creation of an electrochemical half-cell that can be embedded indefinitely into the concrete, providing corrosion rate measurements. The technique uses as an electrode of Copper/Copper Sulphate, a segment of the same steel of the studied structure as working electrode and an auxiliary electrode made from pure graphite, all covered by mortar.

Chapter 2 considers the mechanical model and control equation for analysing the mechanical behaviour of marine risers in working conditions. The control equation has been employs the weighed residual method, while the analysis model has been verified using finite element analysis (FEA). The deformation and stress distribution within the marine riser are acquired. The simulated experimental system is introduced together with the system composition, functions and operational approach of the experimental setup. A tubular sample has been manufactured to simulate the marine riser. The simulated experiments are carried out considering this setup. Finally, the strain and stress values of the specimen are obtained take into account the analysis of the data obtained.

The next Chapter 3 presents a novel beamforming array technique and probability-based diagnostic imaging method to determine acoustic emission (AE) sources in plate-like structures. The technique differs from the common beamforming array techniques as a sensor network instead of a linear sensor array is employed. To reduce uncertainty, avoid the boundary reflection effect and ensure the rationality of the signal superposition, a Hilbert transform–based signal processing is applied before the delay-and-sum algorithm, and a probability-based diagnostic imaging method is developed for AE source localisation. Finite element numerical simulations and Hus-Nielson source tests have been carried out on aluminium plate and a thin-walled cylinder pipe-like structure as part of the development of the proposed method. The results indicate that this method is efficient and capable of visualising the location results highlighted in the probability images.

Chapter 4 describes guided wave experiments and modelling results for pipeline wall thinning. Propagation of guided waves and classification of detected features based on signal amplitude are considered in this chapter. The outcome of this interpretation is a decision as to whether the follow-up inspection is classified as high, medium or low priority, and it requires to be determined. This can be done by analysing the signal amplitude corresponding to a specific level of metal loss. If the resonance can be obtained for specific frequency, the operator can identify the wall thickness loss by interpreting the reflected guided wave energy amplitude. The research presented in this chapter identified the suitable way of deploying this strategy using thickness resonance. Dispersion curves are evaluated and the thickness loss effect on the dispersion curves are evaluated. The experiments carried out prove the effectiveness of the methodology employed.

Chapter 5 analyses the application of the AE technique to detect crack location in steel and concrete structures. An improved approach, modified from the classical crack source location method, is proposed to provide more reliable data in masonry structures. It is introduced a new parameter, the degree of inhomogeneity, ξ, that is employed in the case of the modified method considered herewith. This parameter helps to decrease the unavoidable propagation delay related to refraction due to the layers in masonry material. The modified method is successfully applied to the AE source detection during the Hsu-Nielson source test. Tests are done on the central pier settlement of a model bridge to analyse the crack pattern. The results indicate that the proposed method is a reliable tool for crack source location in masonry structures.

Chapter 6 emphasises the need of a new electromagnetic NDT methodology to provide the basis of the accurate magnetic representation of surface topography. It considers that a concave-shaped feature produces "positive" magnetic flux leakages, and therefore forms a "raised" signal wave, but a bump-shaped feature generates "negative" magnetic fields, leading to a "sunken" signal one. In this case, the corresponding relationships between the wave features and surface topography are presented, and the relevant evaluation system for testing surface topography (concave, bumped and flat features) is built. The methodology is analysed and verified by FEA and experiments. The different dimension parameters, height/depth and width of surface topography, are also studied.

Chapter 7 introduces the metrology and measurement methods applied for NDT investigation of electrical discharges occurring in oil insulation systems of high-voltage devices. The main objective is a detailed and multivariate analysis of physical phenomena associated with generation of electrical partial discharges that occurs in oil insulation of electrical equipment. An important cognitive component is the verification whether the form of partial discharge (PD) has an effect on the energy contribution of the physical phenomena associated with their generation. An analysis of physical phenomena associated with the generation in insulating oil is designed and implemented, where the PD is simulated as follows: 1) a surface system; 2) needle-needle system in insulating oil and 3) needle-needle system in insulating oil with gas bubbles. The optical signals (IR, UV and visible), ultra–high-frequency electromagnetic and high-energy X-ray radiation, acoustic emission and thermal images are also considered. The signals are studied as multi-variant an in the time and frequency domains.

Chapter 8 considers the NDT applicability for evaluating cultural heritage buildings because of the high cultural value of the concerned assets and need to preserve them as intact as possible. Different NDT techniques applied for the evaluation of the condition of historical buildings are presented. The selected techniques concern the measurement of some physical properties of the construction materials measured at the surface such as water absorption, permeability, water content, cohesion, hardness, etc.; the actual condition of the building including stress state, deformation, crack growth, etc., and the in-depth physical properties, including mechanical properties, inner structure of walls, damp location and salt content. Some of these techniques are used for inspection of the buildings at a given time; others can be applied for long periods of time to investigate the degradation of the structural integrity with time. The main objective of this review is to focus on the applications, especially to discuss which information can be supplied and to present published results in each case. Some techniques are very simple and require inexpensive equipment, but others, using more sophisticated technology, require post-acquisition treatments based on more complex physical principles. Some examples of combinations of different techniques are presented.

Chapter 9 discusses the principle and application of two model-based algorithms for processing non-dispersive and dispersive ground penetrating radar (GPR) data over layered medium under monostatic antenna configuration. Both algorithms are selected for their high time resolution capability and reduced computational burden. They allow GPR radar to measure a layer thickness smaller than the fraction of the dominant wavelength. For non-dispersive data, the ESPRIT algorithm is generalized to handle different types of data models employed in experiments and literature. For dispersive data, the proposed adaptation of the MPM algorithm allows recovering the full-time resolution and jointly estimating the time delays and quality factors of a layered medium with reduced bias. Both processing techniques are applied to probe layered roadways for NDT purposes.

The last chapter presents several methods of NDT for induction motors based on the motor current signature analysis (MCSA), magnetic flux and vibration analysis. The method of analysis is a good alternative tool for destructive tests and fault detection in induction motors. Numerical and experimental results demonstrate the effectiveness of the proposed technique. This chapter also presents a model suitable for computer simulation of induction motor in a healthy state and with general asymmetries that can be analysed simultaneously. The model makes it possible to conduct research on different characteristics of engines and outstanding effects produced by the faults.

Fausto Pedro Garcia Marquez
UCLM, Spain

Mayorkinos Papaelias
University of Birmingham, UK

Noor Zaman
King Faisal University, Kingdom of Saudi Arabia

Chapter 1

Remote Monitoring Technique for Evaluation of Corrosion on Reinforced Concrete Structures

Guillermo Roa Rodríguez and Willian Aperador

Additional information is available at the end of the chapter

http://dx.doi.org/10.5772/62314

Abstract

This chapter presents the development of a new remote monitoring technique for the evaluation of corrosion on reinforced concrete structures, which uses embeddable variations of known ASTM standards and telecommunication technologies as a new way to estimate the rate of loss of the steel used as a component of rebar, since such phenomenon is the main cause of deterioration and degradation of the civil infrastructure. The adaptation of the technique was carried out to obtain an electrochemical half-cell that can be embedded indefinitely into the concrete, which provides the measurements corresponding to the corrosive state and allows calculating indirectly the rate of corrosion through the linear polarization resistance. The adaptation is based on a reference electrode of copper/copper sulphate, a segment of the same steel of the studied structure as working electrode and an auxiliary electrode made from pure graphite, all covered by mortar.

Keywords: Corrosion monitoring, Non-destructive testing on reinforced concrete, Corrosion Rate, Corrosion potential, reinforced concrete, Monitoring technique

1. Introduction

The reinforced concrete structures have a limited service life; such durability depends on the concrete resistance to various physical and chemical factors and its ability to protect the embedded steel against the corrosion processes. The referred processes create products of steel corrosion, which generate a volumetric expansion on the embedded reinforcements, causing extremely high stresses inside the concrete, transforming it into a medium susceptible to the generation of cracks from the position of the reinforcement to the surface or between the rebar. Once the cracks have appeared, the oxygen and moisture diffuse directly up to the reinforcement at a greater rate that increments the rate of corrosion [1–3]. Like any natural disaster, earthquakes or severe weather disturbances, corrosion can generate highly risky and costly damage to multiple areas, starting from the automotive, fluid transport systems to the area of infrastructure (bridges, ports, public buildings). As a reference, from 1980 to 2002, the United States suffered 52 major disasters related to climate, which included hurricanes, tornadoes, tropical storms, floods, fires, drought and frost, incurring on losses estimated at US$ 380 billion (US$ 17 billion annually). According to the latest study, the direct cost carried by corrosion of metal structures is US$ 276 billion annually, representing 3.1% of gross domestic product [3,4]. Unlike natural disasters, the phenomenon of corrosion can be monitored and controlled [3,4].

The detection of such corrosive phenomenon is usually performed when the structure clearly shows damage signals which are directly reflected in a rise in cost of corrective maintenance. From the perspective of project managers, there is a marked interest in obtaining information about the states of initiation of degradation processes, to avoid incurring the referred costs. A considerable number of articles and techniques have been published and implemented to perform the monitoring of the corrosive state on reinforced concrete, where some of them not only use a single technique but also use a combination of several evaluation methods that allow to gather information about the condition of the concrete and/or the actual situation of rebar [5,6]. Within those techniques, the standard ASTM C876-91 determines the reinforcement state in terms of probability accordingly to the corrosion potential of rebar (10% of probability of occurrence of the phenomenon, region of uncertainty, 90% of probability of occurrence of the phenomenon) [5,7]. Also a complementary technique could estimate the corrosion rate makes to consider a set of measurements as a more specific evaluation of the corrosion process; according to the aforementioned, the linear polarization resistance method (LPR) rises as the most common procedure, given its ease of interpretation of the results and its high reproducibility [5,8]. Although the mentioned techniques are widely applied, its major constraint relies on their flexibility, since they depend at all times on a hardware operated at the site of study (conducted by a user), which increases the costs due to equipment and travel expenses for the specialized personnel [9].

The purpose of this work is to introduce the design of a new remote monitoring technique for the evaluation of corrosion on reinforced concrete structures, which uses embeddable variations of known ASTM standards and telecommunication technologies as a new way to estimate the rate of loss of the steel used as a component of rebar which also allows to avoid cost overruns related with specialized technicians.

2. Introduction to the electrochemical techniques to assess the corrosive state on reinforced concrete

The corrosion of steel implies a separation of the anodic and cathodic processes that occur simultaneously on the surface; therefore, corrosion of carbon steel inside the concrete acts as a combined electrode over which the anodic and cathodic reactions take part; at the anode (where the corrosion process is carried out) steel is dissolved and oxidized into steel ions. Due to the electronic balance condition, the ions, which have left the steel, are consumed by the cathodic reaction on the steel surface, where the oxygen is reduced and the hydroxide ions are produced. Depending on the spatial distribution of these reactions and the conductivity of the medium, it can be generated into two kinds of corrosion, uniform corrosion and macrocell corrosion. Uniform corrosion occurs when the anodic and cathodic reactions on the same area are coupled to form microcells. The electrode potential of such coupling is called corrosion potential *Ecorr*. In the case of macrocell corrosion, the anodic and cathodic areas are separated; according to the circumstances of the case, it is presented as a considerable drop of the Ohmic potential that permits to separate the potentials of the cathodic and anodic area. In this process, the corrosion potential varies depending on the evaluated position within the limits of the anodic and cathodic corrosion potential [3].

2.1. Half-cell potential measuring on embedded reinforcements on concrete (open circuit)

The corrosion potential is measured as the potential difference according to a reference electrode or half-cell. This measuring is based on the electric and electrolytic continuity between the rebar, the reference electrode and the measuring element (voltmeter), which needs a high input impedance in such a way that the current which flows through the reference electrode does not disrupt the stability of the potential (it is acceptable as an input impedance of 10 MΩ). The connections must to be made such that the reference electrode is over the negative terminal and the reinforcing steel is brought to the positive port of the voltmeter; in this way the readings of the potential will be generally negatives, where the positive measurements are possibly due to passivated steels on dry concrete [3,10].

As previously mentioned, the half-cell potentials correspond to the corrosive level of steel, influenced by the surrounding environment, which is composed by variables such as the type of concrete cover, concrete resistivity, moisture content, oxygen availability and aggressive agents such as chloride ions and carbon dioxide. In corroded steels at contaminated environments, chloride ions can be obtained with potentials within a range between −400 and −600 mV versus an electrode of copper/copper sulphate; furthermore, the passive steel reacts as an oxygen electrode subject to pH variations in the pores of the concrete; hence, potentials on carbonated concrete are less negative. The corrosion potential for passive steels also depends on the oxygen availability and it varies over a great range of voltages [3,10].

The measurement technique is normalized in the standard ASTM C876-91, applicable to the evaluation in situ and is independent of the size of the reinforcement and depth at which it is located inside the concrete paste. This standard allows relating the corrosion potential measured with the probability of corrosion [3,11].

2.2. Calculation of the corrosion rate using the linear polarization resistance method (LPR)

Linear polarization resistance (LPR) is a non-destructive method designed to measure directly the polarization resistance (*Rp*) and indirectly the corrosion rate around the corrosion potential [3].

In order to implement the LPR measurement method, a small disturbance of polarization is applied to the reinforcement steel in the concrete surface, where its response is measured after a proper time delay. Using a potentiostatic method, an overpotential ΔE, between 10 and 30 mV, is supplied on the corrosion potential and after a typical time delay of 30 seconds, the response in current ΔI, is recorded [3,12]. Although the corrosion of steel is an electrochemical process and does not obey the Ohm's law, it has been demonstrated that if the applied polarization does not exceed ±30 mV, the *Rp* value can be determined from the relation between the potential variation and the current variation and, if the polarization can be confined either to a known area *A* or to a small section isolated from the steel, the corrosion current density (i_{corr}) also can be calculated [3,12]. As ASTM G102 indicates, using Eq. 1, the corrosion rate can be directly determined [3,13].

$$Corrosion\,rate = (B \cdot K \cdot W_e) \cdot (d \cdot A \cdot Rp)^{-1} \tag{1}$$

where the value of *B* corresponds to the slopes of Tafel curves, where typically for the steel embedded in concrete they are assumed with values of 25 mV for steel under active conditions of corrosion and 50 mV for passive conditions, W_e is the equivalent weight in equivalent grams units and *d* is the density of the working electrode (rebar). *K* is the constant that defines the units of the corrosion rate, taking values either of 3272 to configure the units in millimeters per year *mmpy* (its units are) or a value of 1.288×10^5 for mils penetration per year *mpy* (its units are *mils* · (*ampere* · *cetimeter* · *year*)) [3,13].

3. Remote monitoring technique for the evaluation of corrosion on reinforced concrete structures

The technique allows to adapt the ASTM standards into a remote mode with embeddable instrumentation with which it is possible to avoid movement of personnel to the measuring zone. The overall design allows both half-cell voltages (open circuit potential) and linear polarization resistance values become available and visualizable to the user through a web server.

3.1. Design of the embeddable cell

The reinforcement steels are typically protected by the alkaline nature of concrete. If that alkalinity is compromised at some point, the corrosion in the steel will initiate if oxygen and moisture combine. The corrosion reaction will promote the anodic and cathodic activity across the rebar, thus generating a corrosion cell by the electric difference between the two mentioned regions [14]. The design of the embeddable cell was made with a configuration of three electrodes (reference electrode, *Re*, working electrode, *Wee*, and auxiliary electrode, *Ae*), framed within the norms ASTM C876 and ASTM G-102.

The embeddable cell was designed with a diameter of 15 and 30 cm height. At its center, it was positioned as a segment of structural steel of 0.5 inches diameter and 15 cm height as the

http://dx.doi.org/10.5772/62314

working electrode. It was set as the embeddable reference electrode of copper/copper sulphate ($Cu/CuSO_4$) 45 mm next to the segment of steel, and finally at 42 mm of distance from the working electrode and 50 mm from the reference electrode it was embed as a pure graphite electrode (10 cm height), acting as an auxiliary electrode. The electrodes were covered by mortar with a water/binder relation for mortar as 0.65, which produces a porous mix that allows the surrounding medium to penetrate through the mortar up to the three-electrode system. This phenomenon assures that the medium inside the embeddable cell is equal to the medium of the whole structure, thus the measurements provided by the cell are reliable and equivalents. Such configuration allows to obtain an electrochemical half-cell that can be embedded indefinitely into the concrete (**Figure 1**).

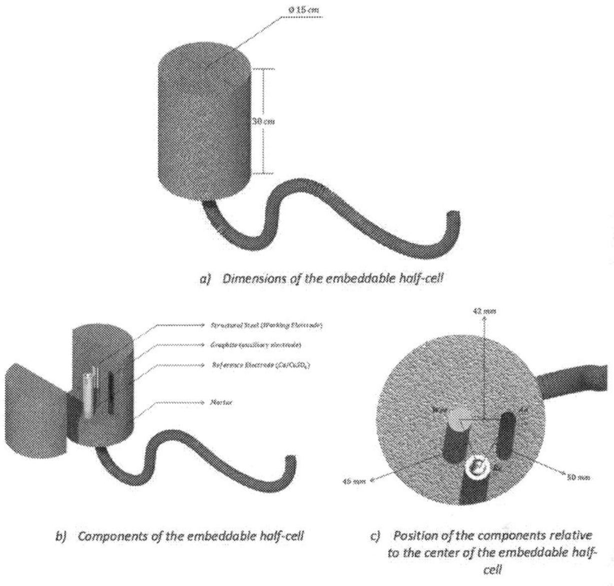

Figure 1. Design of embeddable half-cell.

3.2. Design of the control system

The design of the control system allows the equipment to prepare itself to perform the measurements after receiving a start command (the order is sent via a text message, at the discretion of the user). Once the command is received, the system determines the open circuit potential (half-cell potential) between the working electrode (rebar) and the reference electrode, subsequently, through an integrated circuit for analog to digital conversion (DAC), which guarantees accurate voltage increments up to 75 mV; owing to its resolution of 16 bits, it is supplied as a potential difference between the auxiliary electrode and the working electrode that is varied until it equals the open circuit potential minus 20 mV. To proceed to the start of the voltage sweep, the test is done by varying the half-cell potential 1 mV per second, from −20 mV of the open circuit potential to 20 mV above it [3,14].

With each increase, the half-cell potential and the voltage are measured with a shunt resistance arrangement of 10 Ω, 50 Ω, 100 Ω, 1 KΩ, 5 KΩ and 10 kΩ, which are selected according to the algorithm implemented. The mentioned selection is done to obtain the best measured resolution and also to perform de calculus of the current by Ohm's law. At the end of the voltages sweep, the measured values are used to calculate the polarization resistance and the corrosion rate using Eq. 1. Finally, the system sends the stored information in a format of a *.txt* archive to the GSM/GPRS module, which allows the transmission of information to a web server, where the variables are stored in a proprietary database, which gives the user the ability to view the information through a web application (**Figure 2**) [3,14].

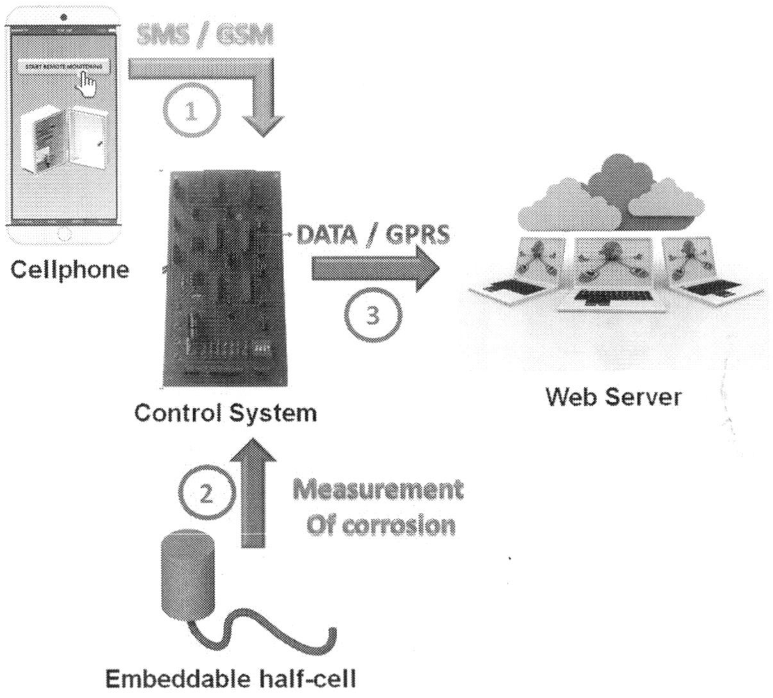

Figure 2. Design of the control system.

Once the test finishes, it is necessary to provide the information available to the user regardless of the place where comes from; for that reason, the design uses a GSM/GPRS board (Global Systems for Mobile Communications/General Packet Radio Service) which provides the ability to use the mobile network inserting a subscriber identity module for a mobile phone (SIM card), allowing to send and receive text messages, with which the remote command of the

http://dx.doi.org/10.5772/62314

equipment is done. The configuration of the module and the control board allows using the file transfer protocol FTP with which it is possible to send the *.txt* file to a web server to query the data generated on the tests.

3.3. Method to assess remotely the corrosive state of rebar

The method to assess remotely the corrosive state of rebar begins with the preparation of the structure to set the embeddable half-cell. First, it is necessary to identify the zone where the structural steel has the major risk of suffering the corrosive phenomenon as a critic representation of the whole structure. Once the position is known, and if it is a new construction in progress, it is elemental to locate the embeddable half-cell parallel to the reinforcement. In contrast, if the place to be measured is an existing construction, after identifying the critic zone, it is mandatory to locate the rebar and next to it bore a hole of 22 cm diameter and 40 cm depth, at which the embeddable half-cell can be placed, parallel to the structural steel. Finally, the hole needs to be backfilled with mortar where its water/binder relation is 0.65 (**Figure 3**).

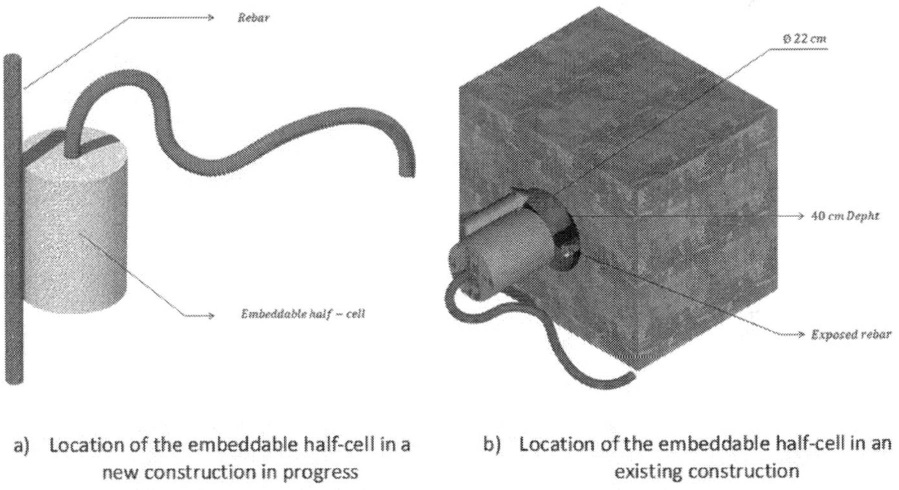

Figure 3. Location of the embeddable half-cell.

After placing the half-cell, we proceed to install the control system to perform the measurements. Once the system receives a text message with the start command, the data acquisition begins where the acquired information corresponds to the half-cell potential and the polarization potential. Then the system sends the data package to a web server through which a developed web supervisory allows consulting the information of the corrosive state (probability of corrosion and corrosion rate) (**Figure 4**).

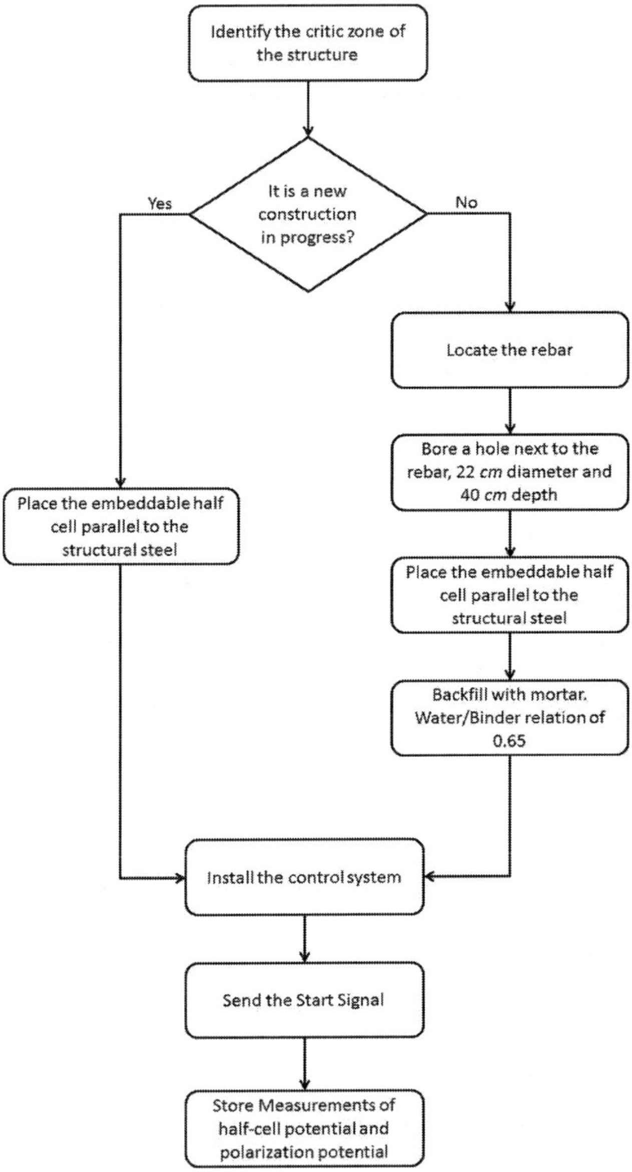

Figure 4. Flowchart for the designed method.

3.4. Corrosion monitoring test on concrete specimen

An embeddable half-cell with the specifications mentioned above was prepared. The probe was allowed to cure for 28 days at a temperature of 25°C at constant moisture (**Figure 5**).

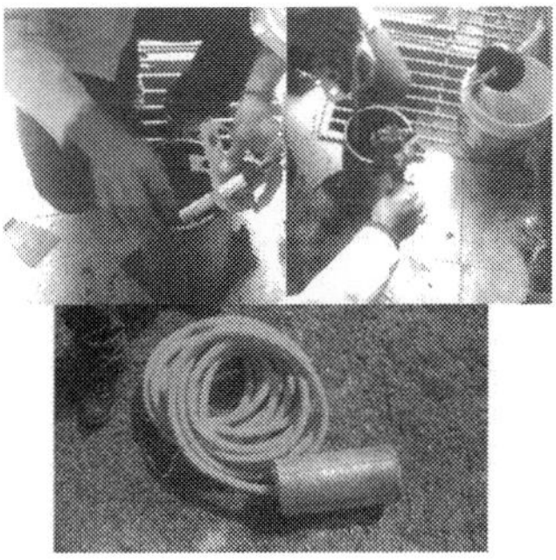

Figure 5. Preparation of the embeddable half-cell.

After the curing period, the designed system and a commercial potentiostat (Gamry PCI4) were used in parallel in order to establish a parameter measurement for error values on the measures. Constant monitoring for 192 hours was conducted in an environment free of aggressive agents. These measurements allowed determining the state of the embeddable half-cell prior to set it into the structure object of study [3].

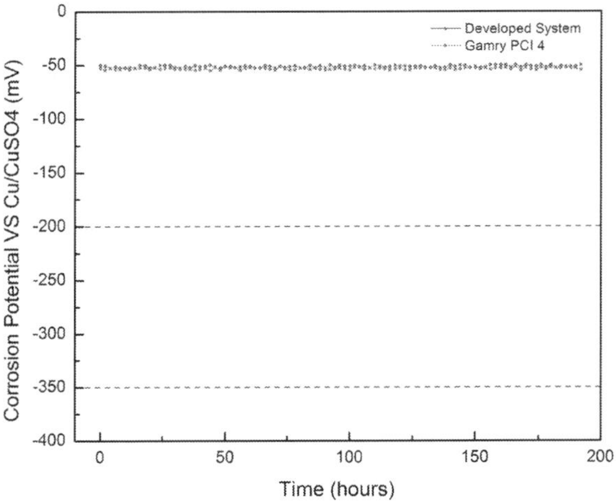

Figure 6. Corrosion potential of embeddable half-cell prior to set it into the structure [3].

As shown on **Figure 6**, the corrosion potentials are constants, allowing to calculate an open circuit potential of −52 mV, which corresponds to a 10% or minor probability of corrosion. Also, the data comparison between the proposed design and the commercial equipment over the total time of the test gives a maximum error of 8.4 and 0.03% as minimum where the average error is 3.62% [3].

The linear polarization resistance value for the test was 13 kΩ which when used in Eq. 1 with the values of the steel segment area, *K*, *We*, density and *B* yields a corrosion density of 0.0045 and a corrosion rate of 526.11 × 10^{-6} *mmpy*, which corresponds to a classification of corrosion as very low, concordant with the obtained corrosion probability, where *B* was assumed as 50 mV because the passive corrosion conditions, *We* takes a value of 27.92 equivalent grams, the density of steel is considered as 7.87 and *K* was assumed as 3272 to establish the units of corrosion rate as millimetre per year *mmpy* [3]. The error values of the linear polarization resistance were found, wherein the maximum gives 6.58%, the minimum error gives 0.019% where the average error is 3.46% (**Figure 7**) [3].

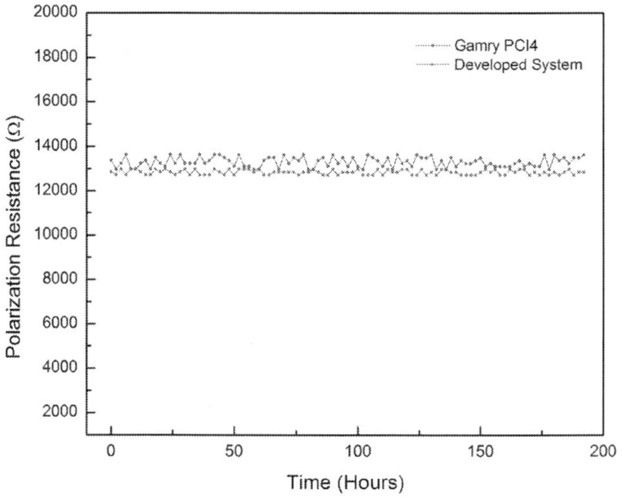

Figure 7. Polarization resistance of embeddable half-cell prior to set it into the structure [3].

After determining the error of the half-cell, it was taken to the measuring site, located in the city of Cajicá in the Department of Cundinamarca in Colombia, specifically at the Acoustical Shell of the Universidad Militar Nueva Granada.

As the place of measuring was a new construction in progress, first it was identified as the zone where the structural steel has a major risk of suffering corrosion, the embeddable half-cell was set parallel to the reinforcement as shown in **Figures 8** and **9** and where the place has a high concentration of chloride ions, as the acoustical shell is located 14 km from a mining area of extraction and refining of salt. Finally, the control system was installed.

http://dx.doi.org/10.5772/62314

Figure 8. Acoustical Shell of the Universidad Militar Nueva Granada.

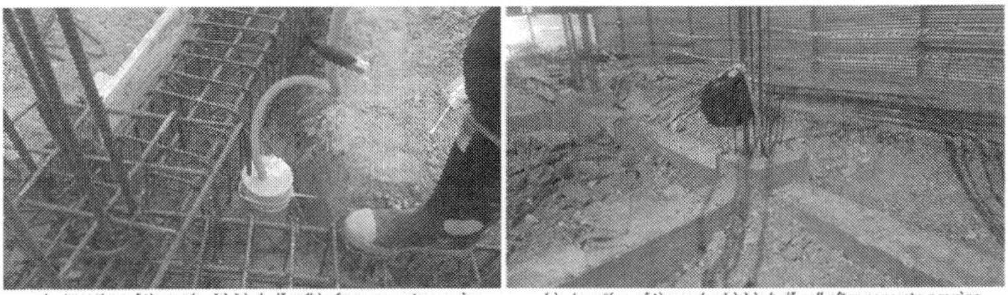

Figure 9. Installation of the embeddable half-cell.

As shown in **Figure 10**, monitoring was carried out each 2 hours during 10 days, reaching a maximum potential of −386m V corresponding to a probability of corrosion just over the limit of 90%. Once the pores of the concrete are saturated owing to the high concentration of CO_2 and chloride ions, the corrosion potential stabilizes over −305m V on the uncertainty region.

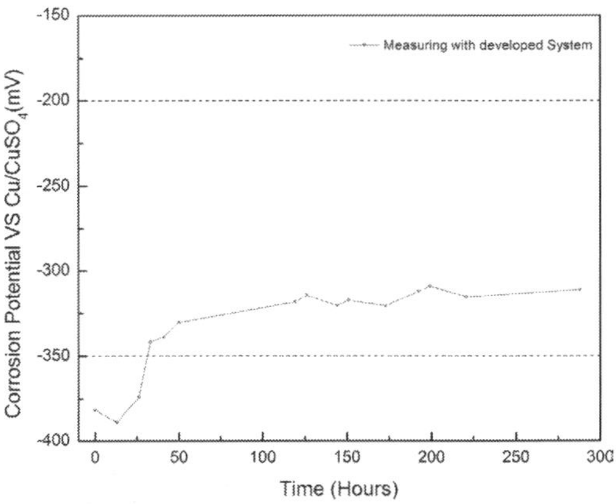

Figure 10. Remote monitoring of corrosion potential [3].

The test of linear polarization resistance was carried out after 10 days as shown in **Figure 11** wherein the average value of the linear polarization resistance was 1.98 kΩ when used in Eq. 1. Together with the values of the steel segment area, *K*, *We*, density and *B* (*B* is assumed to be 26 mV, since the corrosion conditions are active) yield a corrosion density of 0.309 and a corrosion rate of 3.592×10^{-3} *mmpy*, which corresponds to a classification of corrosion as low/moderated, concordant with the obtained corrosion probability, *We* takes a value of 27.92 equivalent grams, the density of steel is considered as 7.87 and *K* was assumed as 3272 to establish the units of corrosion rate as millimetre per year *mmpy*.

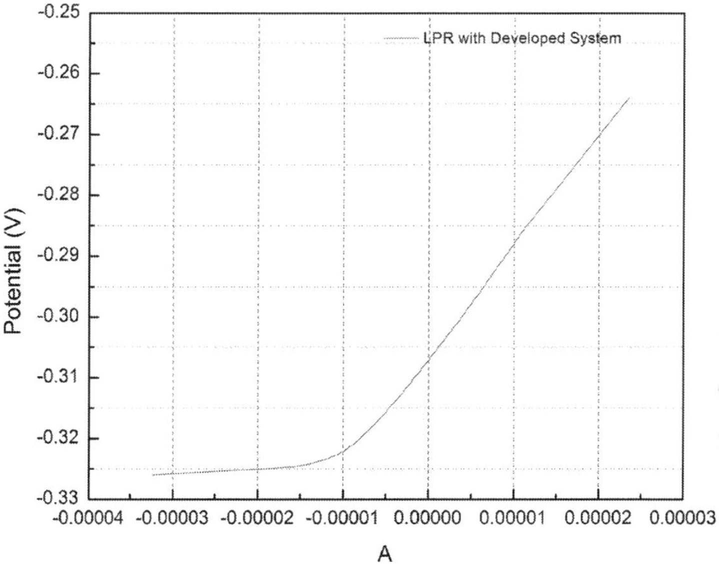

Figure 11. Remote monitoring of linear polarization resistance [3].

4. Conclusions

Through this study, the feasibility of implementing an embeddable remote monitoring technology for determining the corrosive state of reinforced concrete structures subjected to the carbonation phenomenon following the regulations present in the standards ASTM C876 and ASTM G-102 was verified. Such viability was determined from the given performance of the developed system versus commercial equipment used as a pattern, where the mean error on open circuit potential was 3.62% while the mean error on the test for resistance to linear polarization with respect to the calculations from the polarization resistance was 3.46%. Use of the remote measuring technique and the information provided by the developed system allow diagnosing a reinforced concrete structure identifying the corrosive state of the reinforcing steel structure by determining the corrosion probability and the corrosion rate of the material, avoiding any movement of personnel to the measuring site and any destruction of the structure.

Acknowledgements

To my Grandfather Ramon Rodriguez Martinez and my Parents Guillermo Jose Roa Torres and Carmencita Rodriguez Martinez, for their constant support with this research and its funding.

This research was supported by "Vicerrectoría de investigaciones de la Universidad Militar Nueva Granada" under contract ING 1760 validity 2015.

Author details

Guillermo Roa Rodríguez and Willian Aperador

*Address all correspondence to: guillermoroa.rrg@gmail.com

School of Engineering, Universidad Militar Nueva Granada, Bogotá, Colombia

References

[1] Capozucca Roberto. Damage to reinforced concrete due to reinforcement corrosion. Construction and Building Materials. 1995;9(5):295-303. DOI: 10.1016/0950-0618(95)00033-C

[2] Lundgren Karin. Modelling the effect of corrosion on bond in reinforced concrete. Magazine of Concrete Research. 2002;54(3):165-173. DOI: http://dx.doi.org/10.1680/macr.2002.54.3.165

[3] Roa Rodríguez Guillermo; Aperador Chaparro Willian. Remote monitoring system for probabilities and rates of corrosion on reinforced concrete structures. Computers and Concrete. Forthcoming.

[4] Koch G.H.; Brongers M.P.H.; Thompson N.G.; Virmani Y.P.; Payer J.H.; Costs and Preventive Strategies in the United States. Houston, TX: NACE International; 2002. 12 p.

[5] Romano Pedro; Brito Paulo S.D.; Luiz Rodrigues. Monitoring of the degradation of concrete structures in environments containing chloride ions. Construction and Building Materials. 2013;47:827-832. DOI: 10.1016/j.conbuildmat.2013.05.042

[6] Otieno Mike; Beushausen Hans; Alexander Mark. Prediction of Corrosion Rate in RC Structures - A Critical Review. In: Andrade Carmen; Mancini Giuseppe, editors. Modelling of Corroding Concrete Structures. Madrid, Spain: Springer; 2011. pp. 15-37. DOI: 10.1007/978-94-007-0677-4_2

[7] ASTM. ASTM C876-09, standard test method for corrosion potentials of uncoated reinforcing steel in concrete. In: West Conshohocken, PA:2009.

[8] Romano Pedro; Paulo Brito S. D.; Estudo da degradação de estruturas de betão armado expostas ao ataque por sulfatos. Ingenium. 2012;DOI: http://hdl.handle.net/10400.26/4101

[9] Rodríguez Guillermo Roa; Chaparro Willian Aperador; Tobón Emilio Delgado. Diseño de sistema de monitoreo remoto para evaluación de la corrosión en estructuras de concreto reforzado sometidas a ion cloruro . Ingeniería Investigación y Tecnología,. 2015;16(4):565-572.

[10] Elsener B.; Andrade C.; Gulikers J.; Polder. R;Raupach M.; Hall-cell potential measurements—Potential mapping on reinforced concrete structures. Materials and Structures. 2003;36(7):461-471.

[11] Elsener B.; Half-cell potential mapping to assess repair work on RC structures. Construction and Building Materials. 2001;15(2-3):133-139. DOI: 10.1016/S0950-0618(00)00062-3

[12] S.G Millard; Law D; J.H Bungey; Cairns J. Environmental influences on linear polarisation corrosion rate measurement in reinforced concrete. NDT & E International. 2001;34(6):409-417. DOI: 10.1016/S0963-8695(01)00008-1

[13] ASTM. ASTM G102-89(2010), Standard Practice for Calculation of Corrosion Rates and Related Information from Electrochemical Measurements. In: West Conshohocken, PA: ASTM International; 2010.

[14] Roa Rodriguez Guillermo; Aperador Willian; Delgado Tobón Emilio. Desempeño de sonda basada en electrodo de referencia de cobre sulfato de cobre en sistema de monitoreo de corrosión in situ desarrollado para estructuras de concreto reforzado. In: May 19–2, 2014; Fortaleza, Brasil. Fortaleza, Brasil:2014.

Chapter 2

Mechanical Behavior Analysis and Testing of Marine Riser in Deepwater Drilling

Yanbin Wang, Deli Gao and Jun Fang

Additional information is available at the end of the chapter

http://dx.doi.org/10.5772/62315

Abstract

In this chapter, the mechanical model and control equation have been established to analyse the mechanical behaviour of marine riser in working condition. The control equation has been solved by weighted residual method, and the analysis model has been verified by finite element method (FEM) in ABAQUS framework. Based on this, the deformation and stress distribution of the marine riser have been acquired. Then, a simulation experimental system has been introduced, and the system composition, functions and operational approach of the experimental setup have been stated in detail. After that, a tubular sample has been manufactured to simulate the marine riser, and the simulation experiments have been carried out based on this setup, where the experimental procedures, key aspects, difficult points of the experiment and its corresponding solutions have been elaborated. At last, the strain value of the specimen has been measured successfully after the experiment, and the stress state of the specimen has been obtained based on the analysis.

Keywords: Deepwater drilling, Marine riser, mechanical behaviour, Simulation experiment, Stress state testing

1. Introduction

The exploration and development of offshore oil and gas resources are gradually from shallow to deep sea area. The special sea state conditions and engineering problems in deepwater put forward higher request to the drilling technology, and a higher level of drilling equipment and techniques are urgently need to face the challenges.

Generally, the technical challenges in deep offshore drilling include: (1) water depth—huge drilling platform or drilling ship is needed in deepwater drilling. Moreover, the underwater equipment and tools must be precision, intelligent and flexible with higher reliability. The gravity of drilling riser and the complexity of marine environment increase with water depth. Therefore, the drilling vessel must have sufficient bearing capacity and deck space. (2) Sea wave and current—deviation of drilling vessel will occur under the action of sea wave and current, which will result in riser deformation and stress re-distribution. Riser under dynamic lateral force deduced by sea wave and current will generate vortex-induced vibration (VIV), which puts forward higher requirements on fatigue strength design of offshore pipelines. What's more, during drilling operation, riser is needed to disconnect from subsea wellhead and suspended on the drilling vessel, which will cause dynamic compression and even local instability. (3) Narrow drilling mud density window—the undercompaction of deepwater formation causes narrow drilling mud density window, which will result in frequent drilling accidents, such as lost circulation, well kick, borehole collapse and sticking. Besides, the narrow window will bring about the increase of casing layers even unable to drill to the target stratum. Meanwhile, well control problems always exist in narrow window condition [1]. (4) Geological disasters—geological disasters in deepwater drilling mainly include loose submarine soil and shallow flow. Sinking of subsea wellhead, blowout preventer (BOP) and surface conductor will appear if shallow flow is serious. When gas in shallow flow enters the sea water, the density of sea water and the buoyancy of drilling vessel will decrease, which will cause big accident and even rig capsized. (5) Well control—the well-control risk of drilling equipment, submarine tools and hydrate problems is also remarkable [2].

Various tubulars, such as drilling riser, drilling pipe, production tubulars and subsea tubulars, are indispensable for deepwater exploration and development. These pipelines are subjected to kinds of loads under service condition, and how to ensure the security of them is the foundation of efficiently acquiring oil and gas in deepwater. Taking the drilling riser as example, the operating loads subjecting to the riser include internal pressure, external pressure, axial tension, lateral force generated by sea wave and current and top displacement resulted from floating vessel, which can cause kinds of failure models, such as wear, rupture and collapse. Besides theory analysis, obtaining the mechanical behaviour of these tubulars under service loads through experiment is an effective way to guarantee the safe and efficient of offshore drilling operations [3].

However, few literatures have reported the experimental equipment, which can impose external pressure to tubulars. There is a set of experimental system in the University of Rio De Janeiro, Brazil, which can provide 10 MPa water pressure, and the maximum simulated water depth is 1000 m, and the maximum operating depth of the fifth- and sixth-generation drilling platform (ship) is 3000 m. An experimental system named Mechanics Behaviour Simulation Experimental System for Deep Water Drilling and Production Strings has been developed to provide 30 MPa water pressure and to simulate external pressure of 3000 m water depth [4]. Until now, the system can impose external pressure, internal pressure, axial tension/compression load and lateral load to the tubulars, which provide a good platform for exploration and development of offshore oil and gas and can perfect deepwater drilling theory, reduce

http://dx.doi.org/10.5772/62315

corresponding operation risk and improve the safety and economy of deepwater oil and gas development.

2. Theory analysis of riser mechanical behaviour

Marine riser is the key equipment connecting subsea wellhead and floating drilling platform (ship) in deepwater drilling and exploration. The main functions of marine riser are to provide channel for drilling mud in the annulus of riser and drill pipe, support auxiliary lines, guide drilling tools and install the BOP. As riser mechanical behaviour has great impact on deepwater drilling engineering, we take the riser in service as example in this chapter. Then, the mechanical model and control equation have been established. Next, the model validation and sensitivity analysis have been presented. Finally, we get the key parameters to control its mechanical behaviour.

2.1. Mechanical model

The schematic diagram of deepwater drilling riser in service is shown in **Figure 1**.

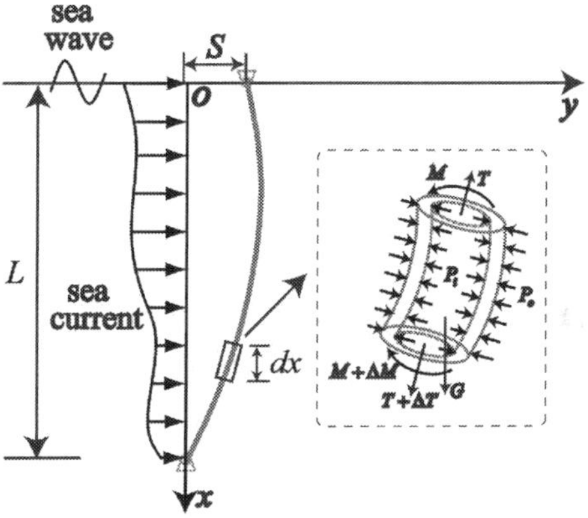

Figure 1. Static analysis model of riser mechanical behaviour.

The analysis model can be regarded as a beam located in the vertical plane and subjected to both non-uniform tension force and lateral force. Both the top and the end of riser are connected with the ball joints, so the boundary conditions are hinge constraints. Take the connection point of drilling ship and riser top as the origin of the coordinate, and the positive direction of x axis is vertical to the bottom of the sea, and the positive direction of y axis is the same as that of the lateral force. Besides, the following assumptions are applied during equation deduction:

- The material of riser is homogeneous, isotropic and linear elastic, and the riser bending stiffness is constant.

- The movement direction of sea wave and current is in the same plane.
- Top tension does not change with time.
- Riser curvature and torsion are small, and the geometric non-linearity is neglected.

The riser differential control equation can be represented as follows [5]:

$$\mathrm{EI}\frac{\mathrm{d}^4 y}{\mathrm{d}x^4} - T(x)\frac{\mathrm{d}^2 y}{\mathrm{d}x^2} - w\frac{\mathrm{d}y}{\mathrm{d}x} = F(x) \tag{1}$$

where EI(x) is the riser flexural rigidity along the x axis (N·m^2); $T(x)$ is the riser tension force distribution along the x axis (N); w is the per unit length weight of riser in sea water (N/m); $F(x)$ is the total lateral force distribution along the x axis (N).

As the top and end of the riser are connected with ball joints, the boundary conditions of Equation 1 can be written as follows:

$$\begin{cases} y|_{x=0} = S; & \theta|_{x=0} = 0 \\ y|_{x=L} = 0; & \theta|_{x=L} = 0 \end{cases} \tag{2}$$

2.2. Load calculations

2.2.1. *Tension force*

The main axial loads subjected on riser include gravity and top tension force, and the axial force of riser section at x below the sea surface can be represented by the following equation:

$$T(x) = T_{\mathrm{top}} - wx \tag{3}$$

where T_{top} is the top tension imposed on riser top end (N).

2.2.2. *Lateral force*

The main lateral force of riser is generated by sea wave and current. If ignoring the dynamic effects, the combined wave and current force can be calculated by the following equation:

$$F(x) = C_{\mathrm{D}}\rho D(v_{\mathrm{w}} + v_{\mathrm{c}})|v_{\mathrm{w}} + v_{\mathrm{c}}| / 2 + C_{\mathrm{m}}\rho\pi D^2 a_{\mathrm{w}} / 4 \tag{4}$$

where C_{D} is the drag force coefficient (a dimensionless quantity); C_{m} is the inertia force coefficient (a dimensionless quantity); ρ is the density of sea water (kg/m^3); D is the outer diameter of riser (m); v_{w} is the horizontal velocity of sea wave particle (m/s); v_{c} is the current velocity (m/s); $a_{\mathrm{w}} = \mathrm{d}wt/\mathrm{d}t$ is the horizontal acceleration of sea wave particle (m/s^2).

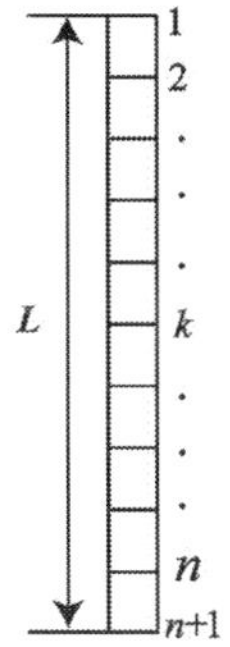

Figure 2. Riser discretisation model.

The sea current velocity under a certain depth of the sea surface can be calculated by the following equation recommend by American Bureau of Shipping:

$$v_c(x) = v_m(x/L)^{1/7} + v_t(x/L) \tag{5}$$

where v_m is the sea surface wind velocity (m/s); v_t is the sea surface tide velocity (m/s).

The linear wave theory is usually chosen to calculate the horizontal velocity of sea wave particle, and the equation is as follows:

$$v_w = \pi H / T_w \exp\left[k(x-L)\right]\cos(ky - \omega_w t) \tag{6}$$

where H is the wave height (m); T_w is the wave period (s); k is the wave number (a dimensionless quantity); ω_w is the wave circular frequency (rad/s).

Generally, the maximum of wave current is always used to calculate the maximum value of riser deformation and stress, and the maximum lateral force can be obtained through substituting Equations 5 and 6 into Equation 4.

2.3. Model solution

Equation 1 is very complex and cannot obtain the analytical solutions. In this chapter, we have solved Equation 1 by weighted residual method.

2.3.1. Residual equation

As shown in **Figure 2**, riser length can be divided into n spans. The first node locates at riser top and is marked with No. 1, the last node locates at riser end and is marked with No. $(n+1)$. So, the whole discrete elements have $(n+1)$ nodes and n elements.

Because Equation 1 is a fourth-order differential equation, the shape function in weighted residual method can be written as follows [6]:

$$y_i = C_{i,0} + C_{i,1}x + C_{i,2}x^2 + C_{i,3}x^3 + C_{i,4}x^4 + C_{i,5}x^5,\ \ x \in [0, l_i],\ (i = 1, 2, \cdots, n+1) \tag{7}$$

where y_i is the displacement of discrete nodes; l_i is the length of each span; C_{ij} (j = 0, 1, 2, 3, 4, 5) is the undetermined coefficients.

The derivative forms of the shape functions are as follows:

$$\begin{cases} y_i' = C_{i,1} + 2C_{i,2}x + 3C_{i,3}x^2 + 4C_{i,4}x^3 + 5C_{i,5}x^4 \\ y_i'' = 2C_{i,2} + 6C_{i,3}x + 12C_{i,4}x^2 + 20C_{i,5}x^3 \\ y_i''' = 6C_{i,3} + 24C_{i,4}x + 60C_{i,5}x^2 \\ y_i^{(4)} = 24C_{i,4} + 120C_{i,5}x \end{cases},\ \ x \in [0, l_i],\ (i = 1, 2, \cdots, n+1) \tag{8}$$

Substituting Equations 7 and 8 into Equation 1, the residual equation can be written as follows:

$$R_i(x) = \mathrm{EI}_i y_i^{(4)}(x) - T_i(x) y_i''(x) - w_i y_i'(x) - F(x) \tag{9}$$

As the shape function determined by Equation 7 is fifth order, there are six undetermined coefficients C_{ij} (j = 0, 1, 2, 3, 4, 5) in each shape function. Besides, there are n discrete elements in the system, so there are $6n$ equations that are needed to solve the problem. Generally, two collocation points are needed in weighted residual method. If $l_i/3$ and $2l_i/3$ are chosen as the collocation points, the residual value should be equal to 0 as follows:

$$\begin{aligned} R_i\left(\frac{1}{3}l_i\right) &= \mathrm{EI}_i\left(24C_{i,4} + 40C_{i,5}l_i\right) - T_i\left(\frac{1}{3}l_i\right)\left(2C_{i,2} + 2C_{i,3}l_i + \frac{4}{3}C_{i,4}l_i^2 + \frac{20}{27}C_{i,5}l_i^3\right) \\ &- w_i\left(C_{i,1} + \frac{2}{3}C_{i,2}l_i + \frac{1}{3}C_{i,3}l_i^2 + \frac{4}{27}C_{i,4}l_i^3 + \frac{5}{81}C_{i,5}l_i^4\right) - F_i\left(\frac{1}{3}l_i\right) = 0,\ \ (i = 1, 2, ..., n) \end{aligned} \tag{10}$$

$$\begin{aligned} R_i\left(\frac{2}{3}l_i\right) &= \mathrm{EI}_i\left(24C_{i,4} + 80C_{i,5}l_i\right) - T_i\left(\frac{2}{3}l_i\right)\left(2C_{i,2} + 4C_{i,3}l_i + \frac{16}{3}C_{i,4}l_i^2 + \frac{160}{27}C_{i,5}l_i^3\right) \\ &- w_i\left(C_{i,1} + \frac{4}{3}C_{i,2}l_i + \frac{4}{3}C_{i,3}l_i^2 + \frac{32}{27}C_{i,4}l_i^3 + \frac{80}{81}C_{i,5}l_i^4\right) - F_i\left(\frac{2}{3}l_i\right) = 0,\ \ (i = 1, 2, ..., n) \end{aligned} \tag{11}$$

$2n$ equations can be obtained through Equations 10 and 11, other $4n$ equations are needed to list according to the boundary conditions and continuity conditions.

2.3.2. Continuity conditions

The continuity of geometry and physical properties of riser guarantees the continuity of displacement, deflection angle, bending moment and shear force on the adjacent discrete element as follows:

$$\begin{cases} x_i(l_i) = x_{i+1}(0) \\ \theta_i(l_i) = \theta_{i+1}(0) \\ M_i(l_i) = M_{i+1}(0) \\ Q_i(l_i) = Q_{i+1}(0) \end{cases}, (i = 2,3,\cdots,n) \tag{12}$$

According to the relationship among displacement, deflection angle, bending moment and shear force in mechanics of materials, the following equations can be listed after substituting Equations 7 and 8 into Equation 12 as follows:

$$\begin{cases} C_{i,0} + C_{i,1}l_i + C_{i,2}l_i^2 + C_{i,3}l_i^3 + C_{i,4}l_i^4 + C_{i,5}l_i^5 = C_{i+1,0} \\ C_{i,1} + 2C_{i,2}l_i + 3C_{i,3}l_i^2 + 4C_{i,4}l_i^3 + 5C_{i,5}l_i^4 = C_{i+1,1} \\ \mathrm{EI}_i\left(2C_{i,2} + 6C_{i,3}l_i + 12C_{i,4}l_i^2 + 20C_{i,5}l_i^3\right) = 2\mathrm{EI}_{i+1}C_{i+1,2} \\ \mathrm{EI}_i\left(6C_{i,3} + 24C_{i,4}l_i + 60C_{i,5}l_i^2\right) = 6\mathrm{EI}_{i+1}C_{i+1,3} \end{cases}, \quad (i = 2,3,\cdots,n) \tag{13}$$

$4(n–1)$ equations can be obtained according to Equation 13, and $6n–4$ equations can be listed through adding $2n$ equation as mentioned above. Other 4 equations can be acquired by boundary conditions.

2.3.3. Boundary conditions

The boundary conditions of the system are shown in Equation 2. Substituting the shape function and its derivative forms into Equation 2, the boundary conditions can be represented by the following equation:

$$\begin{cases} C_{1,0} = S; \quad C_{1,1} = 0 \\ C_{n+1,0} + C_{n+1,1}l_{n+1} + C_{n+1,2}l_{n+1}^2 + C_{n+1,3}l_{n+1}^3 + C_{n+1,4}l_{n+1}^4 + C_{n+1,5}l_{n+1}^5 = 0 \\ C_{n+1,1} + 2C_{n+1,2}l_{n+1} + 3C_{n+1,3}l_{n+1}^2 + 4C_{n+1,4}l_{n+1}^3 + 5C_{n+1,5}l_{n+1}^4 = 0 \end{cases} \tag{14}$$

4 equations can be obtained according to Equation 14, and all of the $6n$ equations have been obtained finally. In other words, the $6n$ undetermined coefficients and the displacement of the riser system can be determined unique. Once the deformation has been obtained, the internal force and strength can be discussed later.

2.4. Model validation

Taken a real deepwater drilling operation as example, the water depth is 1500 m, the riser outer diameter is 533.4 mm, the riser wall thickness is 15.875 mm, the top tension is 1.3 G, the top displacement of floating vessel is 30 m, the steel density is 7850 kg/m^3, the elastic modulus of steel is 206 GPa, the drilling fluid density is 1200 kg/m^3, the sea water density is 1030 kg/m^3, the wave height is 6.5 m, the wave period is 8 s, the sea surface wind velocity is 2 m/s, velocity of sea tide is 0.5 m/s, the drag force coefficient is 0.8 and the inertia force coefficient is 1.5.

Numerical simulations have been conducted with two-dimensional finite element model using ABAQUS software. The models are completed in the ABAQUS version 6.12-3 software. The computer used is DELL OPTIPLEX-390, and the RAM is 4 GB. Results of analysis calculated by the theoretical method and the numerical simulations are shown in **Figures 3** and **4**.

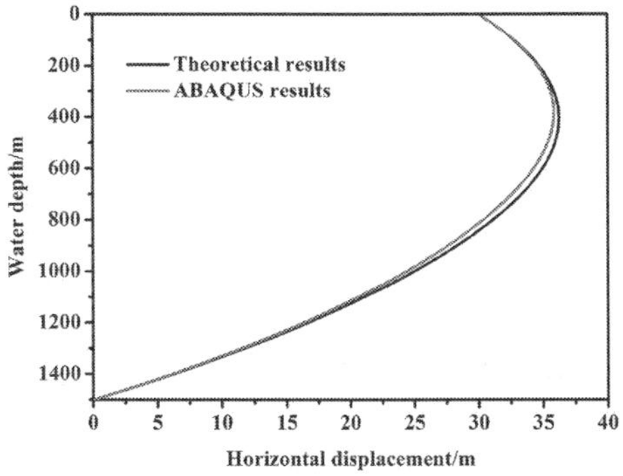

Figure 3. Horizontal displacement.

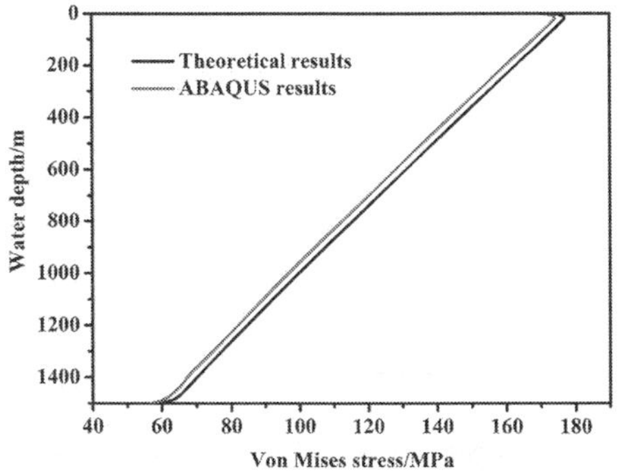

Figure 4. Von Mises stress.

As shown in **Figures 3** and **4**, good consistency has been shown between the calculation results, which validate the availability of the theoretical analysis. In particular, in regard to horizontal displacement, the maximum horizontal displacement calculated by theoretical analysis and ABAQUS software is 36.3 and 35.9 m, which locates at 406 and 383 m below the sea surface, and the maximum calculation error is 1.1% and 5.6%, respectively. However, in regard to stress, the Von Mises stress of the maximum value is 177 and 174 MPa, and both of them appear at 16 m below the sea surface.

3. Experimental system

3.1. System compositions

The experimental system is named mechanics behaviour simulation experimental system for deep water drilling and production strings, which has four sub-systems, namely mechanical structure, hydraulic power and control system and data acquisition system. The main system configuration is shown in **Figure 5** [7].

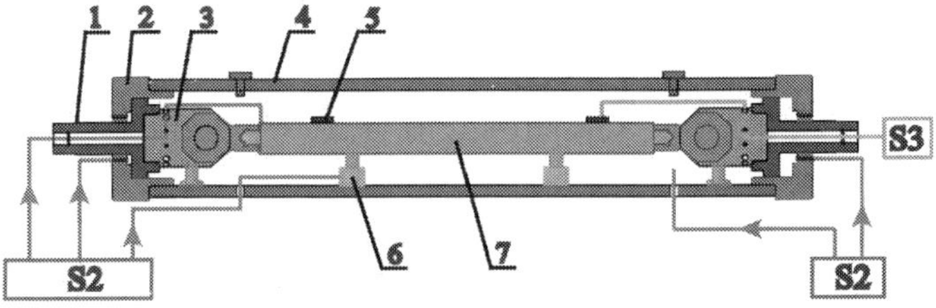

Figure 5. System compositions (1) axial piston; (2) end clamp; (3) connection joint; (4) main cylinder; (5) strain gauge; (6) hydraulic cylinder; (7) tubular specimen; (S2) hydraulic power and control system; (S3) data acquisition system.

The mechanical structure, which is the main body of the experimental equipment, is consist of the main cylinder, two axial pistons, two end clamps and two connection joints. The hydraulic power and control system includes proportion booster cylinder, servo booster cylinder and two hydraulic cylinders to apply transverse force. The main parameters of mechanical structure and hydraulic power and control system are shown in **Tables 1** and **2**.

Main cylinder				End clamp		
O.D. (mm)	I.D. (mm)	Length (mm)	P_w (MPa)	O.D. (mm)	I.D. (mm)	Stroke (mm)
1500	1200	8000	30	1000	400	220

Note: O.D. = outer diameter; I.D. = inner diameter; P_w = pressure-withstanding value.

Table 1. Main structure parameters of the experimental system.

	D_{ip} (mm)	D_p (mm)	Stroke (mm)	V_{max} (m/s)	γ	P_{max} (MPa)
C_1	180	120	700	0.5	2.25	56
C_2	180	120	700	1.0	1.65	46
C_3	140	80	100	–	–	25

Note: C_1, C_2, C_3 are proportion booster cylinder, servo booster cylinder and transverse hydraulic cylinders. D_{ip} = inner diameter of piston cylinder; D_p = outer diameter of piston rod; V_{max} = maximum output speed; γ = Booster ratio; P_{max} = maximum output pressure.

Table 2. Main parameters of the hydraulic control system.

Besides, the data acquisition system is consist of strain gauges and instruments for displacement, strain, pressure and flow rate measurement.

3.2. System working principle

3.2.1. Determination of simulated condition

Generally, we do small size simulation experiment, but in some cases, we need to conduct full size simulation experiment. So, the first matter is to decide which kind of the experiment is. Then, the experimental samples are needed to manufacture. However, no matter which kind of experiment is, we must know the loads subjected to the samples before the experiment. So, the simulated conditions are needed to clearly confirm. Next, the internal pressure, external pressure, axial force and lateral force subjected on the samples are in need for precise calculation . Finally, we must program the load spectrum and upload into the control system to be called during the experiment. It should be pointed out that, if the experiment is not a destructive test, we must limit the load to ensure the safety and integrity of the sample.

3.2.2. Sample installation and load applying

Once the load spectrum is accomplished, the tubular sample is needed to install into the main cylinder. The installation process should be based on the following procedures: first, the strain gauges should be pasted on the sample, and the test leads should connect to the data acquisition system, then connect the sample with one of the connection joint, the axial piston and the end clamp successively, then put the two lateral hydraulic cylinders into the main cylinder, then connect the hydraulic lines with the hydraulic system and then install the left axial piston and end clamp. When conducting the experiment, the hydraulic power system receives the instructions (pressure and displacement) from the control host, and apply internal pressure, external pressure, axial load and lateral force to the tubular sample. The tubular sample generates displacement and strain under the simulation loads. The displacement and strain signals transmit into the data acquisition system through the teat leads. And finally, the strain and stress state of the sample can be obtained by analysing the data.

http://dx.doi.org/10.5772/62315

3.2.3. *Pressure relief*

Pressure relief is necessary after the experiment. To ensure the safety of the specimen in pressure unloading, all kinds of loads are needed to be unloaded in proportion.

3.3. System main functions

3.3.1. *External pressure-withstanding test*

When the external pressure is generated by water pressure in the main cylinder, the axial piston will be pushed towards axially, which will result in axial tension at the tubular sample. In some cases, the axial tension may exceed the maximum allowable load and result in sample damage. So, when external pressure-withstanding tests are done in addition to the water pressure in the main cylinder, the pressure of water must be pumped into the annulus between the end clam and the axial piston to balance the axial force generated by water pressure in the main cylinder. Besides, these two kinds of water pressure must be pumped proportionally.

3.3.2. *Internal pressure-withstanding test*

Similarly, when the internal pressure is generated by water pressure in the tubular sample, the axial tension will also be appeared. So, when internal pressure-withstanding tests are done, the pressure of water in the sample and the annulus between the end clam and the axial piston must be pumped concurrently and proportionally.

3.3.3. *Combination test under external, internal and lateral forces*

If we want to study the mechanics performance of tubular under combined internal pressure, external pressure and lateral force, the pressure of water pumped in the main cylinder, tubular sample and the annulus between the end clam and the axial piston must be loaded concurrently and in proportion.

3.3.4. *Sealing performance test*

Many devices used in the deepwater engineering require good sealing performance. Since the experiment equipment can provide 30 MPa pressure, it is a good platform to conduct sealing performance test for some special devices.

3.4. System performance test

Through system debugging, the hydraulic power system has presented good performance in receiving and executing the instructions from the control host. Besides, the hydraulic servo valves have excellent performance in regulation of pressure and flow rate according to the feedback signals. No phenomenon of leaking water and oil has been occurred. All of the technical indicators have been reached in the laboratory. After completion of the debugging, the internal and external pressure loading tests have been completed, and the actual value collected by the data acquisition system and the target pressure value is shown in **Figure 6**.

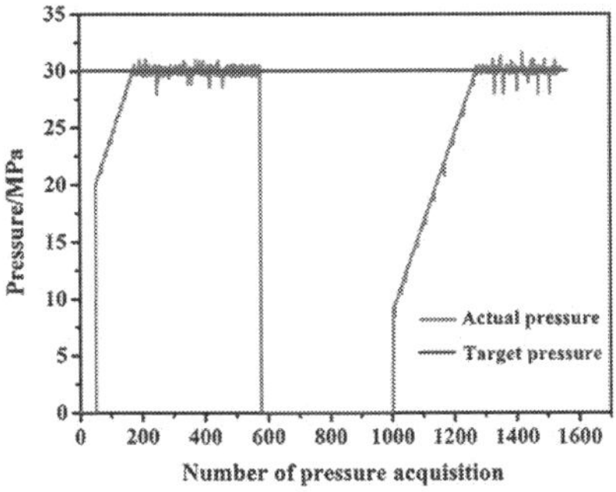

Figure 6. System performance test curve.

As shown in **Figure 6**, the pressure can rapidly response and remain unchanged for long time period. After pressure relief, the reload performance is still excellent. During pressure maintenance phase, the maximum pressure is 31.61 MPa, the minimum pressure is 28.15 MPa, and the average pressure is 29.97 MPa, which indicate that the pressure control system has outstanding performance and can carry out well in accordance with the experimental expectations.

4. Simulation experiments

4.1. Theoretical analysis

As shown in **Figure 1**, the marine drilling riser in working status is subjected to internal pressure, external pressure, axial tension and bending moment induced by drilling fluid, sea water, top tension and lateral force, respectively. If we choose micro-unit from riser outer wall surface, the stress state is shown in **Figure 7**.

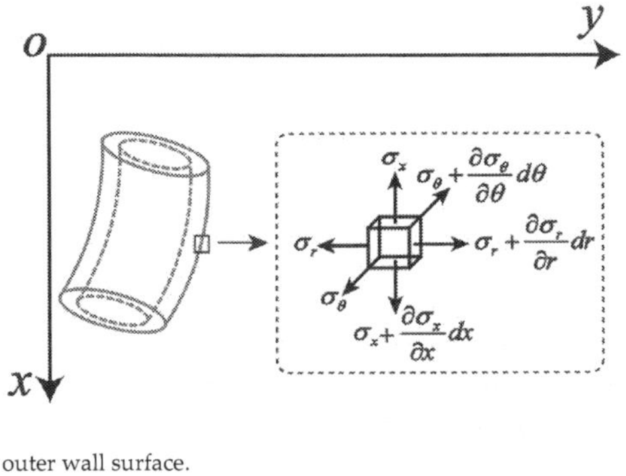

Figure 7. Stress state of riser outer wall surface.

As shown in **Figure 7**, the micro-unit is under the stress state of axial stress, radial stress and hoop stress, and all of the stresses are principal stress. The relationship between the stress and the loads is satisfied by the following equation:

$$\begin{cases} \sigma_{\mathrm{r}} = \dfrac{p_{\mathrm{i}}a^2 - p_{\mathrm{o}}b^2}{b^2 - a^2} - \dfrac{(p_{\mathrm{i}} - p_{\mathrm{o}})a^2b^2}{(b^2 - a^2)r^2} \\ \sigma_{\theta} = \dfrac{p_{\mathrm{i}}a^2 - p_{\mathrm{o}}b^2}{b^2 - a^2} + \dfrac{(p_{\mathrm{i}} - p_{\mathrm{o}})a^2b^2}{(b^2 - a^2)r^2} \\ \sigma_{\mathrm{x}} = \dfrac{F_{\mathrm{a}}}{S_{\mathrm{a}}} \pm \dfrac{M}{W} \end{cases} \tag{15}$$

where σ_θ is the hoop stress (Pa); σ_r is the radial stress (Pa); σ_x is the axial stress (Pa); a and b are the inner diameter and outer diameter of riser, respectively (m); p_i and p_o are the internal and external pressures, respectively (Pa); F_a is the axial tension (N); S_a is the cross-sectional area (m^2); M is the bending moment (N·m); W is the bending modulus (m^3); "+" is selected at the side under tension stress generated by bending moment, whereas "–" is applied at the side under compressive stress generated by bending moment.

Once the stress state has been obtained, the strain value of the micro-unit can be calculated by the generalized Hooke's law as follows:

$$\begin{cases} \varepsilon_r = \dfrac{1}{E}\left[\sigma_r - \mu\left(\sigma_\theta + \sigma_x\right)\right] \\ \varepsilon_\theta = \dfrac{1}{E}\left[\sigma_\theta - \mu\left(\sigma_r + \sigma_x\right)\right] \\ \varepsilon_x = \dfrac{1}{E}\left[\sigma_z - \mu\left(\sigma_\theta + \sigma_r\right)\right] \end{cases} \tag{16}$$

where ε_r, ε_θ and ε_x are the radial strain, hoop strain and axial strain, respectively, and E and μ are elastic modulus and Poisson's ratio, respectively.

As shown in **Figure 3**, the riser has the maximum horizontal displacement at about $x = 400$ m. Therefore, we select the riser here as the simulation object, and the theoretical calculation results of loads, stress and deformation are shown in **Table 3**.

P_i (MPa)	p_o (MPa)	F_a (N)	M (N·m)	σ_x (MPa)	σ_r/ (MPa)	σ_θ/ (MPa)	ε_x (με)	ε_r (με)	ε_θ (με)
4.8	4.0	3.08e6	1.78e4	124.8	–4	6.8	602	–211	–143

Table 3. Theoretical calculation results of riser at $x = 400$ m.

What calls for special attention is that ε_x and ε_θ can be obtained after experiment, while ε_r is difficult to be measured directly. However, if we use the theoretical value of radial stress, σ_x, ε_x and ε_θ are known in Equation 16, and σ_r, σ_θ and ε_r can be solved. So, finally, the stress state strength check can be determined by σ_x, σ_r and σ_θ.

4.2. Simulation experiment method

The four-point bending method has been adopted in the simulation experiment, and the mechanical analysis model of the tubular sample is shown in **Figure 8** [8].

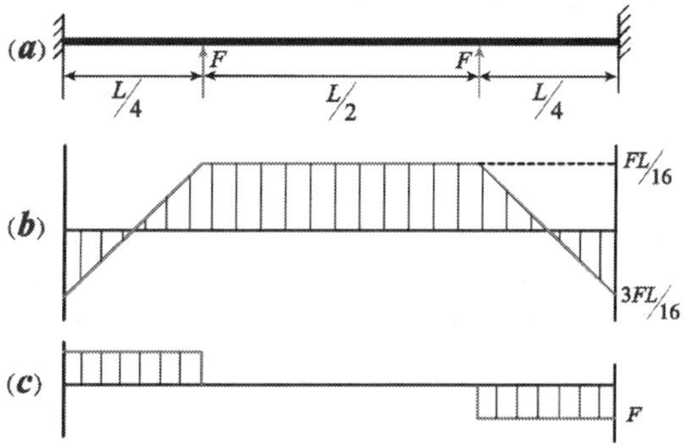

Figure 8. Simulation experiment method: (a) mechanical analysis model; (b) bending moment diagram of tubular sample and (c) shearing force diagram of tubular sample.

Because the two ends of the tubular sample are fixed with the connection joints, this constrain can be regarded as a fixed end. The force outputted by two lateral hydraulic cylinders pushes the sample upward. If both of the two outputted forces equal to F, then the bending moment between $L/2$ is constant and equals to $FL/16$. So, the tubular sample between $L/2$ is the real simulation object.

4.3. Key problems

4.3.1. Insulation problem between strain gauges and tubular sample

Because the strain gauges and its welding wires are working in high water pressure, the insulation between strain gauges and tubular sample must be done to avoid the high water pressure infiltrating into the strain gauge base. Otherwise, the strain gauge will separate from the sample, which will result in rapid drop of insulation resistance and even fall off of the strain gauges. All of these have strong impact on the accuracy of the measurement data. Besides, if the welding wires are exposed to high water pressure directly, the wires would be conductive with water, which will also result in distortion or even failure of the measurement. Therefore, the exposed metal parts in welding wires must be isolated from the water. Generally, the chemical coating method is used in strain measurement under high water pressure. Before the experiment, the sealing protection measures must be adopted after the strain gauges are pasted on the outer wall of the tubular sample.

The measure taken to insulate the strain gauges is shown in **Figure 9**.

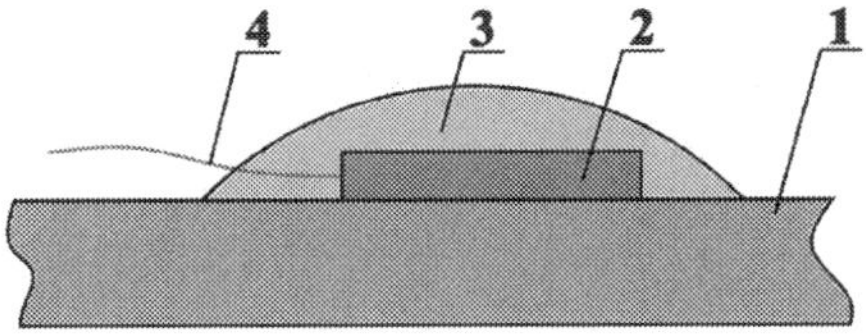

Figure 9. Sealing protection measures of strain gauges: (1) outer wall of the tubular sample; (2) strain gauge; (3) sealing glue and (4) test wires.

4.3.2. Sealing problems between test wires and the main cylinder

As the test leads connect the strain gauges and the strain acquisition instrument, which is put outside of the main cylinder, there must be satisfactory sealing problems between the test wires and the main cylinder. If the sealing measures do not conform to the requirements, the water leakage would happen during pumping water pressure, which would cause pressure fluctuations in the main cylinder and unstable strain signals. The key to solve this problem is to design reasonable sealing plugs. Design principles of the sealing plug are sealing easy and effective, large capacity of test leads, not complicated machinery manufacturing and installa-

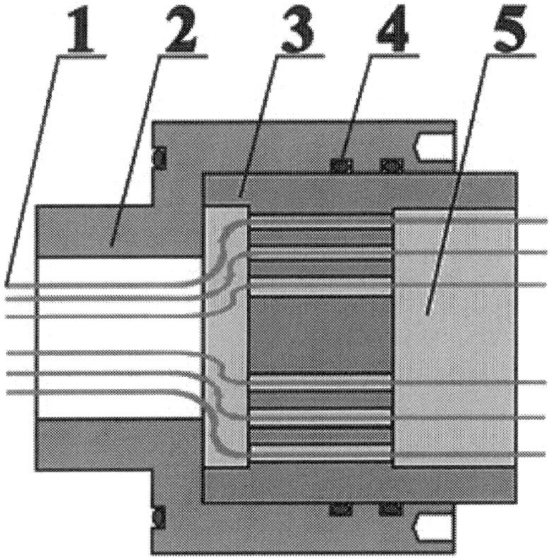

Figure 10. Sealing plug: (1) test wires; (2) sealing plug body; (3) sealing plug core; (4) seal rings and (5) sealing glue.

tion and the possibility to be reused. The comprehensive cleaning of residue and oil stain is necessary and important to ensure that the sealing performance of the plugs after manufacture is completed. In this experiment, sealing glue is used to inject into the plugs to seal them. Moreover, the sealing performance test must be conducted to ensure that the plugs meet the requirements of the pressure in the experiment. Otherwise, the sealing plugs need to be re-made. The sealing measure taken to seal the test wires and the main cylinder is shown in **Figure 10**.

4.4. Tubular sample

The tubular sample is a steel pipe, with a length of 6000 m, outer diameter of 520 mm, inner diameter of 496 mm (wall thickness is 12 mm) and Elastic modulus and Poisson's ratio of 206 GPa and 0.3, respectively.

4.5. Pressure loading scheme

According to **Table 3**, the bending moment subjected to the sample is 1.78e4 N·m, so the force outputted by the lateral hydraulic cylinders is 18.5 MPa based on the experiment equipment shown in **Table 2**. So, there are three parameters needed to be controlled, which are the internal pressure (4.8 MPa), the external pressure (4.0 MPa) and the lateral hydraulic pressure (18.5 MPa). Equal proportion loading scheme has been selected in the experiment, as shown in **Figure 11**.

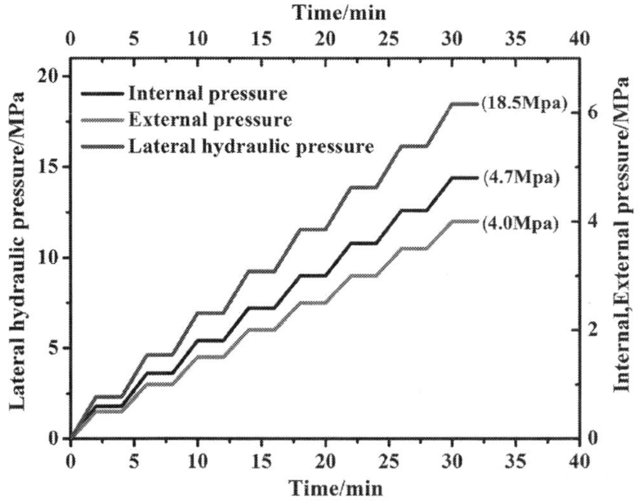

Figure 11. Pressure loading scheme.

As shown in **Figure 11**, the loading scheme of "pressurization–stabilization–pressurization–stabilization..." is adopted. The displacement, pressure and flow rate are collected at all the experimental time, while the strain of the sample is only collected during pressure stabilization period. In each load period, the pressurization duration is 2 min, and the stabilization duration is 2 min. So, the duration of each load period is 4 min. After each load period, the increases of internal pressure, external pressure and lateral hydraulic pressure are about 0.6, 0.5 and 2.31 MPa, and the three kinds of pressure will eventually achieve 4.7, 4.0 and 18.5 MPa, respectively, after 8 load periods. Then, the stress state of the tubular sample reaches that of the marine riser at x = 400 m. To acquire more valid data, the duration of stabilization can be extended as expected after 32 min.

4.6. Strain gauges pasted scheme

As shown in **Figure 8**, four-point bending method is used in the experiment, and the tubular sample between $L/2$ is the real simulation object. So, the strain gauges are all pasted on this section. The whole scheme of pasting strain gauges is shown in **Figure 12**.

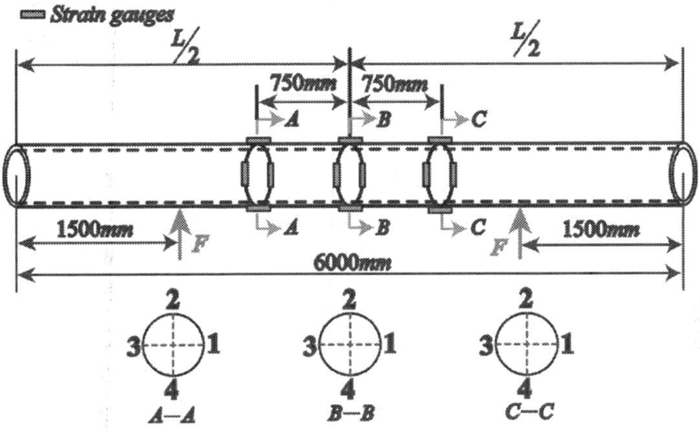

Figure 12. Strain gauges pasted scheme of the whole sample.

To acquire more valid data on different section, three sections (A–A, B–B and C–C) have been chosen to paste the strain gauges. One section locates the centre of the sample, and the other two sections symmetrically locate 750 mm from the centre of the sample. On each section, there are four points (1, 2, 3, 4), and the circumference interval of the four points is 90°, as shown in **Figure 12**. To measure the axial strain and hoop strain simultaneously, there are two strain gauges that are pasted at each point, of which one pasting direction is coincidence with the one of the mother lines of the tubular sample, and the other pasting direction is perpendicular to the prior strain gauge. The pasting scheme at each point is shown in **Figure 13**.

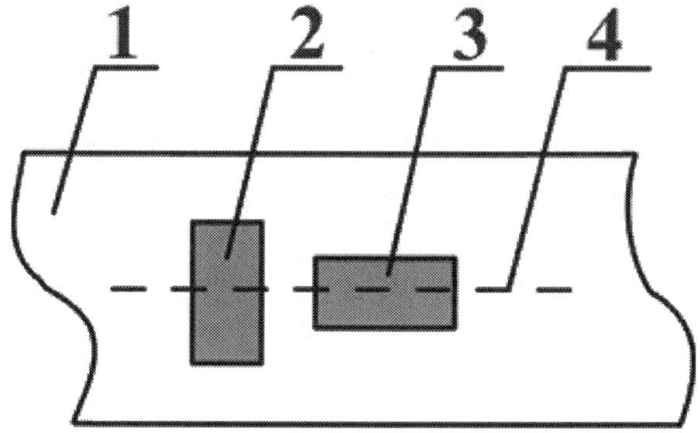

Figure 13. Pasting scheme at each point.

Therefore, 8 strain gauges are pasted at each section, and 24 strain gauges are pasted at the whole sample, of which 12 strain gauges measure the axial strain, and other 12 strain gauges measure the hoop strain. From theoretical perspective, the hoop strains measured at each point are equal, whereas the axial strains measured at point are unequal (Point 2 > Point 1 = Point 3 > Point 4).

4.7. Pressure acquisition

According to the experiment purpose, the three kinds of pressure (internal pressure, external pressure and lateral hydraulic pressure) are measured real time. The actual measured value of the pressure during stabilization period is shown in **Figure 14**.

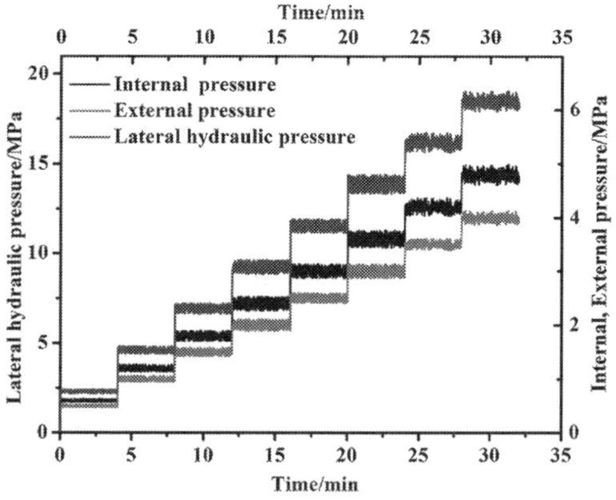

Figure 14. Measured pressure value.

As shown in **Figure 14**, good feedback adjustment has been revealed. Although the actual value has fluctuated during pressure loading, they generally increase in the forms of steps. Finally, all of the pressure values have reached the experiment expectation with high consistency, and the average values of the internal pressure, external pressure and lateral hydraulic pressure are 4.8, 4.1 and 18.5 MPa, respectively. Besides, during the stabilization duration, the strain data have been measured successfully.

4.8. Strain acquisition

The average strain values of the tubular sample under 4.8, 4.1 and 18.5 MPa are shown in **Table 4**.

Measure points	Strain value		Measure points	Strain value		Measure points	Strain value	
	ε_x (με)	ε_θ/ (με)		ε_x (με)	ε_θ/ (με)		ε_x/ (με)	ε_θ/ (με)
A-1	609	−147	B-1	603	−144	C-1	598	−150
A-2	639	−149	B-2	633	−133	C-2	622	−150
A-3	608	−134	B-3	606	−137	C-3	601	−134
A-4	587	−149	B-4	578	142	C-4	570	−151

Table 4. Strain data measured by the gauges.

As shown in **Table 4**, all of the strain gauges have collected the strain data during the experiment, which indicates that the key problems (insulation problem between strain gauges and tubular sample and sealing problems between the test wires and the main cylinder) involved in the experiment have been solved effectively, and the strain of the tubular sample can be measured by the experiment method as mentioned above.

4.9. Data analysis

Due to the influence of the bending moment, the axial strain at the same section is different. The average value of the axial strain at Points 1, 2, 3 and 4 are 604, 631, 605 and 578 με, respectively, which verifies the above analysis (Point 2 > Point 1 = Point 3 > Point 4). However, if we take the axial strain (29 με) induced by the bending moment (1.78 × 104 N·m) into consideration, the axial point of the four points can be represented by the following: Point 2 = Point 1 + 29 με = Point 3 + 29 με and Point 1 = Point 3 = Point 4 + 29με. The hoop strain is approximately equal at the 12 measure points, and both of the theoretical value and the average value of the measured value are −143 and −143 με, which verifies the accuracy of the experiment method. The theoretical and measured axial strains are shown in **Figure 15**.

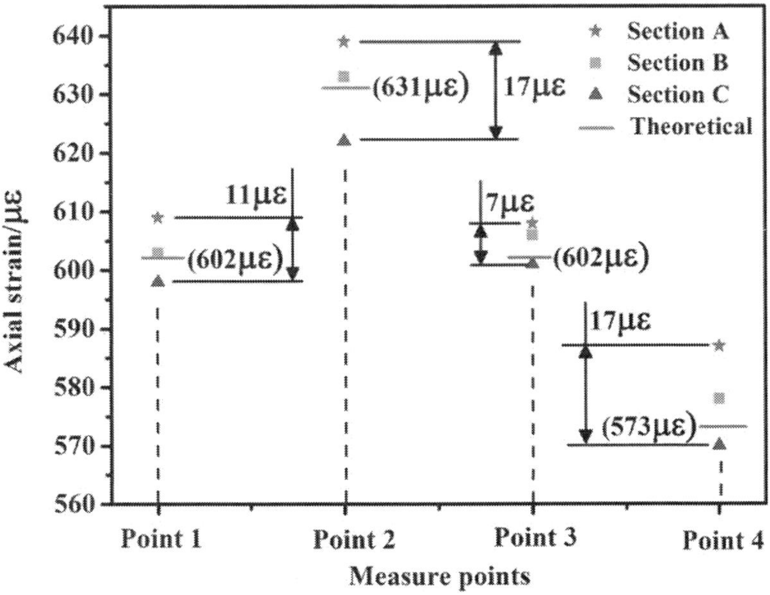

Figure 15. Theoretical and measured value of axial strains.

As shown in **Figure 15**, the maximum differences of the four measure points at the three sections are 11, 17, 7 and 17 με. If we take the average of the three sections at the same point, the statistical results can be as presented in **Table 5**.

Items	Point 1	Point 2	Point 3	Point 4
Theoretical value (με)	602	631	602	573
Average of the measured value (με)	603	631	605	578
Error (%)	0.2	0	0.5	0.9
Variance ($\mu\varepsilon^2$)	22	50	18	86

Table 5. Statistical results of the measure axial strain.

4.10. Stress state analysis

Because the radial strain cannot be measured directly, the stress state can be determined by the theoretical axial stress, the measured hoop strain and axial strain according to Equation 16. After analysis, the stress state of the experimental sample under the eight stabilization phases is shown in **Figure 16**.

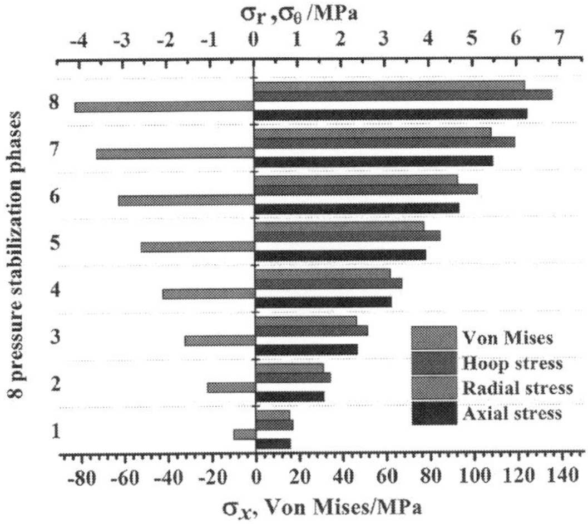

Figure 16. Stress state analysis.

As shown in **Figure 16**, with the increases of internal pressure, external pressure, axial tension and bending moment, the hoop stress, radial stress, axial stress and the Von Mises stress increase gradually. When the loads' state reaches the simulation condition required in the experiment, the Von Mises stress is 121.9 MPa, while the theoretical value is 123.8 MPa. The measurement error is only 1.5%, which verifies the feasibility and validity of the simulation experiment method.

5. Conclusion

Internal pressure, external pressure, axial tension and bending moment are typically loads subjecting on the tubulars in offshore drilling engineering. The mechanical model and control equations have been established to analyse the deformation and stress distribution of the marine riser, and the analysis method has been verified by finite element method (FEM) in the ABAQUS framework.

The experimental system has the ability to provide kinds of loads to simulate the mechanical behaviour of offshore tubulars, which exhibits a good platform to studying the mechanics of tubulars under complicated stress state and can perfect the deepwater drilling from theoretical viewpoint and field practice.

Good solutions must be proposed to solve the key problems in the simulation experiment, and thus, the technique of strain test under high water pressure can be used to measure the strain data effectively. Besides, excellent pressure loading scheme and the strain acquisition scheme are necessary to acquire precise measurement results.

Acknowledgements

The authors gratefully acknowledge the financial support from the Natural Science Foundation of China (NSFC, 51521063, U1262201)

Author details

Yanbin Wang*, Deli Gao and Jun Fang

*Address all correspondence to: wyb576219861@126.com

MOE Key Laboratory of Petroleum Engineering, China University of Petroleum, Beijing, China

References

[1] Gao Deli. (2011).Optimized Design and Control Techniques for Drilling & Completion of Complex-Structure Wells. China University of Petroleum Press, Beijing, China, ISBN 978-7-5636-3598-6.

[2] Lu Baoping. (2014). Key Technologies and Equipment for Deepwater Drilling. China Petrochemical Press, Beijing, China, ISBN 978-7-5114-2695-6.

[3] Robello Samuel & Gao Deli. (2013). Horizontal Drilling Engineering: Theory, Methods and Applications. Sigma Quadrant Publisher, Texas, USA, ISBN: 978-0-615-83770-3.

[4] Wang Yanbin, Gao Deli & Fang Jun. (2014). Development of Mechanics Behavior Simulation Experimental System for Deep Water Drilling and Production String. Oil Field Equipment, Vol. 43, No. 4, pp: 26–29.

[5] Wang Yanbin, Gao Deli & Fang Jun. (2014). Static analysis of deep-water marine riser subjected to both axial and lateral forces in its installation. Journal of Natural Gas Science and Engineering, Vol. 19, pp: 84–90.

[6] Gao Deli & Zhang Hui. (2012). Mechanical Analysis of Tubes in Deepwater Drilling Operation without Riser. Science & Technology Review, Vol. 30, No. 4, pp: 37–42.

[7] Fang Jun, Wang Yanbin & Gao Deli. (2013). Test Method for the Force Deformation of Deepwater Riser. China Petroleum Machinery, Vol. 41, No. 12, pp: 53–57.

[8] Liu Hongwen. (2011). Mechanics of Materials. Higher Education Press, Beijing, China, ISBN: 978-7-0403-0895-2.

Chapter 3

Plate-Like Structure Damage Acoustic Emission Beamforming Array Technique and Probability-Based Diagnostic Imaging Method

Dongsheng Li, Mengdao Jin and Quanming Feng

Additional information is available at the end of the chapter

http://dx.doi.org/10.5772/62413

Abstract

A novel beamforming array technique and probability-based diagnostic imaging method are proposed to determine the acoustic emission (AE) source in plate-like structures. The technique that differs from common beamforming array techniques, in particular a sensor network, is used instead of a linear sensor array, to highlight information on the AE source location in one coordinate system as energy distribution. To reduce the uncertainty, avoid the boundary reflection effect, and ensure the rationality of the signal superposition, a Hilbert transform-based signal processing is applied before the delay-and-sum algorithm and a probability-based diagnostic imaging method is developed for AE source localization. The finite element numerical simulation method and the pencil-lead-broken experiment on aluminum plate are also conducted, and a thin-walled cylinder pipe-like structure is also tested by the pencil-lead-broken experiment to develop the application of the proposed method in various fields. The results indicate that this method is efficient and capable of visually showing the localization results highlighted in the probability images.

Keywords: acoustic emission, source localization, beamforming, probability-based diagnosticimaging, plate-like structure, health monitoring

1. Introduction

Plate-like structures are widely used in a variety of fields, such as material, civil, aerospace, and other industries [1–5]. Confirming the health condition of these plate-like structures is a very meaningful assignment for engineers. One of the various non-destructive techniques, acoustic emission (AE) monitoring is arguably based on the simplest of physical concepts: the release of transient elastic waves in solid materials as a result of a localized source of damage [2–4]. The process of locating the source of these acoustic waves by capturing the related propagating acoustic signals through various sensors or arrays and properly analyzing them is commonly known as the AE source localization technique [3–8]. Common passive monitoring techniques are currently applied for AE source localization in isotropic and anisotropic structures, particularly in two-dimensional plate-like structures [9,10].

The techniques for AE source localization, such as triangulation technique, modal AE, and optimization-based technique, are currently performed by accurately determining the exact time of arrival (TOA) or time difference of arrival (TDOA) [8]. An obvious disadvantage of these methods is that the precision of the time difference measurement has a great effect on the location precision. To avoid this problem, McLaskey et al. [11] introduced a novel beamforming method of AE analysis, which is particularly suited for field applications on large plate-like reinforced concrete structures, such as walls and bridge decks. The advantage of the beamforming technique is that it does not require the TOA or TDOA of a specific wave mode; the technique can therefore handle noisy signals if the noise is a Gaussian white noise. He et al. [12] applied the beamforming technique for source localization in a thin steel plate through finite element (FE) simulation and experiment. Nakatani et al. [13] studied AE source localization on an anisotropic structure and proved that the technique is simple but effective.

Beamforming has been successfully used in source localization by many researchers, but its limitation as a technique is that it is a velocity-dependent method [11]. As such, its localization accuracy and precision are significantly influenced by the propagation characteristics. The propagation characteristics of an AE signal in plate-like structures are complex because of the signal's repeated reflection in the interface. The wave speed is affected by many factors, such as frequencies, material properties whether isotropic or not, multi-mode, and the dispersion in the propagation process. Xiao et al. [14] proposed a novel beamforming method with two uniform linear arrays in determining the AE source location using an unknown wave speed to reduce the effect of the factors (mentioned earlier) on AE beamforming. However, the accurate two-dimensional position cannot be shown at the same time in one coordinate system, because the localization coordinates are correct along the direction parallel to the array but not along the direction perpendicular to the array.

2. Principle of Lamb wave propagation

The Lamb waves are used to interrogate the whole thickness of the plate-like structures. Either the surface defects or the internal damages can be detected.

2.1. The dispersion characteristic of Lamb wave

Since Lamb waves are waves of plane strain that occur in a free plate, only displacements through the thickness (x_3 direction) and in the direction of wave propagation (x_1 direction) are taken into consideration (**Figure 1**).

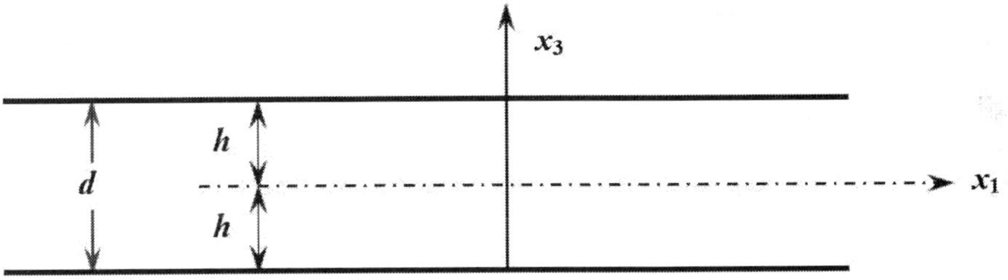

Figure 1. Geometric construction sketching of free plate.

Eq. (1) is for the symmetric Lamb wave mode (S_i, $i = 0,1,2,3 \cdots$.) in which the particle motion of the plate occurs symmetrically about the symmetric axis of the plate. Eq. (2) is for the anti-symmetric Lamb wave mode (A_i, $i = 0,1,2,3 \cdots$.) in which the particle motion of the plate occurs asymmetrically about the symmetric axis of the plate [15,16].

$$\frac{\tan(qh)}{\tan(ph)} = -\frac{4k^2 pq}{(q^2 - k^2)^2}, \tag{1}$$

$$\frac{\tan(qh)}{\tan(ph)} = -\frac{(q^2 - k^2)^2}{4k^2 pq}, \tag{2}$$

where the parameters p and q are defined as follows:

$$p^2 = \frac{f^2}{c_L^2} - k^2, q^2 = \frac{f^2}{c_T^2} - k^2, h = \frac{d}{2},$$

Where c_L and c_T are the propagation velocities of longitudinal wave and shear wave, respectively; and wavenumber $= \frac{f}{c_P}$.

The phase velocity dispersion curves can be solved by Eqs. (1) and (2). And, the group velocity dispersion curves can be plotted from the phase velocity dispersion curves using Eq. (3). The group velocity and phase velocity dispersion curves for an aluminum plate are shown in **Figure 2**(**a** and **b**).

$$c_g = \frac{c_P^2}{c_P^2 - fd\dfrac{\partial c_P}{\partial(fd)}}, \tag{3}$$

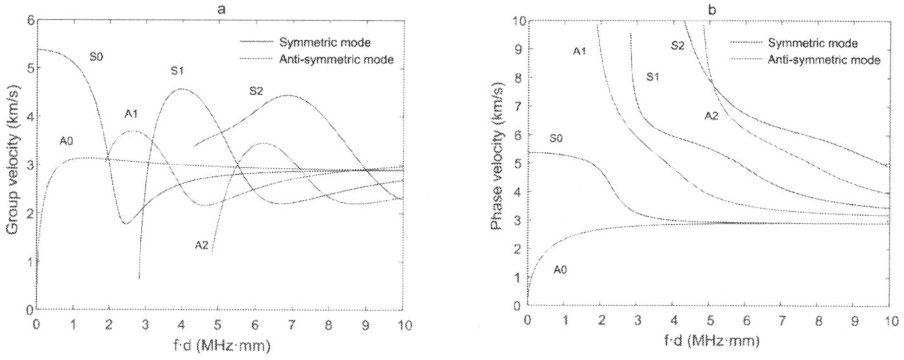

Figure 2. The dispersion curves for an aluminum plate: (a) group velocity dispersion curves and (b) phase velocity dispersion curves.

2.2. Finite element simulation

Numerical simulation is another main method aside from experimental validation that is used in various fields. Modeling the guided waves conveniently provides modeled data with a high signal-to-noise ratio, which can be used to develop useful advanced signal processing and analysis techniques. A number of numerical methods have been applied to analyze the propagation of elastic waves, such as the finite element method (FEM), finite difference method, boundary element method, and finite strip elements (FSE), all of which have their respective merits and defects. FEM is best understood from its practical application, known as finite element analysis (FEA). FEA is a computational tool for performing engineering analysis. It includes the use of mesh generation techniques for dividing a complex problem into small elements, which in turn represent different areas in the physical system; the technique also uses a software program coded with the FEM algorithm. This study simulates the guided wave signal propagation with FEA in the general three-dimensional domain using the commercial package ABAQUS, which has some advantages and is effective. It easily records the solution results of every element or node by time step and extracts results. In addition, it provides sophisticated pre- and post-processing functions and is user-friendly. Many scholars have demonstrated the possibility of using FEM to solve the guided wave formation and guided wave signal propagation in plate specimens.

2.2.1. Explicit dynamic analysis

The explicit dynamics analysis procedure, ABAQUS/Explicit, is based upon the implementation of an explicit integration rule together with the use of diagonal ("lumped") element mass matrices. The equations of motion for the body are integrated using the explicit central-difference integration rule.

$$\dot{u}_{\left(i+\frac{1}{2}\right)} = \dot{u}_{\left(i-\frac{1}{2}\right)} + \frac{\Delta t_{(i+1)} + \Delta t_{(i)}}{2}\ddot{u}_{(i)}, \tag{4}$$

$$u_{(i+1)} = u_{(i)} + \Delta t_{(i+1)}\dot{u}_{\left(i+\frac{1}{2}\right)}, \tag{5}$$

Where u is a degree of freedom (a displacement or rotation component) and the subscript i refers to the increment number in an explicit dynamics step.

The explicit procedure integrates through time by using many small time increments. The central-difference operator is conditionally stable, in order to avoid numerical instability, there are more than 20 points per cycle of the highest frequency. Therefore, the time step can be expressed as follows [17]:

$$\Delta t = \frac{1}{20 f_{max}}, \tag{6}$$

where f_{max} is the highest frequency of interest.

The size of the finite element is depended on the smallest wavelength. In reference [17], it is recommended that 20 nodes per wavelength be used. So that, it can be expressed as

$$l_e = \frac{\lambda_{min}}{20}, \tag{7}$$

where l_e is the element size and λ_{min} is the shortest wavelength of interest.

In the present FEM analysis,$\Delta\, t$ = 0.1 μs and l_e = 1 mm. The plate was discretized using brick elements C3D8R, and the mesh is shown in **Figure 3**. The physical properties of the aluminum plate areas follows: E = 71 GPa, ν = 0.33, ρ = 2700 kg/m^3.

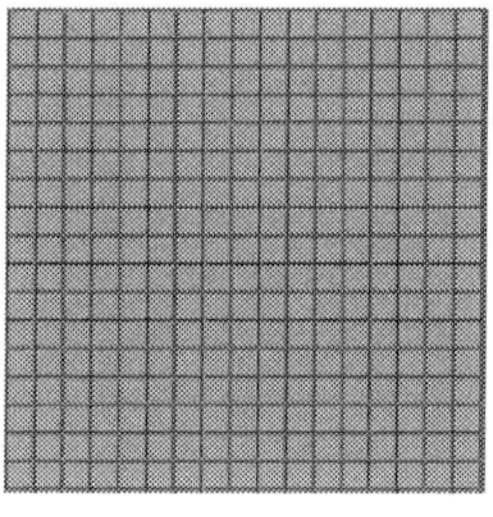

Figure 3. The structured meshing technique.

2.2.2. *Excitation signal*

In the finite element modeling, the excitation signal is a smoothed tone burst obtained by filtering a five-cycle pure tone burst of central frequency f = 150 kHz through a Hanning window [**Figure 4**(**a** and **b**)].The Hanning window is described by the following equation [18]:

$$h(t)=0.5\cdot\left[1-cos\left(\frac{2\pi t}{T}\right)\right],\quad t\in[0,T] \tag{8}$$

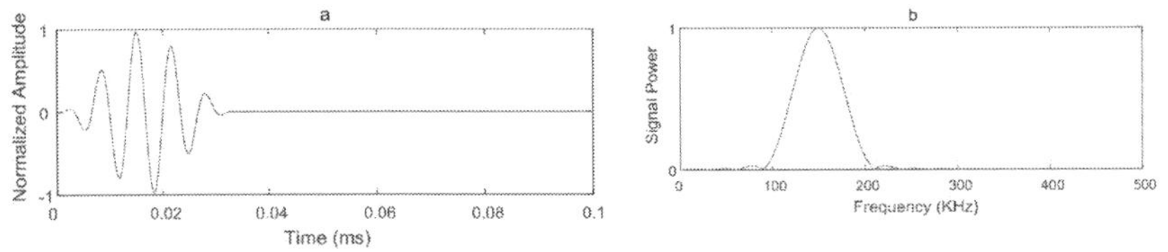

Figure 4. Five-cycle Hanning window-modulated sinusoid tone bursts in: (a) time domain and (b) frequency domain after Fast Fourier Transform.

The number of counts *(N)* in the tone bursts matches the length of the Hanning window:

$$T=\frac{N}{f} \tag{9}$$

The smoothed tone burst is governed by the following equation:

$$x(t)=h(t)\cdot sin(2\pi ft),\quad t\in[0,T] \tag{10}$$

2.2.3. *Selective excitation of Lamb wave modes*

Dual piezoelectric (PZT) actuators are symmetrically mounted on the top and bottom surfaces of a plate, and the frequencies of excitation signals are controlled to a range so as to generate the fundamental symmetric mode S_0 and anti-symmetric mode A_0 only. This mode selection technique can be expressed as:

1. When either of the dual PZTs is energized, both the symmetric mode S_0 and the anti-symmetric mode A_0 will be generated simultaneously.

2. When dual PZTs are in-phase energized, the symmetric mode S_0 with the most signal energy is generated, meanwhile, the anti-symmetric mode A_0 is weaker.

3. When dual PZTs are out-phase energized, the anti-symmetric mode A_0 is generated to dominate the majority of wave energy, and the symmetric mode S_0 is suppressed.

In the FEM, a 1200 mm × 1200 mm aluminum plate with 4 mm thickness was used for the simulation. Because the piezoelectric element C3D8E doesn't exist in ABAQUS/explicit; so we applied 16 self-equilibrating concentrated forces, as shown in **Figure 5.**, to simulate the wave excitation for the round PZT actuators. The mode excitation method is expressed as follows:

When the 16 self-equilibrating concentrated forces are loaded on the top or the bottom of a plate, both S_0 mode and A_0 mode are excited simultaneously.

When the 16 self-equilibrating concentrated forces are symmetrically loaded on the top and the bottom of a plate, the mode S_0 will be excited.

When the 16 self-equilibrating concentrated forces are anti-symmetrically loaded on the top and the bottom of a plate, the mode A_0 will be excited. The excited Lamb wave modes are shown in **Figure 6(a–c)**.

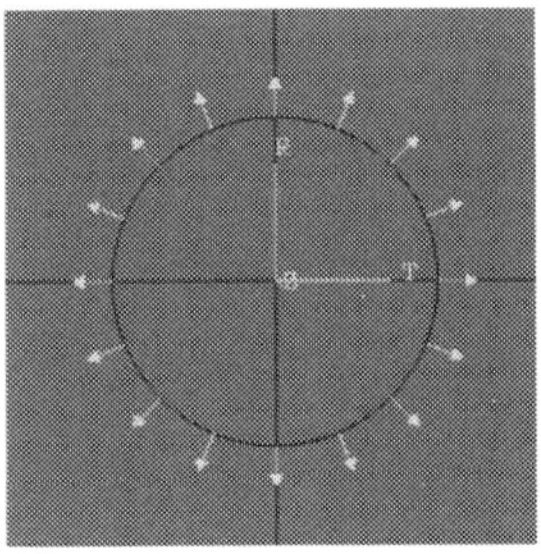

Figure 5. Self-equilibrating concentrated forces for round PZT actuators.

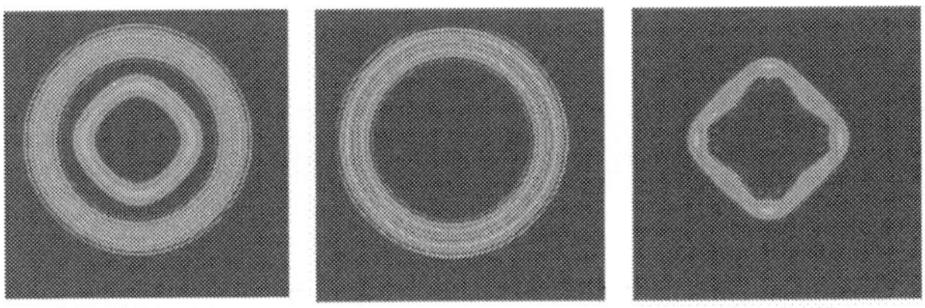

Figure 6. The excited Lamb wave modes.

2.2.4. *Lamb wave scattering simulations*

2.2.4.1. *The interaction of the S_0 mode with a circular hole*

It is well known that mode conversion phenomena will occur when the incident S_0 mode wave interacts with damages; there will be S_0 mode, A_0 mode, and SH0 mode waves in the scattering fields.

When the depth of the hole is through thickness, the scattering S_0 mode and SH0 mode can be captured.

When the depth is 75% of the thickness, the scattering S_0 mode, A_0 mode, and SH0 mode can be captured.

The snapshots of S_0 mode interaction with a circular hole are shown in **Figure 7** and **Figure 8**, respectively.

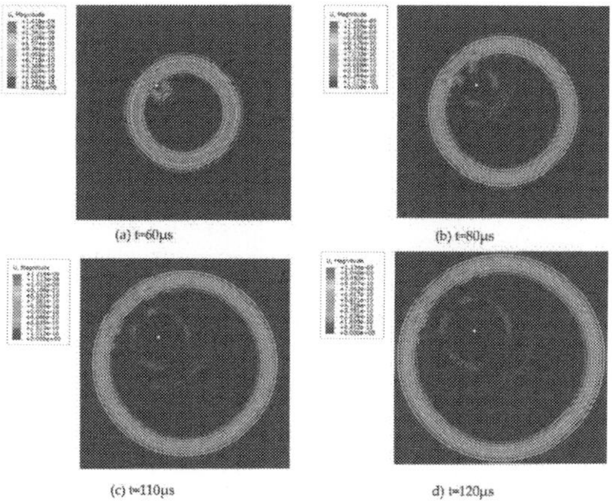

Figure 7. The scattering field of a through-thickness hole.

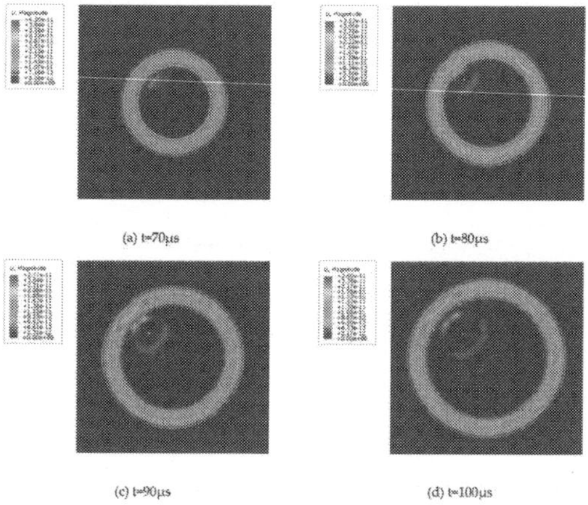

Figure 8. The scattering field of a hole with the depth is 75% of the thickness.

2.2.4.2. The interaction of the S_0 mode with a crack

When the incident S_0 mode wave interacts with a crack, the crack-induced scattering wave components are same as a circular hole. The snapshots of S_0 mode interaction with a crack are shown in **Figure 9**.

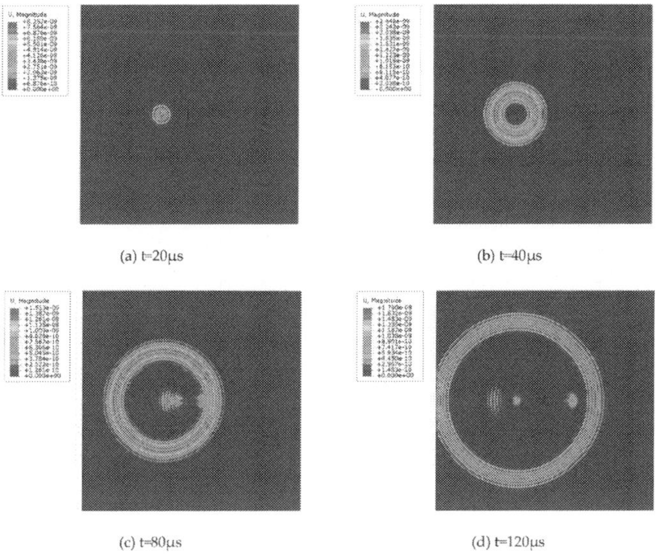

Figure 9. The scattering field of a through-thickness crack.

3. Beamforming algorithm for damage identification

3.1. Principle of the beamforming network method

The beamforming technique is a method based on the delay-and-sum algorithm. **Figure 10** shows that an acoustic event causes an AE source to generate an acoustic wave, which is received by M sensors. Given the difference between the distance of the sensors and that of the sound source, the arrival time must also differ. Although different degrees of attenuation exist on the wave energy, waveform parameters, such as frequency, duration, and the peak of the same sound source remain almost same. The beamforming technique adjusts different sensor waveform signals to the same datum point by time delay to ensure that the entire waveform signal is adjusted to the same wave front. The sensors network output of incident waves in an isotropic plate is expressed by Eq. (11). The variable $b(\vec{r}, t)$ is the network output with the delay-and-sum algorithm, after applying time delays $\Delta_m(\vec{r})$ to the recorded signals and multiplying them with the weight factor as shown further. According to **Figure 1**, the $\Delta_m(\vec{r})$ can be obtained by Eq. (12).

$$b(\vec{r},t)=\frac{1}{M}\Sigma_{m=1}^{M}w_m x_m(t-\Delta_m(\vec{r})), \tag{11}$$

$$\Delta_m(\vec{r})=\frac{|\vec{r}|-|\vec{r}-\vec{r}_m|}{c}, \tag{12}$$

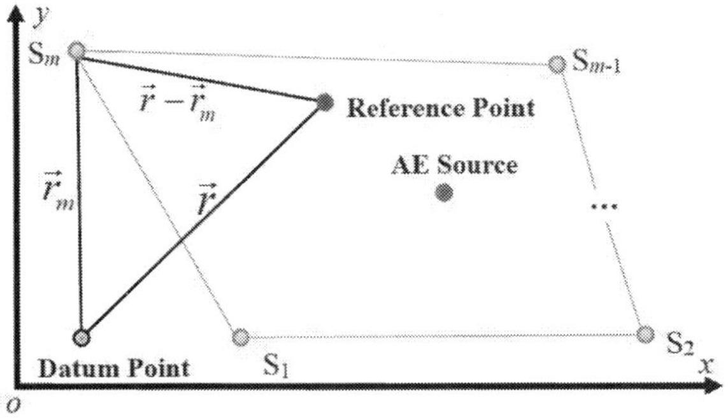

Figure 10. Principle illustration of the beamforming network method.

Here the variable $x_m(t)$ represents the measured signal acquired by the m_{th} ($m = 1, 2,\ldots, M$) sensor S_m. $\vec{r}$ represents the distance of a reference point (also called the focal point) from the previously designated datum point, and $\vec{r}_m$ represents the distance of a reference point from the m_{th} sensor. $\Delta_m(\vec{r})$ is the individual time delay for the m_{th} sensor, and c is the propagation velocity of the acoustic wave. Different reference points (or focal points) have different time delays $\Delta_m(\vec{r})$. By adjusting time delays $\Delta_m(\vec{r})$, the signals associated with the spherical waves emitted from the sound source focus are aligned in time before they are summed. When the reference point and the acoustic source point coincide, the signals are aligned at the same wave front, and the energy output of the sensor network is at its maximum. Beamforming technology compares the network output results of each reference point and displays them into the position-energy curve after normalized processing. The position of the maximal energy distribution is the location of the sound source, and all the weight factors w_m may be taken as 1.

3.2. Hilbert transform-based signal processing

Given the propagation of the elastic wave in materials, the reflection and scattering of signals result in mode conversion, which goes against waveform processing. Research findings have shown that an energy distribution curve with pure signal superposition is not continuous, as shown in **Figure 11(a)**. Beamforming technology needs only the necessary amplitude information and does not require the full waveform information. Thus, to reduce the uncertainty, avoid the boundary reflection effect, ensure the rationality of the superposition, and obtain a smooth curve, the envelopes of the signals obtained using Hilbert transform (HT)-based signal processing $A_m(t)$ (i.e., the first wave packet containing the peak information is intercepted) are used in this study instead of the signal variable $x_m(t)$. The parameter B given in Eq. (15) attains the maximum value when the reference point coincides with the AE source point. The HT series given in Eqs. (13) and (14) have the same amplitude and frequency content as the original data and include phase information, which depends on the phase of the original data, as shown in **Figure 11(b)**. The same source results demonstrate that the HT-based signal processing $A_m(t)$ is reasonable and capable of obtaining the energy distribution information of the AE source, as shown in **Figure 11(c)**.

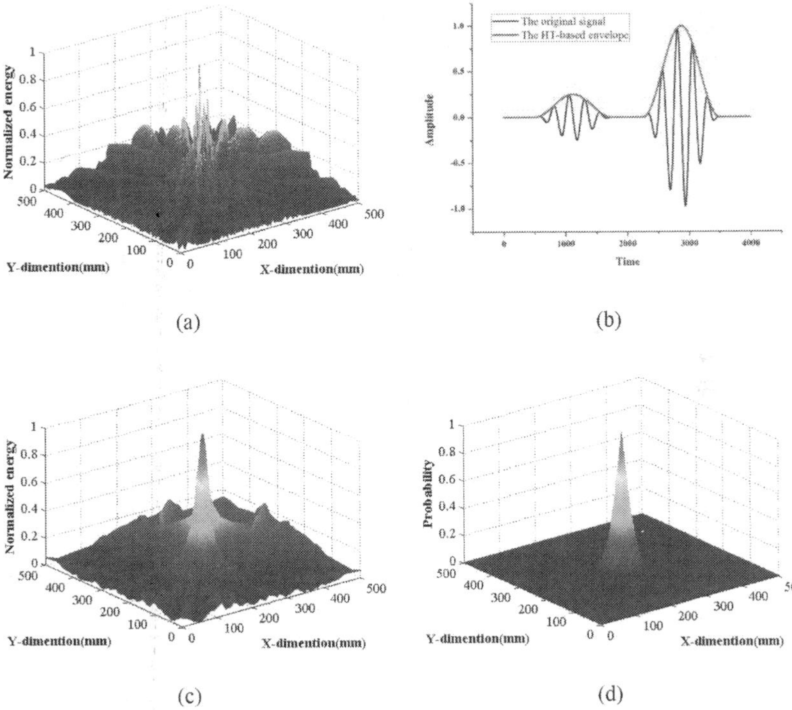

Figure 11. Localization results in multiple ways. (a) The normalized energy distribution with original signals $x_m(t)$; (b) the original signals and the HT-based envelopes; (c) the normalized energy distribution with HT-based envelopes $A_m(t)$; and (d) the probability distribution.

$$\hat{x}_m(t)=\int_{-\infty}^{+\infty}\frac{x_m(\tau)}{\pi(t-\tau)}d\tau\ , \tag{13}$$

$$A_m(t)=\sqrt{x_m^2(t)+\hat{x}_m^2(t)}\ , \tag{14}$$

$$B(\vec{r},t)=\frac{1}{M}\Sigma_{m=1}^{M}w_m A_m(t-\Delta_m(\vec{r}))\ , \tag{15}$$

3.3. Probability-based diagnostic imaging

Given the existence of errors and uncertainty, the real location of the sound source is likely to be where the energy is larger (near the maximum). Thus, a probability diagnostic imaging technology is introduced to explain the results from the perspective of probability theory. After determining the energy distribution, the structure under detection is meshed using L × K nodes for probability-based diagnostic imaging; this technique is an emerging technique for monitoring the health status of a structure and identifying damages and has attracted much attention in recent years [19, 20]. It uses an easily interpretable image whose pixels exclusively correspond to the spatial points to describe the health status of the structure. The field value of a pixel stands for the probability of the presence of damage at the point of the structure corresponding to the pixel, and the degree of probability is often calibrated through different gray scales. In this study, the normalized energy distribution can be regarded as the probability of the AE source location. When the normalized energy distribution focuses on a peak under the correct circumstances, it is similar to the Gaussian distribution. Thus, a cumulative distribution function, $F(z)$, is introduced to facilitate this procedure of defining the field value asfollows:

$$F(z)=\int_{-\infty}^{z}f(z)dz\ , \tag{16}$$

$$f(z)=\frac{1}{\sigma\sqrt{2\pi}}exp\left(-\frac{z^2}{2\sigma^2}\right), \tag{17}$$

$$z=\sqrt{(x_i-x_p)^2+(y_i+y_p)^2}\ , \tag{18}$$

where $f(z)$ is the Gaussian distribution function, which represents the probability density of the presence of the AE source at the mesh node $(x_i, y_j)(I=1, 2, \ldots, L; j=1, 2, \ldots, K)$. The variable z is the distance of the arbitrary node (x_i, y_j) to the node (x_p, y_p), whose peak point is in the normalized energy distribution, and σ is the standard variance. Thus, the field value at node (x_i, y_j), which is the probability of the AE source $I(x_i, y_j)$ is

$$I(x_i, y_j) = 1 - [F(z) - F(-z)], \tag{19}$$

The normalized energy and probability distributions of AE localization results are shown in **Figure 11(d)**.

4. FE model of AE simulation

A sketch of the FE model with the geometric configuration 800 mm × 600 mm × 4 mm is shown in **Figure 12**. The homogeneous isotropic aluminum plate is characterized by the following: Young's modulus E = 71 GPa, Poisson ratio v = 0.33, and mass density ϱ = 2700 Kg/m^3. In this model, damping is not considered, the boundaries are free, and a total of eight sensors (S1–S8) form a rectangular network. The AE source position, datum point, and their location are marked, and the three AE sources are named as cases A, B, and C. All the cases are considered both away from and close to the boundary. All of the sensors are decorated (geometrically) evenly. A Hanning-windowed five-cycle sinusoidal tone burst is used as the transient excitation to simulate the AE wave. The frequency of the excitation signal is 400 KHz. A same off-plane displacement with a peak of 0.1 μ m is loaded on multiple cases, and the time histories of the samples of the sensor nodes are captured. The element size and integration time step in FEA process are designated as 2 mm and 0.1 μs, respectively, to avoid numerical instability [19].

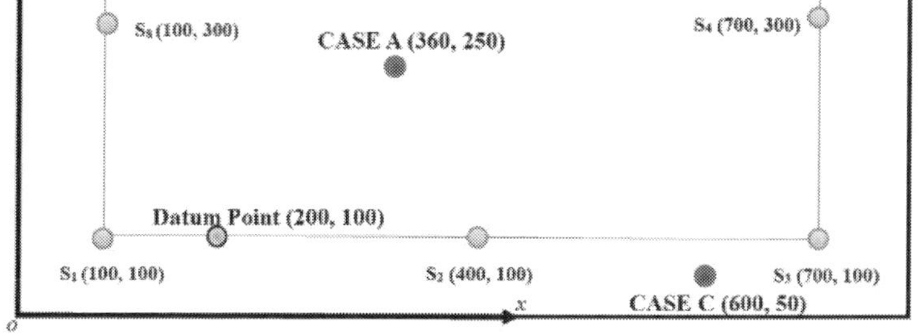

Figure 12. Setup of FE model of AE simulation configuration with a rectangular network.

The AE wave is a common form of the Lamb wave propagation in plate-like structures. In most cases, multiple Lamb modes exist simultaneously, and the dispersive properties of such modes are not identical even for the same mode in different frequency ranges [21]. Different mode waves travel with different groups and phase velocities. As a consequence, source localization strategies must be conducted based on a certain wave propagation mode. The fundamental symmetric mode S_0 wave and the fundamental asymmetric mode A_0 wave are popularly used in plate-like structures. The localization results are presented using S_0 mode wave, given that the propagation velocity of the S_0 mode is faster than that of the A_0 mode and considering the interception of the first wave packet analyzed in this study. The dispersion curve in **Figure 13** shows that the propagation speed of the S_0 mode Lamb wave with a frequency of 400 KHz in 4 mm unconstrained aluminum plate is about 4600 m/s.

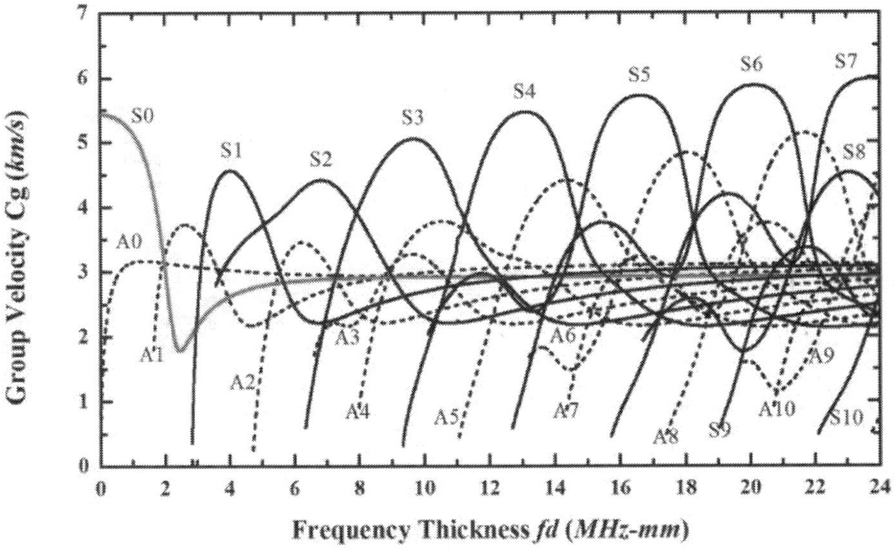

Figure 13. Dispersion curves of Lamb wave in the free isotropic aluminum plate.

4.1. Localization results of the numerical simulation

To verify the validity of the positioning, three cases of real sound sources were tested using the delay-and-sum algorithm with the precise wave speed. Their real coordinates were as follows: case A (360 mm, 250 mm), case B (750 mm, 550 mm), and case C (600 mm, 50 mm). The localization results for the rectangular sensor network are exhibited in **Table 1**.

Multiple case	Normalized energy distribution with HT	Probability distribution

Case A

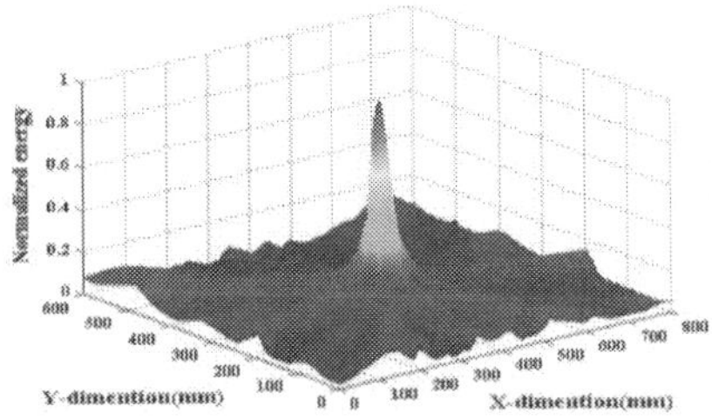

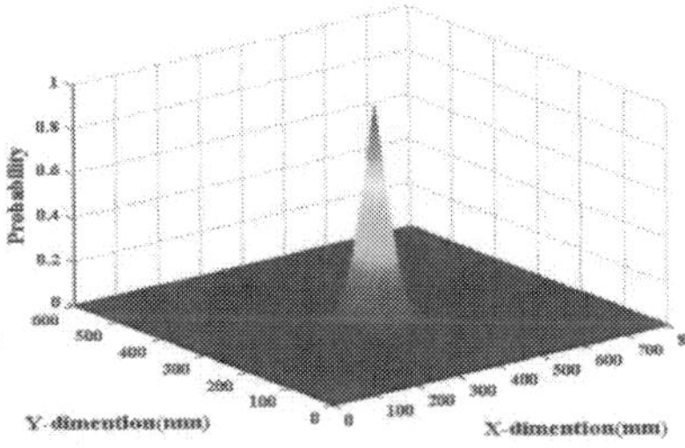

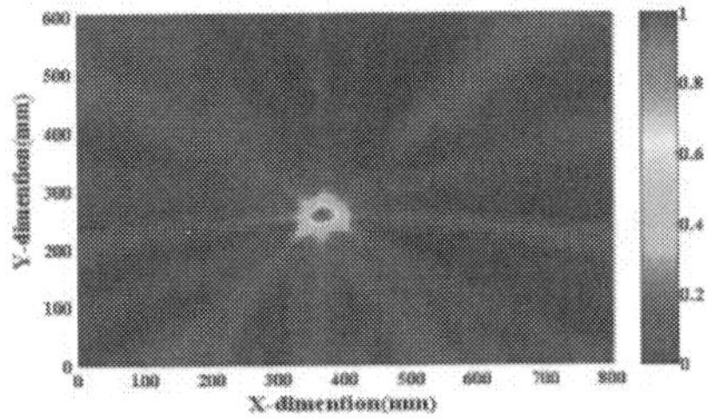

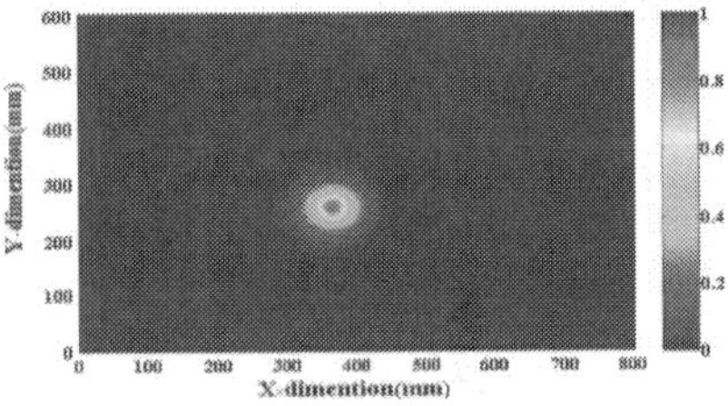

Case B

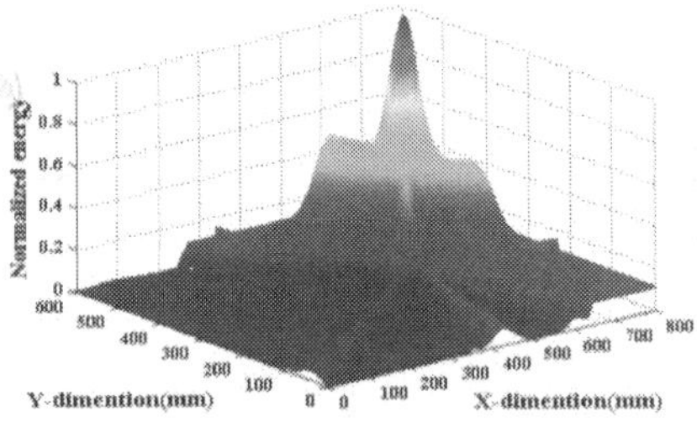

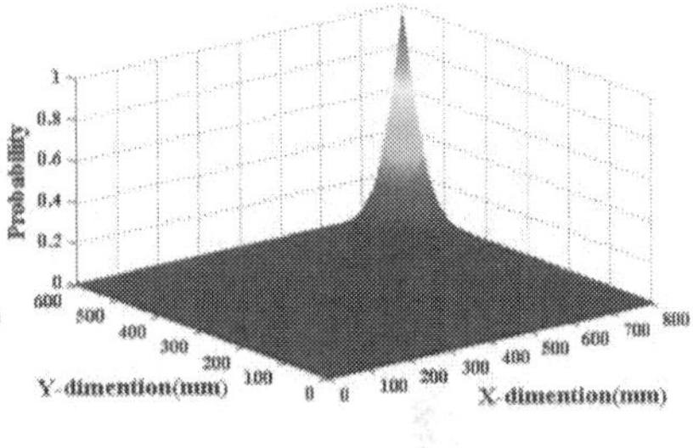

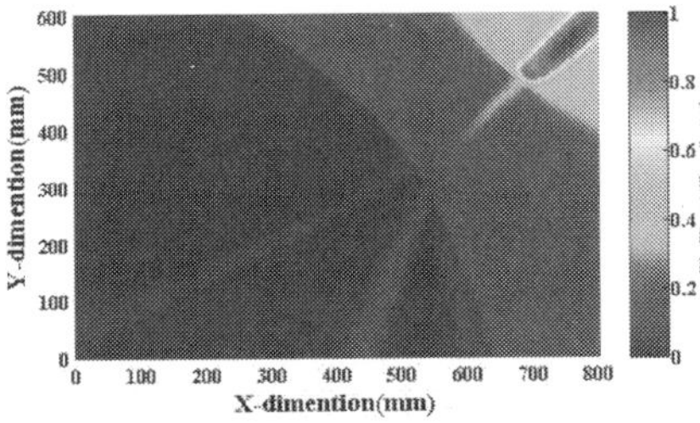

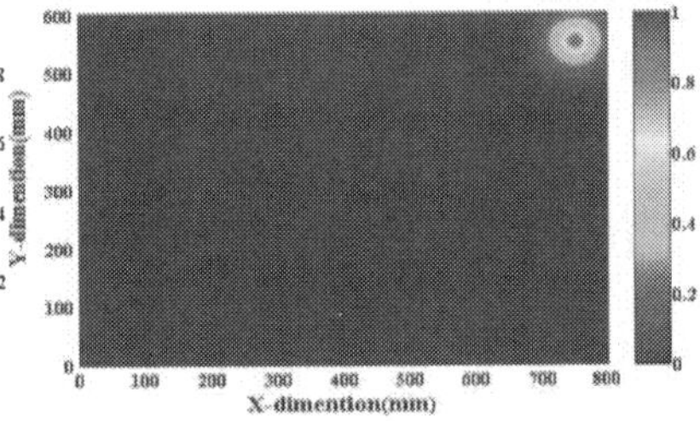

Multiple case	Normalized energy distribution with HT	Probability distribution
Case C	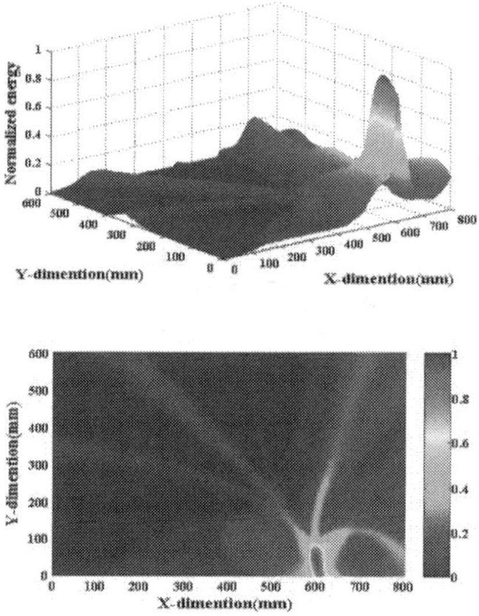	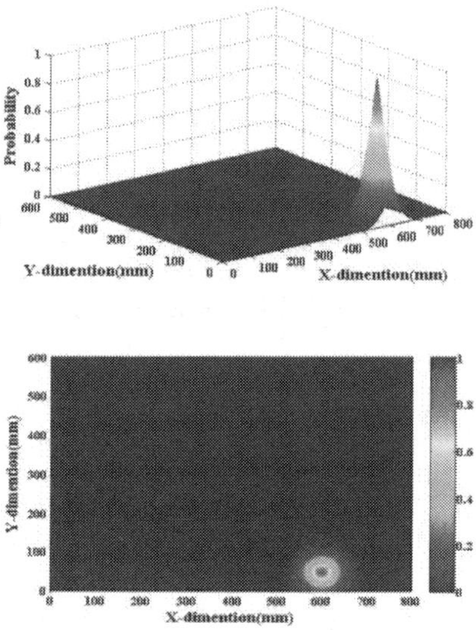

Table 1. Multiple cases of the source localization for aluminum plate based on beamforming method with rectangular sensor network in wave velocity $c = 4600$ m/s

The results show that under the true known wave speed, the beamforming array technique and probability-based diagnostic imaging method can confirm the location of the AE source and produce accurate positioning results, as highlighted in the floor plan image. The case away from boundaries can evidently work much better than the cases close to boundaries, given the same wave speed. The results also prove that HT-based signal processing and probability-based diagnostic imaging are reasonable and effective.

4.2. The influence of the wave speed

The localization results in **Table 1** show that the beamforming technique can effectively confirm the location of the AE source and produce accurate positioning results, as highlighted in the floor plan image. However, it is a wave-velocity-dependent method; a clear limiting factor of this technique is the need to obtain knowledge on its wave propagation speed. For common, uniform, and thick plate materials, we generally find out the dispersion curve to determine the required wave velocity, following the theory of wave propagation. However, given the structure size, material properties, or performance degradation factors, ensuring the dispersive curves of the Lamb wave is not easy, so the sound source localization of the plate structure in the case of unknown wave velocity must be discussed.

http://dx.doi.org/10.5772/62413

According to beamforming theory, the largest impact of wave velocity lies on the length of delay. Under a real wave velocity, signals can be achieved through accurate time delay adjustment to the same wave front, the superposition of energy concentrated, and the maximum output reached. In calculating a wave velocity that is far from the true wave velocity, the energy is unable to concentrate and the peak isn't the maximum. This finding can be proven by **Figure 14**.

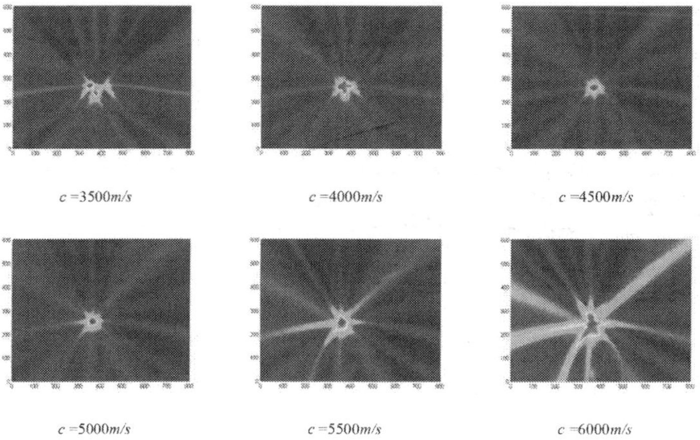

Figure 14. The localization results of case A under different wave speed.

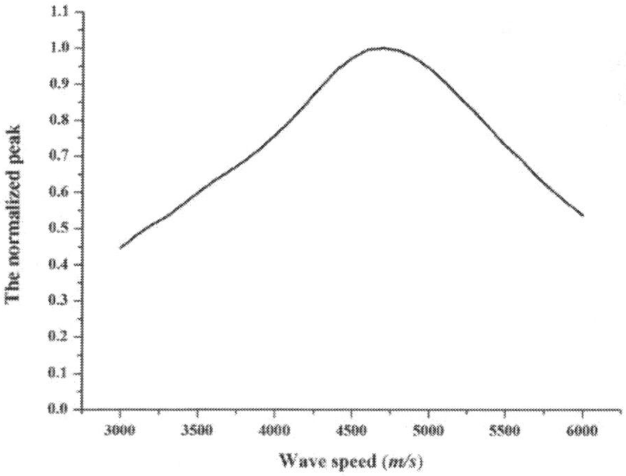

Figure 15. The normalized energy peak curves of FE model in different wave speeds.

The larger the deviation from the real wave velocity, the higher the divergence of the energy; the smaller the deviation from the real wave velocity, the more concentrated the stack power. The normalized energy peak curves of the FE model under different wave speeds are shown in **Figure 15**. When the wave speed is 4700 m/s, the peak of the energy is at its maximum, so that the real wave velocity is around 4700 m/s. At this rate, the localization can also be confirmed by comparing the energy distribution or peak value under the different wave speeds, with no need for the known real wave velocity.

5. Experimental verification

To further verify the effectiveness of AE source localization using beamforming network technology for plate-like structures, a pencil-lead-broken experiment on an aluminum plate and a steel cylinder experiment are implemented.

5.1. Aluminum plate experiment

The size of the aluminum plate adopted in the experiment was 1000 mm × 1000 mm × 4 mm. It was a homogeneous isotropic aluminum plate with the following characteristics: Young's modulus E = 71 GPa, Poisson ratio v = 0.33, and mass density ρ = 2700 Kg/m^3. A rectangular network with eight AE sensors was built, with each of the sensors placed 100 mm away from the closed boundary. An AE acquisition system from Physical Acoustics Corporation was applied in the experiment, and the sampling frequency used was 1 M/s. The pencil-lead-broken experiment was used to simulate the AE source (**Figure 16**). Multiple cases of the source location were set in Cases A (500 mm, 500 mm), B (200 mm, 500 mm), and C (700 mm, 300 mm).

Figure 16. Aluminum plate experiment system.

According to the chapter 3. 1 mentioned method, the normalized energy peak curves of experimental aluminum plate under different wave speeds were shown in **Figure 17**. The AE propagation speed was about 5200 m/s. The localization results for the rectangular sensor network are exhibited in **Table 2**.

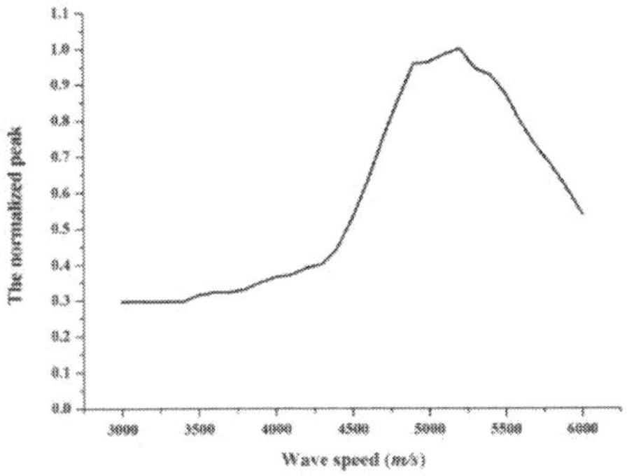

Figure 17. The normalized energy peak curves of 4 mm aluminum plate.

Multiple case	Normalized energy distribution with HT	Probability distribution
Case A	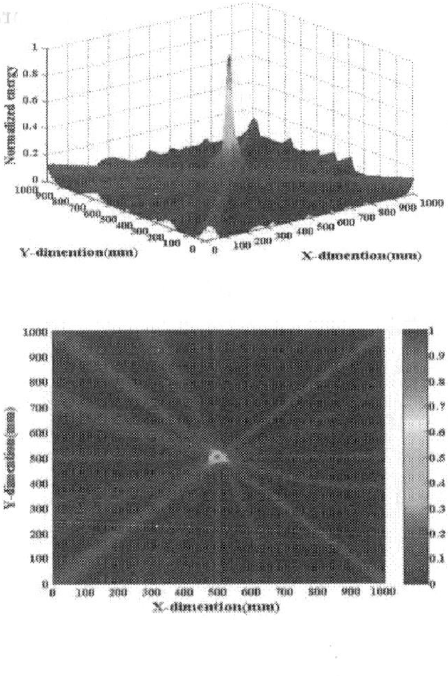	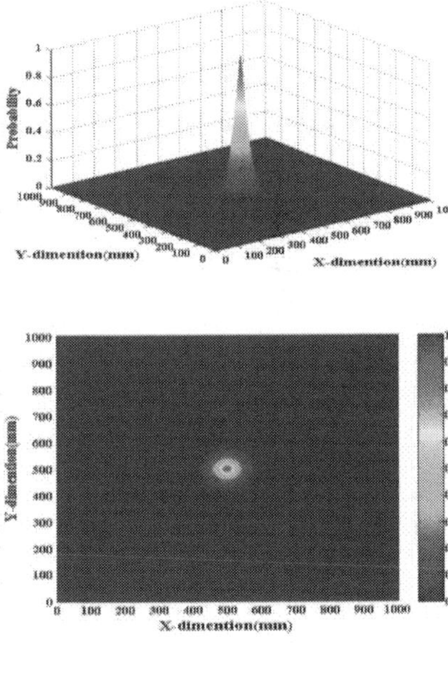

<table>
<tr><th>Multiple case</th><th>Normalized energy distribution with HT</th><th>Probability distribution</th></tr>
<tr><td>Case B</td><td>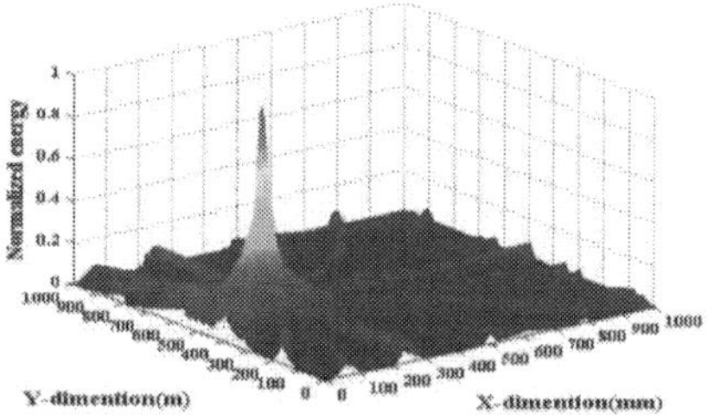

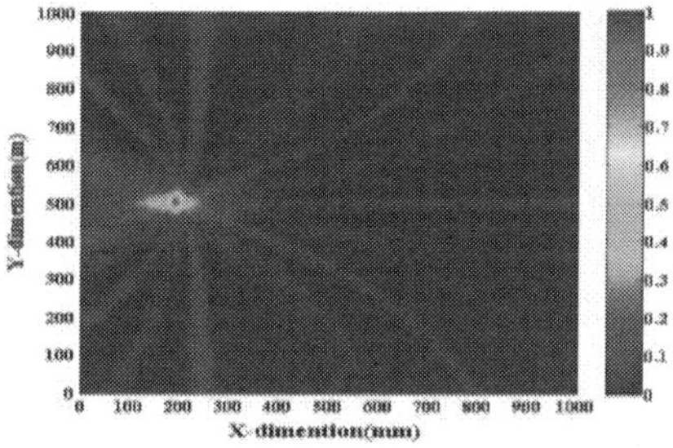
</td><td>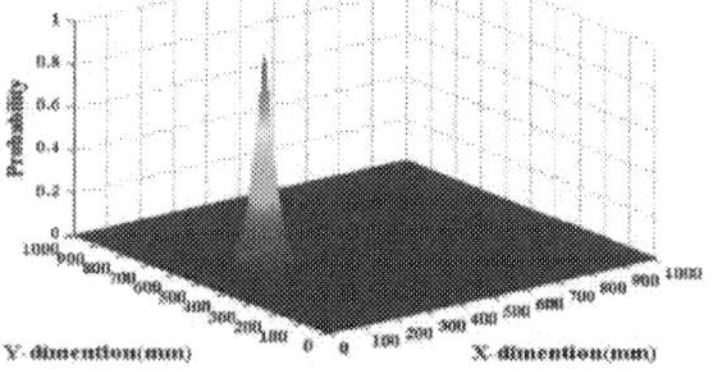

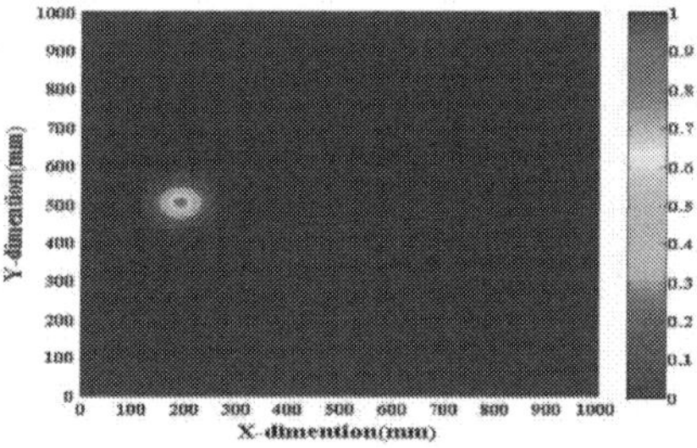
</td></tr>
<tr><td>Case C</td><td>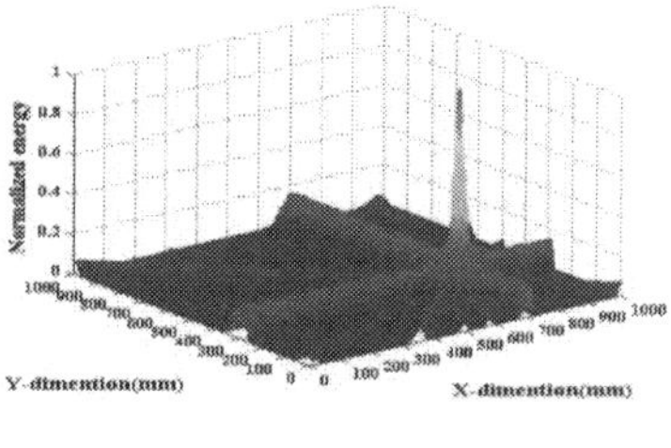

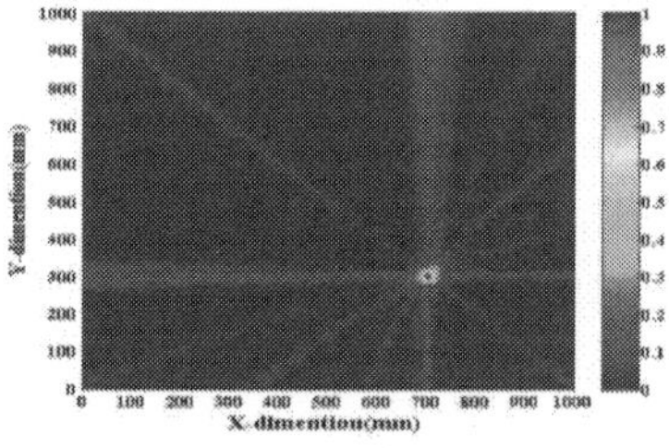
</td><td>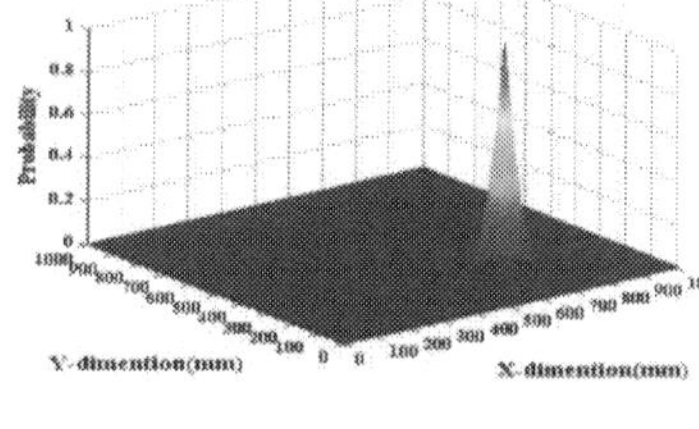

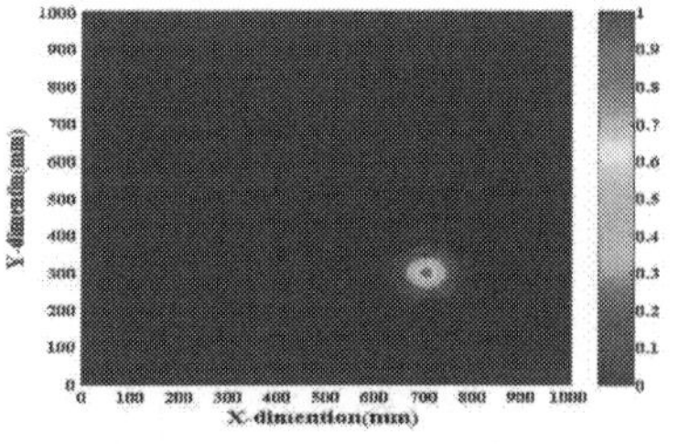
</td></tr>
</table>

Table 2. Multiple cases of the source localization for aluminum plate based on beamforming method with rectangular sensor network in wave velocity c =5200m/s

5.2. Thin-walled steel cylinder experiment

For a thin-walled cylinder with a large radius, the propagation characteristics of the Lamb wave are the same as those of plate-like structure [22]. Thus, this study constructed a sensor network in conformity to a plate on a steel thin-walled cylinder to verify the effectiveness of the beamforming method for AE source localization. The cylinder had an inner radius, external radius, and height of 102, 110, and 700 mm, respectively, and was characterized by the following: Young's modulus $E = 209$ GPa, Poisson ratio $v = 0.3$, and mass density $\rho = 7800$ Kg/m^3. The sensors were arranged, such that the circles with the heights of 100 and 600 mm were stepped by 90° (**Figure 18**). The chosen generatrix was 0°, and the X-axis was transformed into a degree measure. The AE source position was designed at (90°, 350 mm); according to the normalized energy peak curves of the 8 mm steel cylinder in **Figure 19**, the wave velocity was 4900 m/s. Given the strong symmetry of the cylinder, some of the sensors could be used more than once, so that beamforming network algorithm was achieved at a step of 90°. The localization result and energy and probability distribution are exhibited in **Table 3**. It was verified that the algorithm can be applied to AE source localization for thin-walled steel cylinders.

Figure 18. Thin-walled cylinder pipe-like structure experimental system.

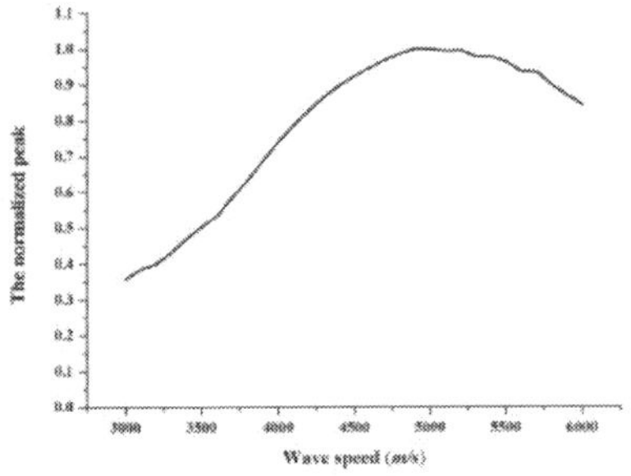

Figure 19. The normalized energy peak curves of 8 mm steel cylinder with different wave speeds.

Normalized energy distribution	Probability distribution

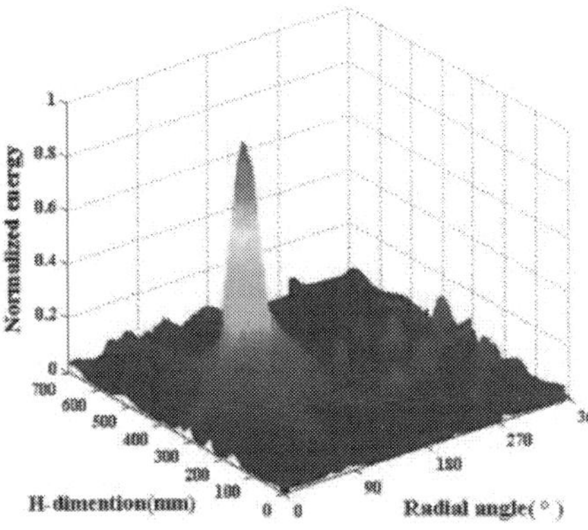

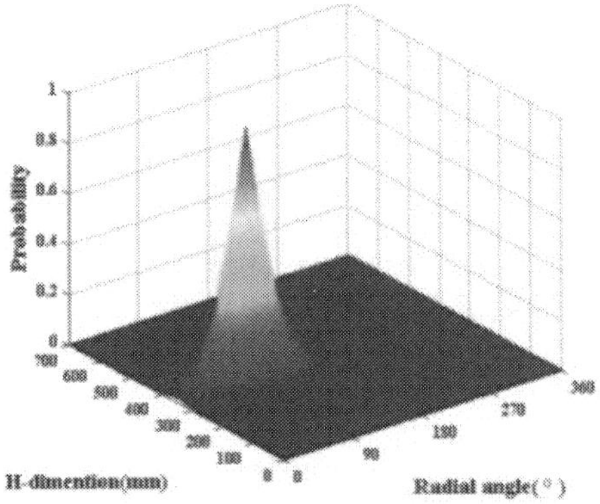

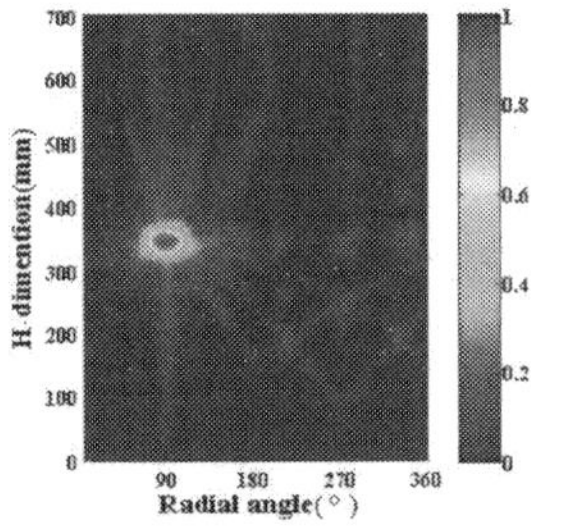

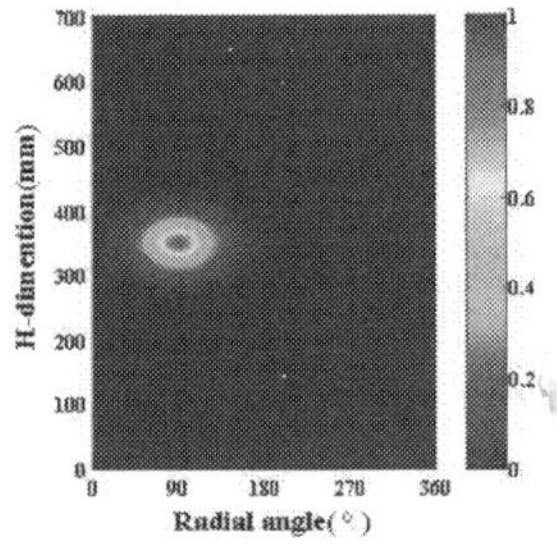

Table 3. The AE source localization for thin-walled steel cylinder based on beamforming method with rectangular sensor network in wave velocity $c = 4900$ m/s.

6. Conclusion

This study investigates AE source localization in plate-like structures using the beamforming network method and probability-based diagnostic imaging. Some conclusions based on the analysis of the localization results are drawn below.

(1) The beamforming technique is effective for AE source localization without the TOA or TDOA. Unlike other localization techniques, it is simple because it avoids the acquirement of detailed information on signals. Through the beamforming network algorithm, the AE source location information is highlighted in one coordinate system as energy distribution.

(2) An HT-based signal processing method was developed. The testing results demonstrate that the HT-based signal processing method is reasonable and capable of obtaining the energy

distribution information of the AE source. Moreover, a probability diagnostic imaging technology is introduced to improve localization accuracy. The field value of a pixel stands for the probability of the presence of damage at the point of the structure corresponding to the pixel, and the degree of probability is often calibrated through different gray scales.

(3) When the wave speed is known, the beamforming technique can accurately determine the source localization, no matter how far away from or close it is to the boundaries of the structure. Even if the wave velocity is unknown, the localization can also be confirmed by comparing the energy distribution or peak values under different wave speeds, as proven accurately in this study.

(4) The algorithm can also be applied to AE source localization in thin-walled steel cylinders pipe-like structure. However, its applicability in long axial length structures is yet to be verified.

Acknowledgements

The authors are grateful for the financial support from National Natural Science Foundation of China (NSFC) under Grant numbers 51278083, 51478079, and Science and Technology Project of Fujian Province (Grant number 2012Y01010047).

Author details

Dongsheng Li*, Mengdao Jin and Quanming Feng

*Address all correspondence to: lidongsheng@dlut.edu.cn

School of Civil Engineering, Dalian University of Technology, China

References

[1] A. Peter, A. Fehervary. Evaluation of acoustic emission from pressure vessels with planar. Theoretical and Applied Fracture Mechanics. 1986;5:17–22.

[2] P. Pellionisz, P. Szüc. Acoustic emission monitoring of pressure vessels. International Journal of Pressure Vessels and Piping.1993;55:287–294.

[3] D. Aljets, A. Chong, S. Wilcox, K. Holford. Acoustic emission source location on large plate-like structures using a local triangular sensor array. Mechanical Systems and Signal Processing. 2012;30:91–102.

[4] T. Kundu, H. Nakatani, N. Takeda. Acoustic source localization in anisotropic plates. Ultrasonics. 2012;52:740–746.

[5] A. Tobias. Acoustic emission source location in two dimensions by an array of three sensors. Nondestructive Testing. 1976;9:9–12.

[6] B. Castagnede, W. Sachse, K.Y. Kim. Location of point like acoustic emission sources in anisotropic plates. Journal of the Acoustical Society of America. 1989;86:1161–1171.

[7] T. Kundu, S. Das, S.A. Martin, K.V. Jata. Locating point of impact in anisotropic fiber reinforced composite plates. Ultrasonics. 2008;48(3):193–201.

[8] T. Kundu. Acoustic source localization. Ultrasonics. 2014;54:25–38.

[9] Archana Nair, C.S. Cai. Acoustic emission monitoring of bridges: review and case studies. Engineering Structures. 2010;32(6):1704–1714.

[10] Marwa Abdelrahman, Mohamed K. ElBatanouny, Paul H. Ziehl. Acoustic emission based damage assessment method for prestressed concrete structures: modified index of damage. Engineering Structures. 2014;60: 258–264.

[11] G.C. McLaskey, S.D. Glaser, C.U. Grosse. Beamforming array techniques for acoustic emission monitoring of large concrete structures. Journal of Sound and Vibration. 2010;329(12):2384–2394.

[12] T. He, Q. Pan, Y. Liu, X. Liu, D. Hu. Near-field beamforming analysis for acoustic emission source localization. Ultrasonics. 2012;52(5):587–592.

[13] H. Nakatani, T. Hajzargarbashi, K. Ito, et al. Locating point of impact on an anisotropic cylindrical surface using acoustic beamforming technique. In: In 4th Asia-Pacific Workshop on Structural Health Monitoring; Melbourne,Australia,December 5–7, 2012.

[14] D.H. Xiao, T. He, Q. Pan, et.al. A novel acoustic emission beamforming method with two uniform linear arrays on plate-like structures. Ultrasonics. 2014;54:737–745.

[15] J. Rose. Ultrasonic waves in solid media. Cambridge: Cambridge University Press; 1999.

[16] Young-Han Kim, Dae-Hyun Kim, Jung-Ho Han, Chun-Gon Kim. Damage assessment in layered composites using spectral analysis and Lamb wave. Composites: Part B. 2007;38:800–809.

[17] Friedrich Moser, Laurence J. Jacobs, Jianmin Qu. Modeling elastic wave propagation in waveguides with the finite element method. NDT & E International. 1999;32(4):225–234.

[18] Buli Xu,Victor Giurgiutiu. Single mode tuning effects on lamb wave time reversal with piezoelectric wafer active sensors for structural health monitoring. Journal of Nondestructive Evaluation. 2007;26:123–134.

[19] T.R.Hay, R.L.Royer, H.Gao, et al. A comparison of embedded sensor Lamb wave ultrasonic tomography approaches for material loss detection. Smart Materials and Structures. 2006;15:946–951.

[20] C. Zhou, Z.Q. Su, L.Cheng. Quantitative evaluation of orientation-specific damage using elastic waves and probability-based diagnostic imaging. Mechanical Systems and Signal Processing. 2011;25:2135–2156.

[21] H.K. Peng, G. Meng, F.C. Li. Modeling of wave propagation in plate structures using three-dimensional spectral element method for damage detection. Journal of Sound and Vibration. 2009;320:942–954.

[22] M.G. Silk, K.F. Bainton. The propagation in metal tubing of ultrasonic wave modes equivalent to Lamb waves. Ultrasonics. 1979;17:11–19.

Chapter 4

Use of Guided Wave Thickness Resonance for Monitoring Pipeline Wall Thinning Using an Internal PIG

Ángela Angulo, Slim Soua and Tat-Hean Gan

Additional information is available at the end of the chapter

http://dx.doi.org/10.5772/62985

Abstract

Experimental data and modelling results for pipeline wall thinning confirm a classification of Guided Wave (GW) propagation and detected features based on signal amplitude. This interpretation leads to a decision on a follow up inspection based on High, Medium or Low priority. The severity of defects must be determined; achievable by examining the signal amplitude as a function of metal loss. Specifically, if resonance can be obtained at a particular frequency, the operator can identify the wall thickness loss through the reflected GW energy amplitude. Previous research presented in this chapter identified a suitable strategy to deploy this thickness resonance technique, starting from dispersion curves (DC) development, to the analysis of the thickness loss effect on the DC, and experiments that prove the effectiveness of the methodology.

Keywords: long-range ultrasonic, pipe thinning monitoring, corrosion, high-frequency-guided waves, thickness resonance technique

1. Introduction

Around 0.5 million kilometres of buried oil pipelines in Europe carry hazardous fluids [1] often at high pressure and temperature. In Europe alone, up to four million gallons of oil per year are leaked into the environment due to corrosion and mechanical damage. Pipeline spills of hazardous fluids into the environment outnumber all other sources (e.g. tanker spills in oceans, etc.) combined. The pipeline network in Europe is increasing at a rate of about 1000 km per year [2]. With this rapid expansion and the existing ageing pipeline (>30 years old), there is a growing growing challenge to maintain its structural integrity. Finding and repairing pipe damage before catastrophic failure occurs, particularly as buildings encroach on pipeline sites, is crucial. Existing inspection using 'intelligent pigs' to find potentially harmful damage and repair or replacement by digging trenches to expose the pipe is very difficult at river, rail and road crossings. Yet, a typical 100-km pipeline might cross six rivers, four railways and three motorways as well as numerous other road crossings [3]. Between 1990 and 2000, there were an average of 75 pipe leakage incidents releasing oil onto land, into rivers and into underground water[4]. To address this issue, pressure is being put on pipeline operators to find new inspection technologies, which provide early warning of pipes at risk of failure [5]. In Brussels, European Union officials have urged 15 government members to begin applying new inspection, re-welding, and repair rules and technology to stem this pollution [6].

Oil and gas transmission lines are generally owned by specialist distribution companies. They usually have only a small technical and engineering base so most of their maintenance activities are outsourced. This is particularly true in the areas of inspection, non-destructive testing (NDT), machining and welding [7]. Estimates put these markets as growing annually by 10% [8]. If current methods of ultrasonic NDT were deployed, aiming to achieve full volume coverage, the position of conventional ultrasonic probes, for example, compression or electromagnetic acoustic transducer (EMAT) probes, would have to be adjusted every few millimetres to keep them in contact or in close proximity with the walls through long lengths of pipeline making it extremely difficult to achieve in practice with a small robot.

Most pipelines are buried in the ground and, once buried, every effort is made to leave them there. Excavation is not only expensive but the site conditions under which any repairs, re-welding or replacements are made are so poor that new defects can be introduced. Excavation is even more problematic when the pipeline is buried under rivers, railways and roads, where damage is more prevalent. Inspection and any repairs need to be conducted with the minimum of disruption, avoided by internal inspections using inspection robots (pigs) although repairs necessarily require the pipeline to be excavated.

The pigs occupy the entire cross section of a pipe due to the size of the sensor collar assembly needed to provide 100% volume coverage. Pigs can cope with moderate changes in diameter and moderate bends in the pipeline but there is a large variation in pipe sizes, for example, standard welded steel pipelines for gas/crude/oil product have internal diameters between 150 and 350 mm while larger pipes have an internal diameter from 500 to 1380 mm. Therefore, a matching pig is required for each pipe size, and for larger diameters the pigs tend to be very bulky. About one-third of all pipelines worldwide are considered unpiggable by existing smart pigs because of difficult access (launching and receiving facilities), restrictions due to valves, substantial changes in diameter along the length of the pipeline, small radius bends, dented or collapsed areas, excessive debris or scale build-up, impassable fittings, low operating pressure, low flow or absence of flow and other configuration issues [9].

The most common type of pig uses Magnetic Flux Leakage (MFL) techniques. These techniques are sensitive to changes in the flux of a strong magnetic field induced in the pipe wall as the 'pig' passes through. Sensors detect leakage of the flux from the pipe wall due to changes in the volume of the wall caused by corrosion, cracks, welds, bends and other geometric effects. MFL pigs however cannot distinguish acceptable anomalies from defects, or determine whether the indication is from the external or internal surface. It also cannot determine the circumferential location of the signal. This technique is also limited to pipes not encased by steel conduits, which is often the case for road and rail crossings.

Using Long Range Ultrasonic Testing (LRUT), sensors need to be adjusted only every 50-150 m, the typical attainable propagation range of GW waves in pipelines, , thus facilitating mechanical adaptation. Data volumes are greatly reduced compared to conventional ultrasound. LRUT can also be deployed internally for pipeline inspection [10]. The advantage of LRUT is a data volume orders of magnitude lower than conventional ultrasound to provide 100% pipe volume coverage [11]. Results can potentially be processed faster with less labour required subsequent to the pigging operation. This advantage is achieved at the expense of loss in sensitivity, with minimum detectable defect sizes much larger than are achievable with conventional ultrasound. with conventional ultrasound. However, LRUT allows serious defects needing immediate attention to be detected far more rapidly.

The identification of defects remains difficult and misinterpretation of indications can have disastrous consequences. For example [12], misinterpretation of a dent in a 16″ gasoline line as an innocuous weld manufacturing flaw led to the release of 237,000 gallons of fuel into Whatcom Falls Park, Washington State, USA, which subsequently ignited killing three young boys and landed BP-Amoco with a $3million penalty [13]. To guarantee proper evaluation of all indications detected with the 'intelligent pig', the pipe must be 'dug out', exposed and tested from the outside with more sensitive NDT techniques such as ultrasonics. This is applied manually and are not sufficiently accurate for Fitness for Service (FFS) assessment. Any potentially damaged section of the pipe is therefore replaced with a new section or 'pup-piece', which is welded into the pipeline at either end with butt welds. Certainly, the excavated pipeline must be inspected again before being re-buried.

2. Background

The LRUT or MFL approaches determine and extract damage-sensitive features from the signal using different signal-processing algorithms. A pattern recognition technique is then required to classify the damage and estimate its severity. It is important to note that Structural Health Monitoring (SHM) always involves the use of threshold values to decide on damage presence in the structure. The choice of the threshold is usually application-dependent and typically relies on some false-positive probability estimation [14]. It is assumed that a signal-to-noise ratio of 6 dB is required for detection in order to reduce the number of false indications. With this in mind, the required sensor detection performance can be evaluated [13]. However, it is not an objective assessment as the nature of the defects themselves requires more detailed interpretation. understanding. For this reason, the requirement of the implementation of GW SHM must involve the detailed characterisation and understanding of the response from the various types and sizes of defects[15]. In [16], the authors assess the feasibility of a monitoring process on an aluminium plate and steel pipe, and damage was simulated by changing the boundary conditions of each structure [14]. It is also crucial to identify a formal classification routine that characterise flaw severity using GW. Damage can be incrementally introduced into the structure, and at each depth multimode wave signals can relate the changes in received signal due to mode conversion and scattering from the flaw.

Lamb wave tomography reconstructions can be used by incorporating several different analysis techniques including wavelet-based feature extraction, and formal pattern classification to create a fully-automated analysis scheme designed to locate, size and identify the severity of unknown flaws. Variations of Lamb wave propagation reflect changes in effective thickness and material properties caused by structural flaws as corrosion, fatigue cracks and voids that can then be mapped via a reconstructed tomographic image [17]. In [18], an extension of classical imaging techniques that take advantage of the chirplet-based matching pursuit algorithm was presented. For non-dispersive propagation, an accurate localisation can be obtained. Even if specific low-dispersive modes are injected, mode conversion at discontinuities can generate dispersive modes superimposed on the targeted modes. This effect significantly complicates the measurements and demonstrates the need for pattern recognition algorithms which can be trained through either modelled or experimental data. The fundamental concept of class distribution within each feature space was discussed in [19] and features were correctly identified with respect to their severity. A linear spread of classes would allow new data corresponding to an intermediate flaw depth to fall correctly between classes, in order to identify a feature space where the classes are mostly linearly distributed.

The approach using torsional ultrasonic GW has strong potential for prognostics-based structural health management due to strong correlation and relationship between damage size change and the signal deviation demonstrated using the error function (erf). This process is effective in monitoring crack growth and shows an evaluation of the Probability of Detection (PoD). The Euclidian distance, which is defined essentially as the signal-to-baseline ratio, was used.

3. Numerical modelling analysis

In this section, the theoretical background to the numerical modelling of GW is presented. Elastodynamic equations govern the dynamic behaviour of a homogeneous linear elastic material. The weak formulation is detailed and provides the equations solved numerically to describe the wave propagation in the pipes. The cross section of the pipe is divided in the thickness direction into layered elements, and waves in the propagating direction z are described by the orthogonal function $e^{(i\xi z)}$, where ξ is the wave number. The mth eigenvalue ξm of the Eigen system derived here denotes the wave number of the mth mode. The elastodynamic variational formulation will be described leading to the discretization of the Finite Element formulation [20]:

$$\int_{\Omega} \delta\varepsilon_{ij}.C_{ik}.\varepsilon_{kj}\partial\Omega + \rho\omega^2 \int_{\Omega} \delta U_i.U_i\partial\Omega + \int_{\Sigma} \delta U_i.\left[\sigma_{ij}.n_j\right]\partial\Sigma = 0$$

The proposed formulation allows the description of wave propagation in complex geometries including oval cross sections induced by external pressure acting on offshore pipes. This aspect will be thoroughly described in this section.

The modelling work presented next predicts the propagation of the GW modes in an isotropic structure containing discontinuities. The key point is to decompose the returning signal into the mode content with regard to the wave-number scale and the frequency. This is obtained through the use of two-dimensional (2D) Fourier transform applied to A-scans obtained for discretised locations along the structure's length.

In the stationary regime domain model, it is possible to plot the frequency spectrum of the GW and compare it to the spectrum of a structure containing a discontinuity. The discontinuity considered here was a thinned region of about 20cm length, representing a corroded or metal loss area.

The second part of the modelling work is based on transient lamb wave excitation in the structure with a discontinuity. The choice of the frequency range of operation uses conclusions obtained in part 1 of the work. The aim of this part is to prove how realistic the stationary calculation results are. A Fast Fourier transform is used to extract the mode contents from the A-scans.

The final part of this work is based on the 2D fast Fourier transform, proven to be an efficient tool of mode identification. The content of the obtained A-scans is clearly affected by the discontinuity and more modes are generated dependent on the discontinuity thickness (and therefore the remaining thickness). Finally, structures with different notches depths were tested.

3.1. The stationary regime

It is possible to separately excite the symmetric and the asymmetric waves in a plate; this is achieved based on the following choices:

- To excite an S0 mode, tangential symmetric and normal asymmetric displacements should be excited.
- To excite an A0 mode, normal and tangent displacement excitation should be inverted.

The statement of the modelling problem uses the following assumptions:

- The solution is obtained in the harmonic domain; this supposes the excited wave to have a single frequency.
- The solution is obtained based on the stationary solver using Finite Element Analysis (FEA).
- The meshing is 2500 elements along a 1 m length.

Figure 1 shows the separation of the resulting symmetrical and asymmetrical displacements. It is proved that the resulting S0 displacement mode is normal; however, the resulting A0 displacement mode is tangent. At this stage, the FEA model seems consistent with the initial assumptions.

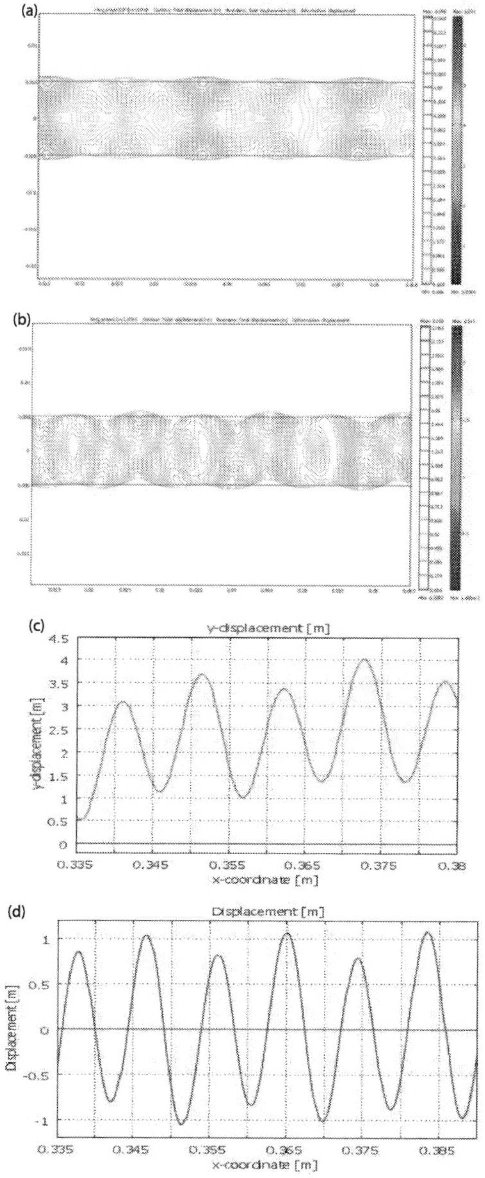

Figure 1. Displacement plots obtained for selective excitation of S0 and A0 modes, respectively. (a) Displacement for S0 mode at 307kHz. (b) Displacement for A0 mode at 307kHz. (c) Displacements along neutral plane for S0 excitation. Lambda=9.7mm. (d) Displacements along neutral plane for A0 excitation. Lambda=8.7mm.

http://dx.doi.org/10.5772/62985

In the next step, a frequency sweep is performed in order to generate a spectrum of the resulting A-scan. The previous excitation models were used and the frequency spectra of the S0 and A0 modes were plotted for a homogeneous plate (constant thickness).

Figure 2 shows the superposition of the frequency spectra for the frequency range [200 - 400kHz] and the displacement resonances. In this specific frequency range, we expect to obtain the natural resonance. The displacement plot shows that the obtained results are in agreement with the previous assumption and the expected modes are obtained.

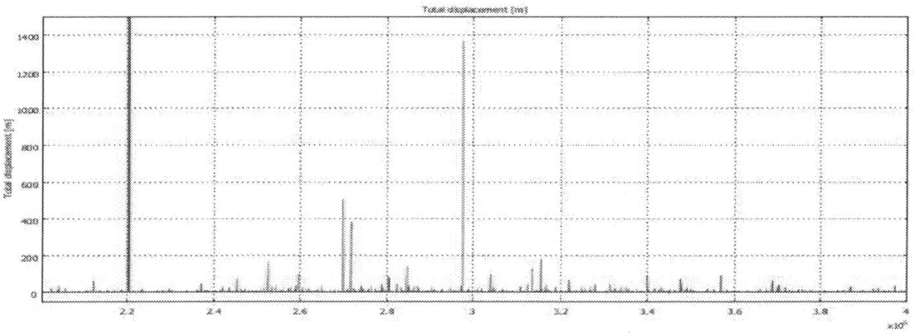

Figure 2. Superposition of spectra obtained for S0 (green) excitation and A0 (red) excitation in a plate.

The next step is to build a spectrum of the plate with a 5mm thickness discontinuity (**Figure 3**) and compare it to the previous spectrum of the defect-free plate (**Figure 4**).

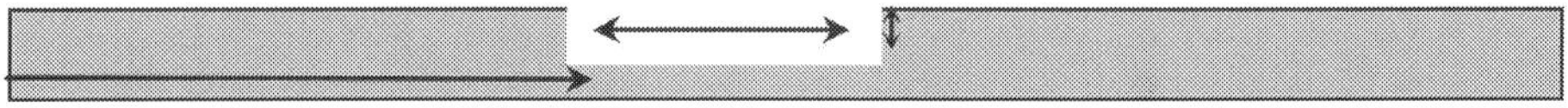

Figure 3. Thickness reduction in a plate of 10 mm.

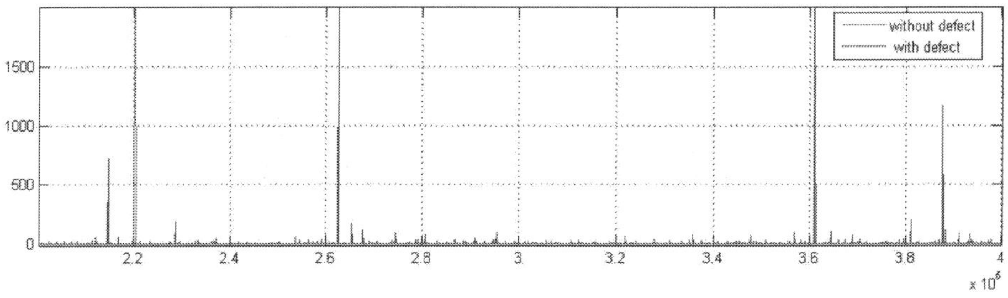

Figure 4. Superposition of spectrum obtained for uniform plate and 'defective' plate.

3.2. Displacement visualisation of resonant frequency for plate with thickness change

Localised thickness reduction was introduced along 10 cm length, 40 cm from the excited edge (**Figure 3**). In the frequency spectrum of the defective plate, compared to the spectrum of the defect-free plate, resonances occurred at 214.8, 262.2, 360.9 and 387.7 kHz. The displacement plots of the excited GW at these frequencies are presented in **Figures 5–8**, the normal and tangential displacement components of an excited pure Lamb (A or S) mode are plotted at the neutral plane of the main plate (e = 10 mm). One of the displacement components is supposed to vanish; the other component is then obtained in the form of the propagating Lamb mode with a constant wave number.

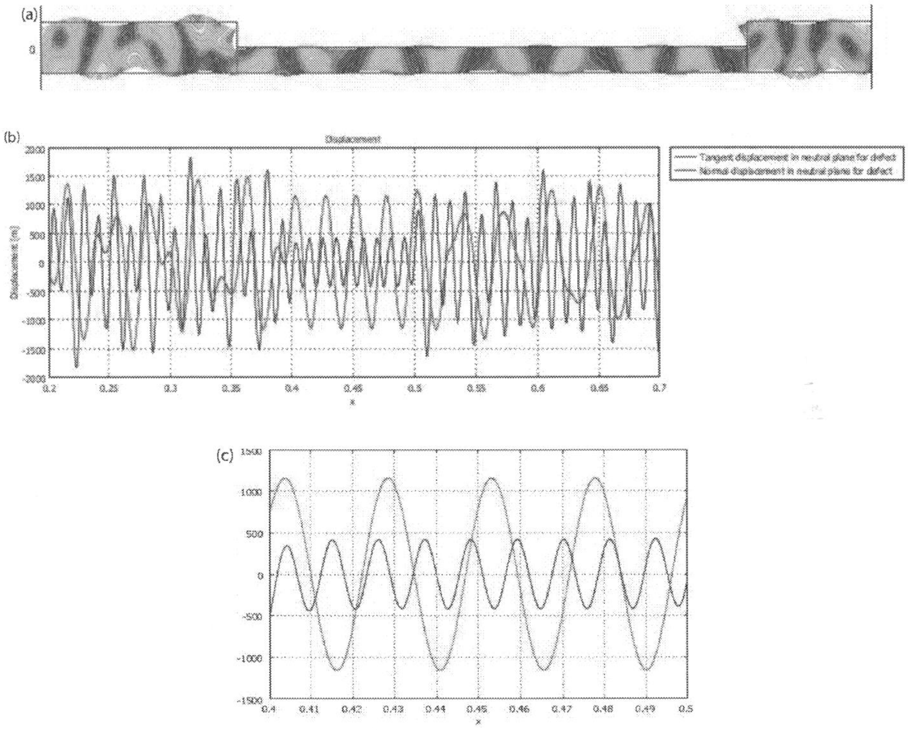

Figure 5. Displacement plots of the excited guided waves at 214.8 kHz. (a) Wave propagation-amplitude plot. (b) Displacement components at y=5 mm. (c) Zoom on the thinner zone.

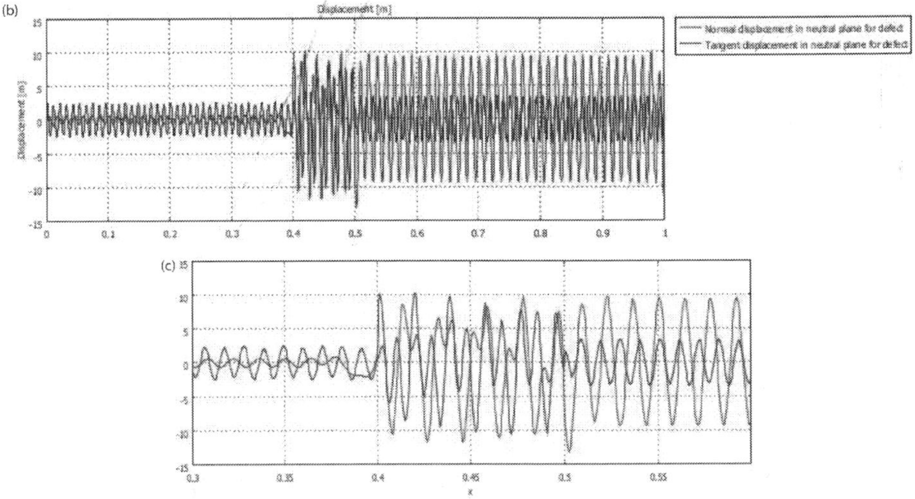

Figure 6. Displacement plots of the excited guided waves at 262.2 kHz. (a) Wave propagation-amplitude plot. (b) Displacement components at y=5 mm. (c) Zoom on the 5mm thickness zone.

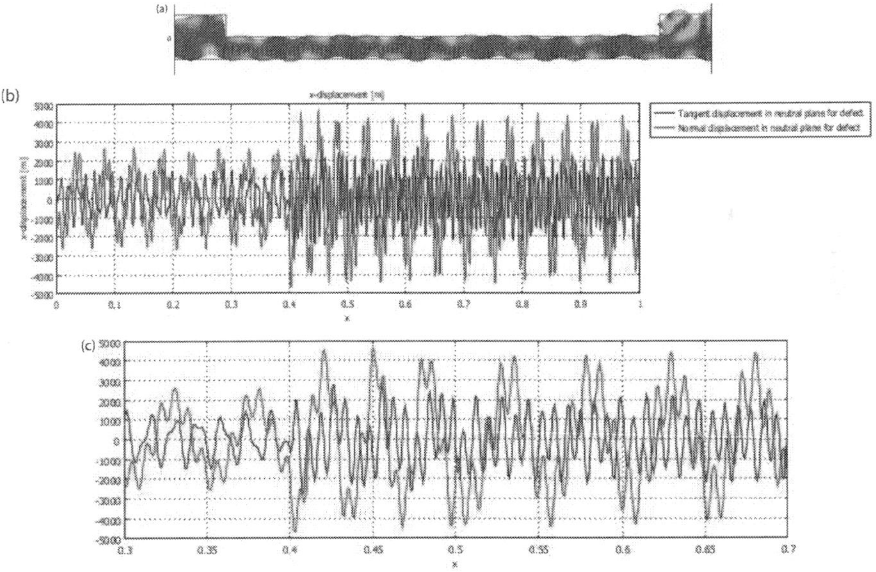

Figure 7. Displacement plots of the excited guided waves at 360.9 kHz. (a) Wave propagation-amplitude plot. (b) Displacement components at y=5 mm. (c) Zoom on the 5 mm thickness zone.

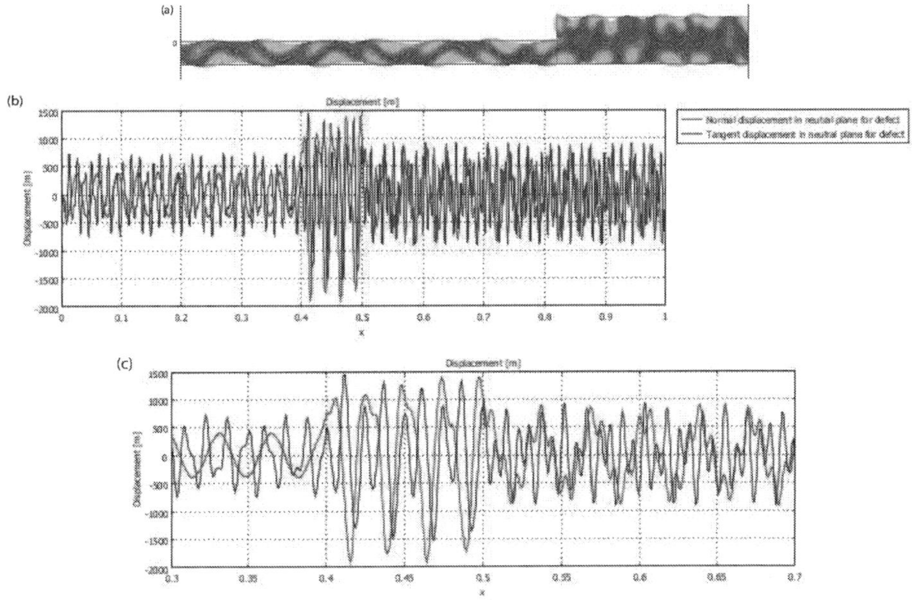

Figure 8. Displacement plots of the excited guided waves at 387.7 kHz. (a) Wave propagation-amplitude plot.(b) Displacement components at y=5 mm. (c) Zoom on the 5 mm thickness zone.

In the case of Lamb wave generation at the thickness reduction part of the plate, both S and A modes can propagate. Their detection is enabled by localising the region where the displacement shows a constant wave number.

The following conclusions may be drawn:

- At the excitation frequency of 214,9 kHz, the notch zone (5-mm thickness) shows a constant wave number. Both symmetric and asymmetric modes are observed in **Figure 9**.
- At the excitation frequency of 262,4 kHz, the main plate (10-mm thickness) shows a constant wave number. Both symmetric and asymmetric modes are observed.
- Both frequencies of 387,6 kHz and 360,9 kHz do not show any Lamb wave generation. Those resonances can be due to the natural frequencies of the plate.

This part of the work provided a good understanding of the behaviour of the defective plate when Lamb waves are generated. The following conclusions may be drawn:

- Although the excitation was designed to be mode-selective, both asymmetric and symmetric waves were excited.
- Each region of the plate with different thickness can excite Lamb wave separately from the rest of the structure.

3.3. Generation of the S0 and A0 modes in the transient regime

Results obtained in the transient regime must be validated. This excitation generates both symmetric and asymmetric modes. For different excitation frequencies using 10 cycle tune burst signal, the Fourier transform of the time response collected in the position at 0.5 m at the neutral plane of the plate was plotted (**Figure 9**).

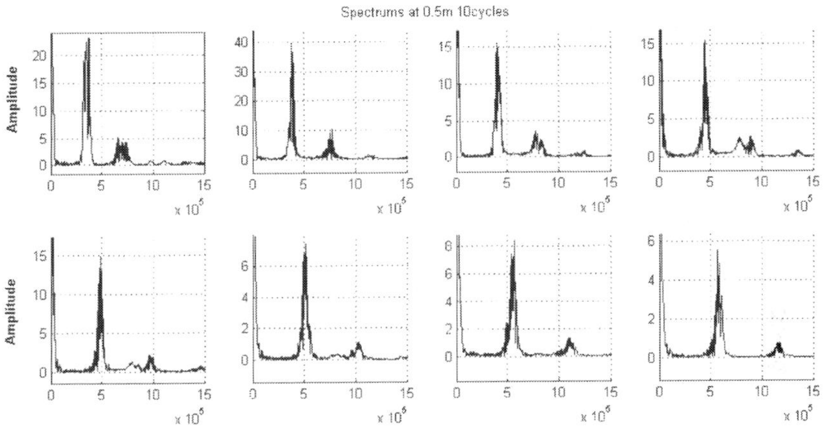

Figure 9. The different excitation frequency spectra.

Figure 9 clearly shows the existence of the two modes S0 and A0. The overlap of the different spectra is presented in **Figure 10** and shows that the maximum amplitude is obtained at a varying frequency, which depends on the excitation frequency.

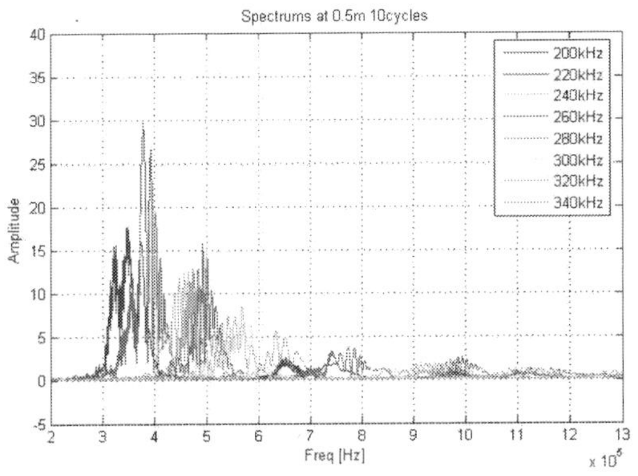

Figure 10. Superposition of the different excitation frequency spectra.

The results show the change in the frequency content introduced by the defect. Additionally, the work needs to be validated with experimental results. A 2D fast Fourier transform (2DFFT) is also used to identify the mode content of the different A-scans and compare them to 'Disperse' results.

3.4. Mode identification 2DFFT

The 2DFFT algorithm transforms a time/space representation into frequency/wave-number representation. It gives information about the mode content and the how dispersive the mode is.

Following on, a plate of 10-mm thickness is considered, excited using single point excitation. **Figure 11(a)** shows the results obtained using the 2DFFT, for the defect-free plate. S0, A0 and A1 modes are generated and this result is consistent with 'Disperse' results.

Figure 11(b) displays the dispersion curves for a plate that presents a thickness reduction of 5 mm, and shows the generation of additional modes S0 and A0.

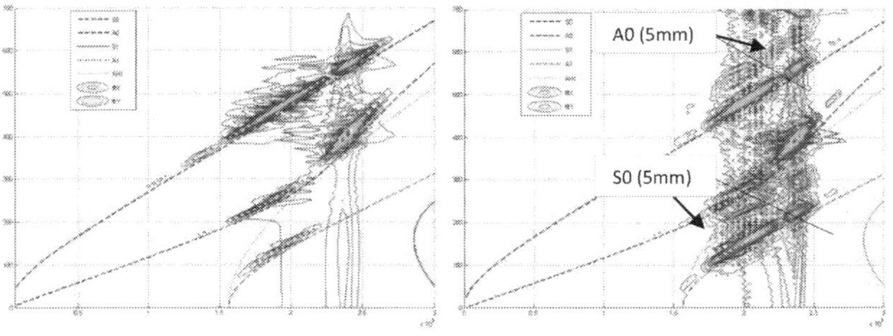

Figure 11. Validation of the 2dfft 2DFFT procedure for a plate of 10-mm thickness. (a) Compared to Disperse results. (b) 2DFFT for a 10-mm plate with 5-mm defect thickness over 20mm length.

The processing procedure is then clearly in good agreement with 'Disperse' and describes well with the problem under study. It was therefore used in the following development.

3.5. Transient modelling of the S0 mode propagation

Consider an excitation using seven-cycle tune-burst signal. Following 'Disperse', the dispersion curve for S0 gives a wavelength of 10 mm at 318,5 kHz and the dispersion curve for A0 gives a wavelength of 10 mm at 279,8 kHz.

The excited frequency bandwidth is about 100 kHz (**Figure 12**). It is noticed that A0, S0, A1 and S1 are generated and are in good agreement with 'Disperse'. AH1 and SH1 are not present in the results because the 2D model does not take the horizontal displacement components into account.

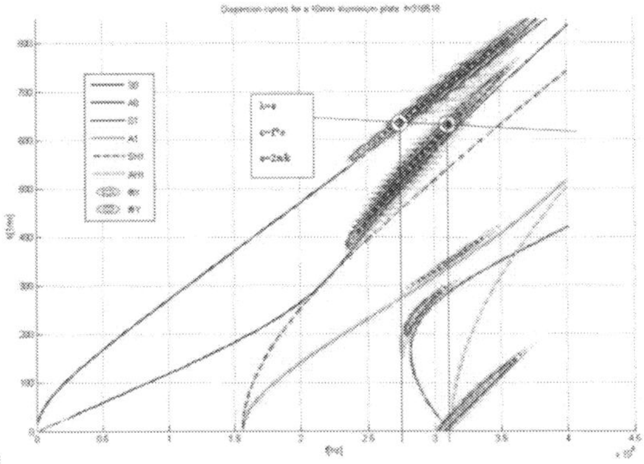

Figure 12. Dispersion curve generation using 2DFFT superimposed with Disperse results for a plate with constant thickness. The dashed lines are obtained from Disperse, and the surfaces plots are obtained from the 2DFFT.

Figures 13 and **14** give an extrusion plot at the frequency of 318,5 kHz and the A-scan at 0.2 m from the excitation location and the extrusion plot of the excited waves in the plate. Consider an excitation using seven-cycle tune-burst signal applied to the plate (10-mm thickness) with the 5-mm-thickness reduction. By applying the 2DFFT on the A-scan results at each position increment (**Figures 15** and **16**), the mode content of the GW generated in the plate is obtained. This is carried out for the frequencies of excitation of 279,8 and 318,5 kHz.

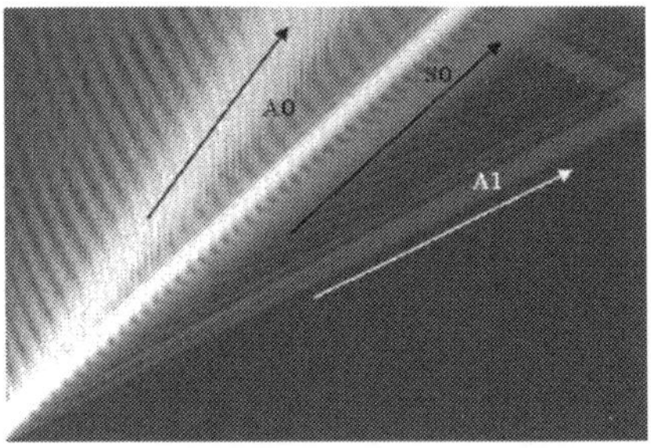

Figure 13. Extrusion of the wave propagation in the plate for the excitation frequency, $f = 318{,}581$kHz.

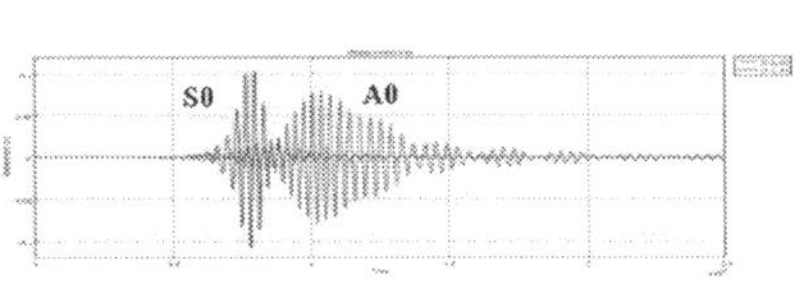

Figure 14. Displacement plot at 0.2 m from the excitation point.

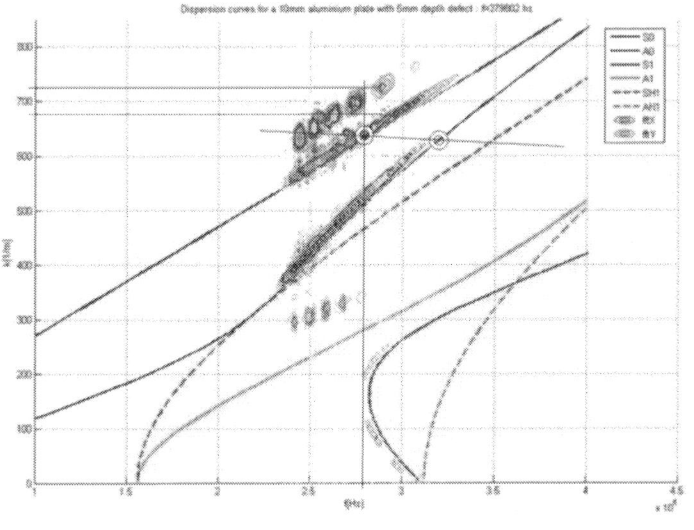

Figure 15. Dispersion curves generation for the defective plate at the excitation frequency 279.8 kHz using 10-cycle tune-burst signal. The dash lines are obtained from Disperse, and the surfaces plots are obtained from the 2DFFT.

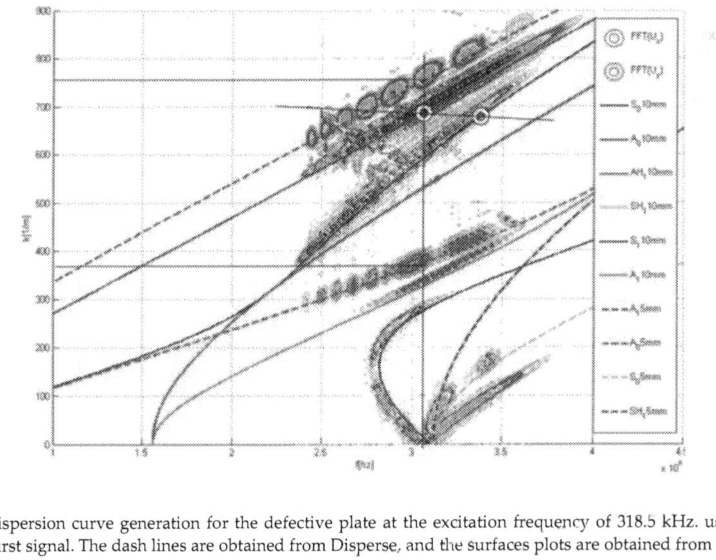

Figure 16. Dispersion curve generation for the defective plate at the excitation frequency of 318.5 kHz. using seven-cycle tune-burst signal. The dash lines are obtained from Disperse, and the surfaces plots are obtained from the 2DFFT.

The displacement extrusion and the A-scan at 0.2 m are presented in **Figures 17** and **18.** Additional modes near the dispersion curve A0 and S0 may be seen. These modes represent the A0 and S0 modes of the part of the plate with 5-mm thickness. The results from 'Disperse' confirm this conclusion.

Figure 17 shows a complex wave generation and wave reflection. The main problem is that both symmetric and asymmetric waves are excited.

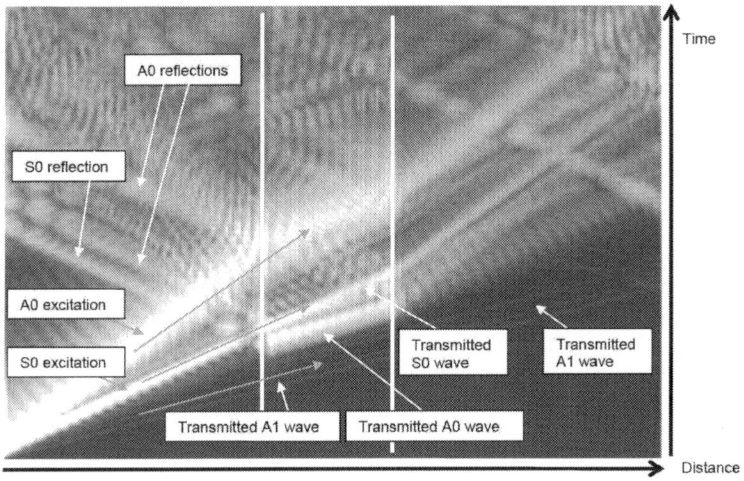

Figure 17. Extrusion of the displacement in the plate for the excitation, f = 318,5 kHz, 5-mm-notch thickness delimited by the white lines.

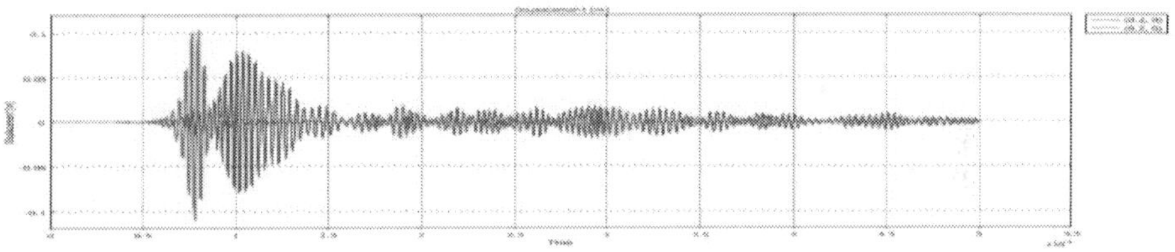

Figure 18. A-scan of the wave propagation in the plate for the excitation at the frequency, f=318.5kHz.

3.6. Mode selection

Using the procedure of mode selection described above, the same excitation was applied to A plate of10mm thickness with a 5mm thickness reduction.

The excited mode is now totally symmetrical, the dispersion curve plot in **Figure 21** is less complex and clearly shows the generation of additional modes due to the thickness reduction zone (**Figures 19** and **20**).

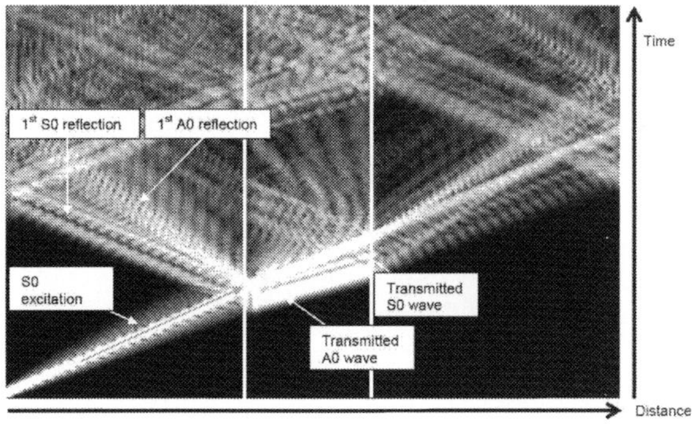

Figure 19. Extrusion of the displacement in the plate for the symmetric excitation at f = 318,5 kHz, 5-mm-notch thickness delimited by the white lines.

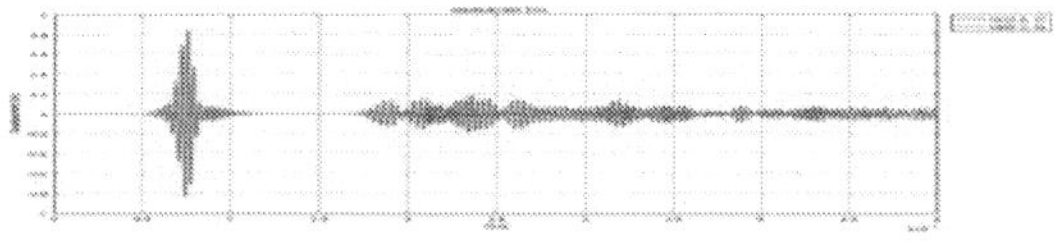

Figure 20. A-scan of the wave propagation in the plate for the symmetric excitation at the frequency, f = 318,581 Hz.

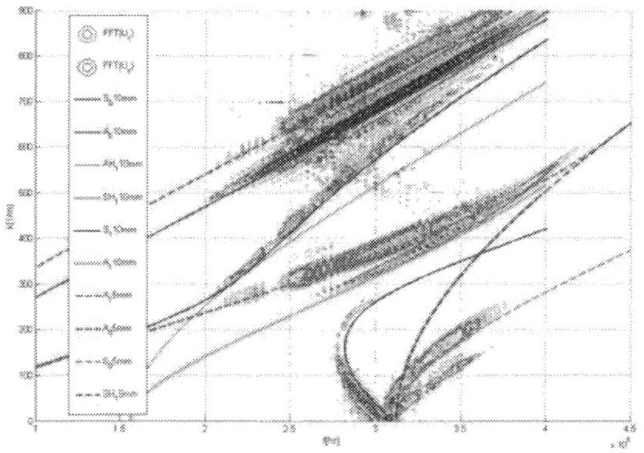

Figure 21. Dispersion curves generation using 2DFFT superimposed with Disperse results.

Figure 21 proves the generation of the additional S0 and A0 modes introduced by the thickness reduction zone. This conclusion is also demonstrated in the A-scan of **Figure 20**. As a result, the defect thickness is directly linked to the generated GW properties. This conclusion will be discussed in further detail in the last part of this work.

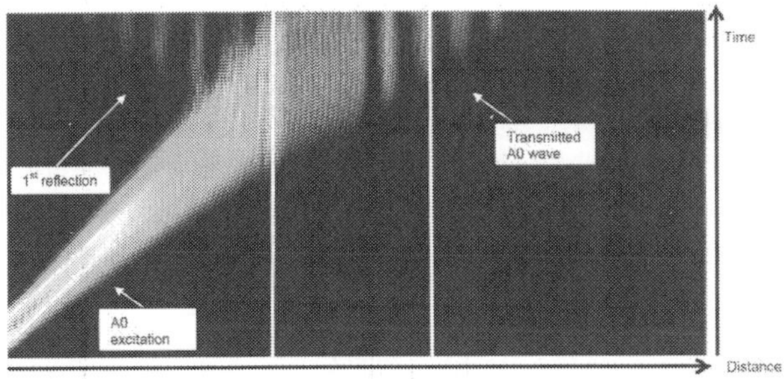

Figure 22. Extrusion of the displacement in the plate for the asymmetric excitation at f = 318,581 Hz, 5-mm-notch thickness delimited by the white lines.

Figure 22 presents the extrusion displacement plot of an asymmetric excitation of the plate with thickness reduction; it appears that this mode of excitation is too dispersive and cannot be used efficiently for inspection.

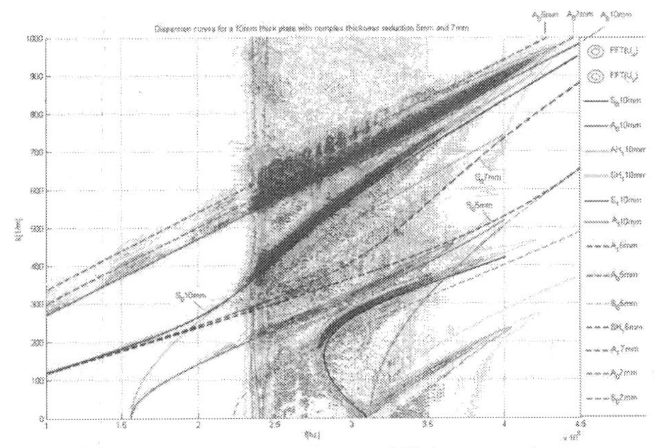

Figure 23. Dispersion curves generation using 2DFFT superposed to Disperse results. The dashed lines are obtained from Disperse, the surface plots are obtained from the 2DFFT.

Here a plate with reduction zones of different thicknesses was examined, which models a corroded area in an aluminium plate. The thickness levels considered here are 10, 7 and 5 mm in 15-cm-length regions as seen in **Figure 24**(top). For the point excitation mode, the extrusion of the displacement in **Figure 24** and the A-scan at 0.2m from the excited edge in **Figure 26**, are presented showing complex wave propagation and diffraction pattern. **Figure 25** presents the dispersion plots which seem to capture the change in thickness. The dispersion curves S0 and A0 are sensitive to the change of thickness from 10mm, 7mm to 5mm. It is clear that this feature relates to the thickness levels in the plate, and each thickness level introduces additional modes with a phase velocity specific to the thickness reduction. From this phenomenon the NDT operator can get an idea of the corroded zone if the A-scan received contains additional echoes. In the practical work these additional echoes can be considered as noise or unknown indications, but following the present analysis it appears that they have significant meaning as indication of the presence of corroded areas or thickness modifications.

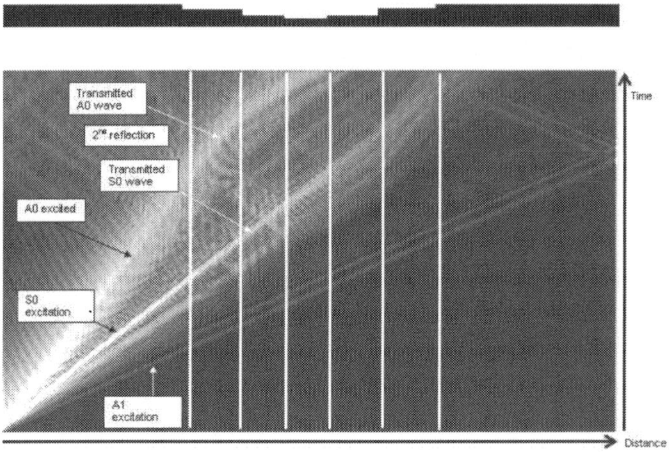

Figure 24. Extrusion of the displacement in the plate for the non-symmetric excitation at f = 318,5 kHz. 5, 7 and 10mm notch thickness delimited by the white lines.

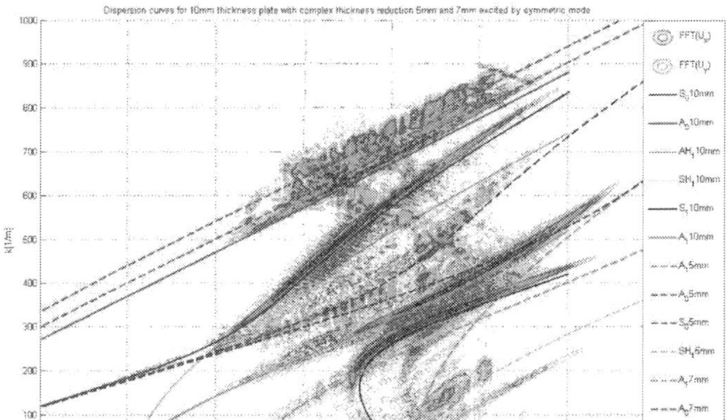

Figure 25. Dispersion curves generation using 2DFFT superimposed with Disperse results.

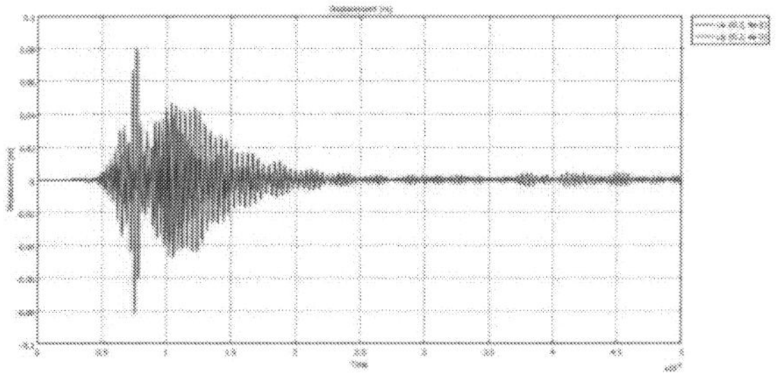

Figure 26. A-scan of the wave propagation in the plate with multi thickness defect for the non-symmetric excitation at the frequency f = 318,5 kHz.

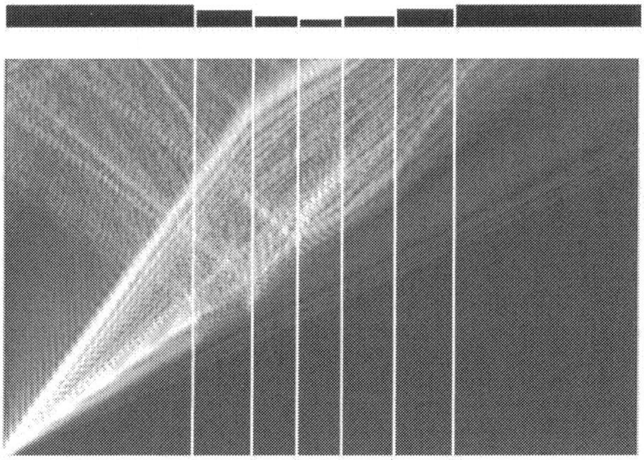

Figure 27. Extrusion of the displacement in the plate with complex thickness reduction zones.

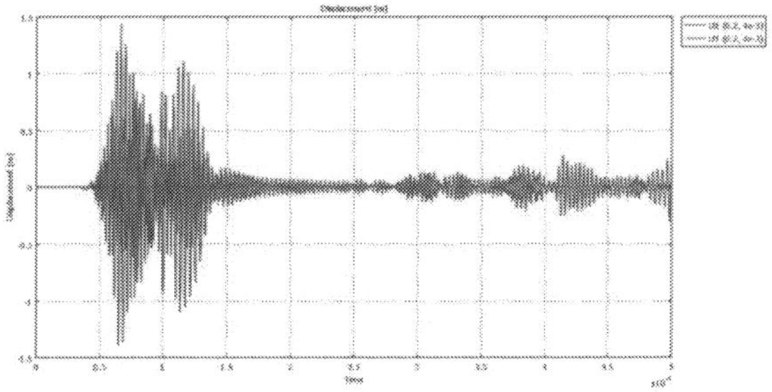

Figure 28. Dispersion curves generation using 2DFFT superimposed with Disperse results.

The final step in this work is to use symmetrical excitation instead of point excitation. The aim is to reduce the complexity of the excited mode which reduces the complexity of the A-scan analysis. The results are given in **Figures 27** and **28**. The echoes that indicate the presence of the corroded zone may be easily located in the A-scan of **Figure 28**.

4. Methodology

In this study the groundwork for the development of a fast, screening tool to detect corrosion defects by evaluating the effects of wall thinning has been presented. To analyse the response of LRU to corrosion, a two-dimensional finite element model was constructed including gradual thinning of the pipe walls. Experimental testing is necessary to validate the modelling results.

There is a wide range of applications where LRU can be applied to detect corrosion. Inspection tools can be used for in-service NDT and SHM. As mentioned at the beginning of this chapter, one of the most relevant uses of LRU is its application to pipeline integrity assessment. The experimental test to be described is based on this principle. Defect growth has been monitored by a novel system designed for this purpose. Specific tooling was designed in order to relate the effectiveness of the methodology (**Figure 29**). The system performs the total volume inspection more accurately than current methods of ultrasonic NDT inspection.

Figure 29. LRU collar for pipeline monitoring.

4.1. Experimental setup

Defects simulating wall thickness loss were introduced in the pipe at distances of approximately 1.2 and 2.4 m from the end of the transducers located at one end of the pipe segment. These defects simulated a through-thickness crack growth and an increased depth of a corroded area. The experimental results are presented in **Figure 30(a)**. The end-of-pipe amplitude reflection is reduced due to the increase of thickness loss at the location of the

http://dx.doi.org/10.5772/62985

simulated corrosion. In other regions of the A-Scans, there are several significant changes in the signal amplitude due to the loss of wall thickness. The most important aspect of the A-scan data is the amplified amplitude change due to a resonance that follows a trend with increasing deviation from the case with no damage. Investigating at the frequency domain of the signals with and without loss of wall thickness (**Figure 30(b)**), there is a greater loss of energy over a wider range of frequencies around 30 kHz compared to the results from the inserted cuts.

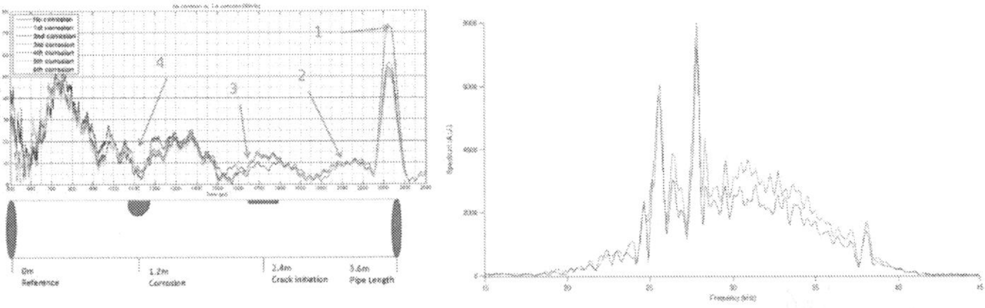

Figure 30. Simulated wall thickness loss in the outside surface of the 10″ pipe. The internal monitoring system being included inside the pipe for trials.

The additional presence of thickness loss on the outer wall of the pipe results in a loss of energy in the received end-of-pipe reflection. In the presence of wall thinning, there are significant changes in the amplitude of the A-scans at several points of the time-base. This is due to the creation of more complex wave behaviour within the pipe as the torsional wave interacts with the inserted feature.

4.2. Results

To validate the technique and the tooling, a crack was simulated with a slot increasing in depth in seven equal increments until the entire wall thickness was penetrated (**Figure 31**, measurements 1–7). A secondary defect simulating wall thickness loss was introduced on the outside wall of the pipe in six steps (**Figure 31**, measurements 8–13).

The results presented in an A-scan (**Figure 31(a)**) shown an end of-pipe reflection, which is reduced in amplitude in the presence of cracking and wall thinning, and as the width of the simulated corrosion is increased. In other regions of the filtered A-scan, where anomalies are present (**Figure 31(a)**), there are several significant increases in signal amplitude due to the loss of wall thickness. The A-scan amplitude changes, following an increasing trend deviation from the defect-free signal.

Investigating at the frequency domain of the signals with and without loss of wall thickness, there is a greater loss of energy over a wider range of frequencies around 30 kHz compared to the results from the inserted cuts and thinning.

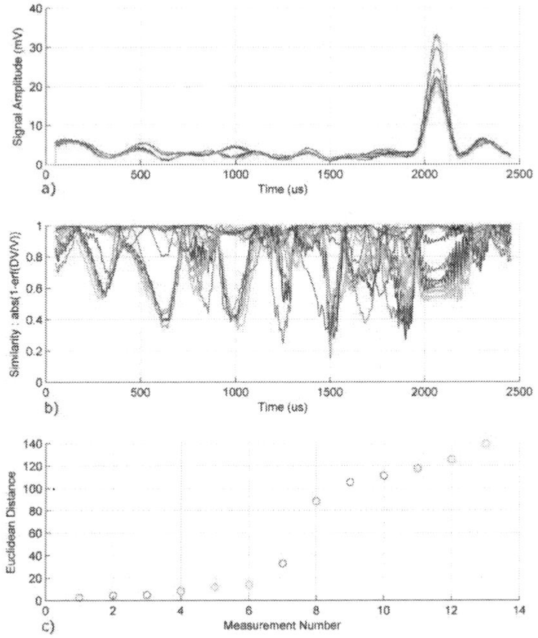

Figure 31. LRU monitoring results. Damage detection (a), similarity (b) and Euclidian distance (c).

Changes on the A-scan's amplitude reveal the presence of wall thickness loss. These different areas are evaluated by analysing the similarity of the signals at a given location.

Similarity is equal to one minus the error function of the rate of change of the A-scan's amplitude:

$$Similarity = 1 - erf\left(x\right) = \frac{2}{\sqrt{\pi}}\int_{0}^{x} e^{-t^2}\,dt, where x = \frac{V - V_0}{V_0}$$

Figure 31(b) shows the correlation between each of the signals acquired in comparison with the flawless pipe reference signal. The similarity is measured from one to zero, with one being equivalent to 100% equivalence of signals. The error is calculated in one dimension; thus, the similarity between two points on the real line is the absolute value of their numerical rate of change difference.

The Euclidean distances were analysed for one dimension, comparing the obtained signals on each measurement against the defect-free pipe signature. The distance between two points in one dimension is simply the absolute value of the difference between their coordinates. Mathematically, this is shown as follows:

http://dx.doi.org/10.5772/62985

$$Euclidean\ distance = \sqrt{\left| p1 - p2 \right|^2}$$

where $p1$ is the first coordinate of the first signal and $p2$ is the first coordinate of the second signal.

In **Figure 31(c)**, the results show an increment in the distance between each signal measured and the reference signal when there is defect growth. The most important aspect considered is that the distance on amplitude change follows a trend with increasing deviation from the defect-free signal. GW allows rapid screening of long lengths of pipe to detect external or internal corrosion. Both large cracks and corrosion are detectable with GW technology. Depending on the position of the crack, the feature can go unnoticed when using only one GW mode. Corrosion can be detectable upwards from as little as 10% of cross section loss under certain conditions. The accuracy of detection is decreased by many factors such as distance, attenuation, scattering, absorption or leakage.

5. Conclusions

This chapter presents results on a 254-mm internal diameter pipe of 10-mm-wall thickness with a 82.9 cm2 Cross sectional Area (CSA). Artificial crack defects showed that significant increases in the signal-to-noise ratio of defect echoes can be obtained with this approach. Corrosion defects can be detected with thinning greater than 10% of wall thickness.

The alternative of using the resonance of LRU to obtain the thickness of each thinned area was investigated. The frequency at which each peak occurs is linked to the actual thickness producing the resonance. The method is to sweep the test frequency over a range which covers the nominal thickness and possible thinning ranges. The amplitude of the spectrum of the output across this range is defined and the existence of peaks determines the remaining wall thickness of the pipe. Different types of damage can be identified due to changes in received signal amplitude of the A-Scans compared to the time baseline.

This technique is then translated into a monitoring strategy that uses an internal monitoring robot (pig) in order to deploy ultrasonic sensors enabling the GW to propagate from inside the pipe and to reach unpiggable areas.

Author details

Ángela Angulo*, Slim Soua and Tat-Hean Gan

*Address all correspondence to: angela.angulo@twi.co.uk

TWI Ltd., Integrity Management Group, Condition and Structural Health Monitoring, Granta Park, Great Abington, Cambridge, UK

References

[1] Western European cross-country oil pipelines, Lyons D. 25yr performance statistics, CONCAWE oil pipelines management group. Report no. 1/02.

[2] Financial Times comment 'Growth in pipelines', May 2002

[3] Non-destructive Testing: An Expanding Market. 2002. Published by Business Communications Company. Web: www.buscom.com.

[4] European industry oil safety performance, statistical summary of reported incidents, Martin D E, Prepared for the CONCAWE, 1999. Safety Management Group

[5] Performance of cross-country oil pipelines in Western Europe, statistical summary of reported spillages, Davis P M. CONCAWE Oil pipelines Management Group, 1999. Special task force on pipeline spillages.

[6] European Commission, Directorate General, Institute for Health and Consumer Protection, 'Physical and Chemical Exposure Unit European Information System on Electromagnetic Fields Exposure and Health Impacts'. Final Report, February, 2005.

[7] Future growth prospects for inspection and repair, Fulop G. www.buscom.com.

[8] Lincoln Electric annual report 2001, www.lincolnelectric.com

[9] In-line inspection of hard-to-pig pipelines, Bukman F., Schmidt R. The Pipeline Pigging Conference, Amsterdam, The Netherlands, June 1997.

[10] PIGWaves – In–line service for internal inspection of unpiggable buried oil pipelines using long range ultrasound guided waves in fifty metre segments, TWI Case Study. www.twi-global.com.

[11] TWI Integrity Management Group, Collaborative Overview, July2014.

[12] Pipeline rupture and subsequent fire in Bellingham, Washington, Accident report NTSB/PAR-02/02, June1999.

[13] PR Newswire Aug 7th 2000. Following is a summary of press releases about Texas, Oklahoma, Louisiana and Arkansas companies and organizations transmitted today by PR Newswire.

[14] Review of Guided-wave Structural Health Monitoring. Ajay Raghavan and Carlos E. S. Cesnik The Shock and Vibration Digest 2007; 39; 91

[15] Strategies for Guided-Wave Structural Health Monitoring, A.J Croxford, P.D Wilcox, B.W Drinkwater, G Konstantinidis. 8 November 2007. DOI: 10.1098/rspa.2007.0048.

[16] A unified approach for the structural health monitoring of waveguides. Xuan Zhu, Piervincenzo Rizzo. Structural Health Monitoring November 2012 vol. 11 no. 6, 629-642.

http://dx.doi.org/10.5772/62985

[17] Ultrasonic Lamb wave tomography in structural health monitoring, Zhao X., Royer R., Owens S., Rose J. Smart Materials and Structures 2011, Volume 20, Issue 10, article id. 105002, 10 pp.

[18] Dispersion-based imaging for structural health monitoring using sparse and compact arrays, N Quaegebeur, P Masson, D Langlois-Demers and P Micheau. January 2011, Smart Materials and Structures, Volume 20, Number 2.

[19] Ultrasonics Classification of flaw severity using pattern recognition for guided wave based structural health monitoring, Corey A. Miller, Mark K. Hinders. Ultrasonics Volume 54, Issue 1, January 2014, Pages 247–258.

[20] Modelling of long range ultrasonic waves in complex structures. Slim Soua, Septimonette Chan, Tat-Hean Gan. TWI Ltd, Granta Park, Great Abington, Cambridge CB21 6AL, UK. BINDT annual conference 2008, 15-18 September 2008, Macclesfield, Cheshire, UK.

Chapter 5

Application of Acoustic Emission Technique in the Monitoring of Masonry Structures

Jie Xu, Qinghua Han and Ying Xu

Additional information is available at the end of the chapter

http://dx.doi.org/10.5772/63093

Abstract

The application of acoustic emission (AE) technique in monitoring the safe condition is a useful technique in steel and concrete structures, whereas its application is restrained in masonry structures due to the layered property. Qualitative and quantitative analyses were investigated in this research to improve the AE application in masonry structures. For quantitative analysis, an improved localization method is proposed to give more reliable crack localization results. In the proposed method, the parameter ξ on the behavior of inhomogeneity of the monitored structure could minimize the unavoidable propagation delay caused by the layers in the masonry structure. The rest results approved the reliability of the proposed method in masonry structures. For qualitative analysis, the parameter analysis, including the cumulative AE event, frequency distribution, time-scaling exponent, and *b*-value, was adopted to monitor one historical church and was approved to be useful.

Keywords: acoustic emission, masonry structures, quantitative analysis, qualitative analysis, crack localization

1. Introduction

Acoustic emission (AE) is the class of phenomena whereby transient elastic waves are generated by the rapid release of energy from a localized source or sources within a material. Clearly, an AE is a stress wave that travels through a material as the result of some sudden release of strain energy. By investigating their origin and characteristics, AE techniques provide an insight into the deterioration processes of a tested object, especially for the monitoring and nondestructive testing of the structural integrity and general quality.

Significant research work has been published in relation to the use of AE sensors for monitoring the health of structures in civil engineering, especially for concrete structures. The most important applications of AE to concrete elements started in the late 1970s, when the original technology developed for metals was modified to suit heterogeneous materials [1,2]. Some fundamental studies with small-scale specimens have shown that, in principle, AE analysis is an effective method for damage assessment [3]. Research from various laboratory loading tests to full-scale models of real structural components was intended to relate observed AE characteristics to failure mechanisms in reinforced or prestressed concrete [4–7]. Some applications were tried to evaluate the structural integrity, load-carrying capacity, or eventual failure for real civil engineering structures such as concrete bridges and so on [8]. Also, a continuous monitoring of a whole structure is applicable (e.g. to detect wire breaks of prestressing tendons) [9]. The development of the corresponding AE equipment, such as wireless monitoring system with microelectromechanical system (MEMS) sensors [10], also furthers the AE research.

As AE gained importance for concrete structures, its application for masonry structures is gradually employed [11]. Huge numbers of ancient masonry structures, such as towers and bridges, are present all over of the world. To preserve this inestimable cultural heritage, a sound safety assessment should taken into account the evolution of damage phenomena [12]. In this respect, the AE monitoring technique can be highly effective. Some researchers [13] have successfully applied AE techniques to monitor masonry buildings, towers, and bridges. In these researches, the interpretation of AE rate has been put forward to monitor the criticality of the ongoing process, and the "*b*-value" analysis was used and proven as a useful method to determine the propagation of the cracks. Also, an ad hoc theory based on fractal concepts for assessing the stability of masonry structures from the data obtained with the AE technique is proposed. In the application [14], a series of multiring brickwork arches have been tested to assess the applicability of the AE technique for masonry arch bridges under static and long-term cyclic loading to study crack propagation and failure mechanisms. A series of short-term creep tests and compressive tests were performed on masonry columns by Verstrynge et al. [15], and they found that the damage accumulation parameter, calculated from the AE results, shows a linear increase in the function of the relative stress level.

However, unlike the conditions in concrete, AE applications in masonry are still at an early stage. The reason is that AE is applicable for homogeneous or quasi-homogeneous materials (e.g. metallic and concrete) with good acoustic transmission. However, its application on heterogeneous materials, such as masonry, is much more problematic, especially in the crack source localization problem. Attempts were made by Carpinteri et al. [16] by involving the location of cracks during an on-site monitoring campaign. Although a large amount of AE events were detected, only a small percentage of them could be located. Damage location in masonry arch bridges was performed by Tomor and Melbourne [17] by simply applying a large set of AE sensors and identifying the sensors at which most damage was detected.

This paper focuses on the applicability of AE technique in the monitoring of masonry structures based on quantitative and qualitative analyses. To address these issues regarding the heterogeneous problem, a series of pencil-lead break tests were carried out on a masonry bridge and a historical church was monitored by the AE technique. The results are discussed in this chapter.

2. Quantitative analysis

Quantitative analysis is also called the signal-based AE technique, and the time series of AE events are recorded along with their waveforms. Quantitative methods allow a more accurate characterization of the fracture process and try to describe the nature of a source using special methods, such as localization and moment tensor analysis. The goal of quantitative AE analysis is to enumerate and explain the source of an AE event. The principle is that AE burst signals within a given time window are detected by more than one sensor and grouped to an AE event that can be associated with an AE source. Source localization method is one of the important ones adopted in this method.

The crack localization of AE sources is important to evaluate the areas of active damage in the monitored structures. The localization issue is normally solved by triangulation methods based on acoustic signal trajectories [18]. Usually, these methods rely on the elasticity modulus, the propagation mode, and signal attenuation caused by the heterogeneity and anisotropy of the material. In this part, the classical localization method was summarized briefly.

2.1. Classical localization method

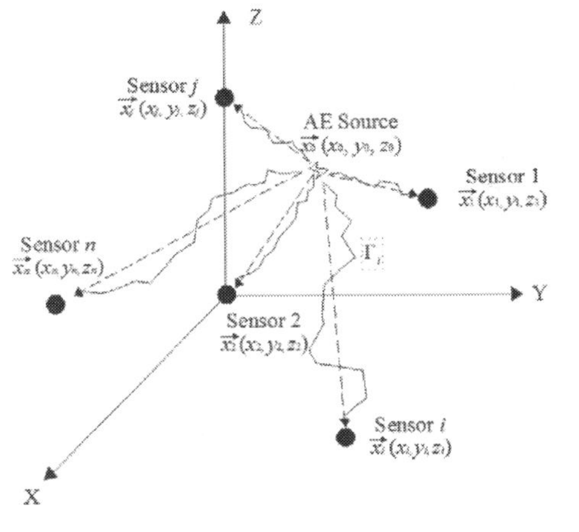

Figure 1. AE source localization by an array of n sensors.

In a theoretical model, with wave propagation velocity v_{Pi}, the arrival time t_i^* at $\vec{x}_i$, unknown coordinates $\vec{x}_0$, and origin crack happening time t_0 can be calculated by an integral along the real path Γ_i, as shown in **Figure 1**:

$$t_i^* = t_0 + \underbrace{\int_{\Gamma_i} (d\Gamma_i / v_{Pi}(r))}_{I}, \tag{1}$$

where $v_{Pi}(r)$ is the field of wave velocity in the monitored structures. If the material is homogeneous, Equation (1) can be simplified as

$$t_i^* = t_0 + \frac{|\vec{x}_0 - \vec{x}_i|}{v_p} = t_0 + \frac{\vec{x}_{i0}}{v_p}. \tag{2}$$

For each sensor i, there will be residual r_i between the observed arrival time t_i and the calculated arrival time t_i^*:

$$r_i = t_i - t_i^*. \tag{3}$$

If t_j is the arrival time at another sensor $\vec{x}_j$, the detected arrival time difference between sensors i and j is adopted. Usually, we have

$$r_i^* = \Delta t_{i1} - \frac{\vec{x}_0 - \vec{x}_i}{v_p} \ (i = 2, \ldots, n), \tag{4}$$

where subscript 1 is related to an arbitrarily chosen reference sensor (e.g. the closest one to the crack source).

If more than four arrival times are available for one event, the equation will be overdetermined. The least-squares method are adopted to minimize the residuals, where the error for (n-1) equations is only the sum of all squared time residuals:

$$\chi^2 = \sum_{i=2}^{n} (r_i^*)^2. \tag{5}$$

Considering Equation (5) is linear, and the issue can be solved iteratively until convergence [19].

2.2. Velocity field of AE-$v_{Pi}(r)$

As shown in Equation (1), the acoustic wave velocity field $v_{Pi}(r)$ is an important point in the calculation. The variation of velocity field $v_{Pi}(r)$ is affected mainly by the material property of monitored structures. If the structure is homogeneous, the wave propagation path from the crack point to the sensor is surely treated to be a straight one, shown in **Figure 2** by dashed line, and the velocity field $v_{P_i}(r)$ is also homogeneous. Thus, the velocity in the structure is a certain value in every path direction, and v_P can be calculated by Equation (5). The metal material is the typical one that can be considered as homogeneous, and Equation (2) can be applied well for it.

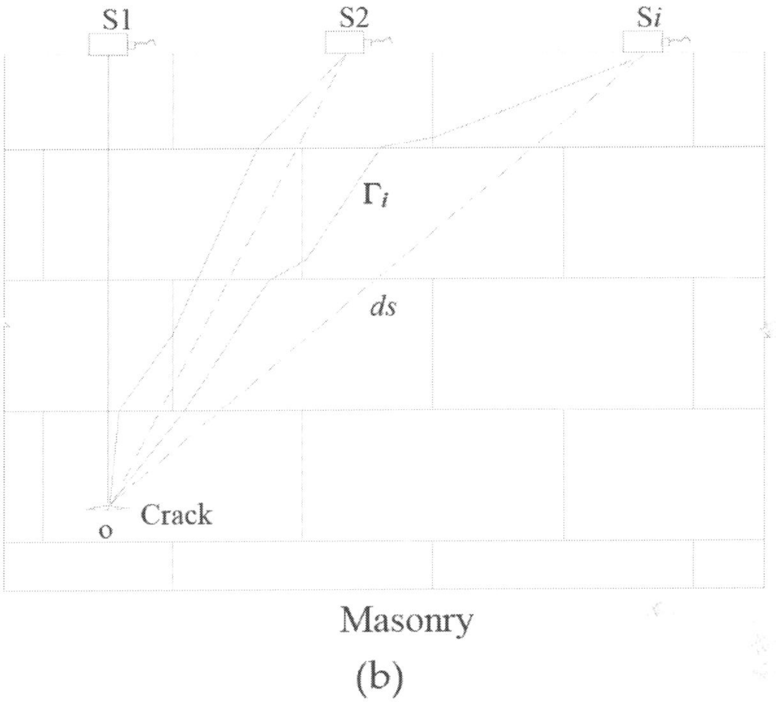

Figure 2. Differences between theoretical wave propagation path and the real path in masonry.

The refraction and reflection of the wave at the layers is unavoidable for masonry materials. Because the reflection phenomenon mainly leads to the decrease of the signal amplitude and of incident energy density, we only take propagation effects related to the refraction of AE wave into consideration. From **Figure 2**, the propagation delays (denoted as *P-D*) are

$$P-D=\frac{\Gamma_i}{v_\Gamma}-\frac{ds}{v_{ds}}=\frac{\int_{\Gamma_i} kdl}{v_{ds}}-\frac{ds}{v_{ds}} \tag{6}$$

where $k=\frac{v_{ds}}{v_\Gamma}$ and v_Γ are the propagation velocity along the real path Γ and v_{ds} is the counterpart along the theoretical path *ds*. Thus, we have

$$P-D=\underbrace{\frac{\int_{\Gamma_i} kdl}{v_{ds}}-\frac{\int_{\Gamma_i} dl}{v_{ds}}}_{Velocity-delay}+\underbrace{\frac{\int_{\Gamma_i} dl}{v_{ds}}-\frac{ds}{v_{ds}}}_{Geometrical-delay} \tag{7}$$

Normally, the difference of velocity in the two different paths for one sensor event is not so obvious and $k = 1.0$. In this case, only the effect of the geometrical delay is considered:

$$P-D=\frac{\int_{\Gamma_i} dl}{v_{ds}}-\frac{ds}{v_{ds}} \tag{8}$$

Generally, the *P-D* value reflects the deviation of the calculated path *ds* from the real wave propagation path Γ_i between the crack point and the sensor. The *P-D* value will increase as the traveling path increases, because the practical situations will be more complicated for propagating through more layers. In this point, the *P-D* value surely exists in masonry structures, which means that the traditional method based on Equation (2) cannot be used here directly.

2.3. Test set-up and velocity field modification

Several pencil-lead break tests based on a two-arch masonry bridge model are conducted in this research to improve the source localization problem in the masonry structure.

2.3.1. Monitoring system of AE

The AE monitoring set-up consisted of six piezoelectric (PZT) sensors and six control units and a PC-based multichannel monitoring system called SAMOS AEwin (Sensor-based Acoustic Multichannel Operating System) that is manufactured by Physical Acoustic Corporation (PAC). Appropriate AE sensor types are important in the fracture monitoring of concrete structures. Because concrete is known to be a highly attenuating material, lower-frequency sensors are suitable for AE studies. The maximum aggregate size in concrete for the present study is 20 mm, and 3500 m/s is normally used as sound velocity in concrete. Using the relation $V=n\lambda$, the frequency of the AE sensor to be used is desirable to be less than 180 kHz, and this frequency is within the range of 100 to 500 kHz. Considering this point, another R15A with highest sensitivity between 50 and 400 kHz (**Figure 3a**) was selected in the test. The gains of the preamplifiers and the acquisition system were set to 20 and 40 dB, respectively. The acquisition threshold was set to 40 dB to ensure a high signal-to-noise ratio to avoid background noise. Vacuum grease LR (high vacuum silicon grease) was used as a coupling agent to fix the sensors on the opposite surfaces of the concrete. The band-pass analog filter is set to between 20 kHz and 2 MHz, and the sampling frequency is set to 10 MHz.

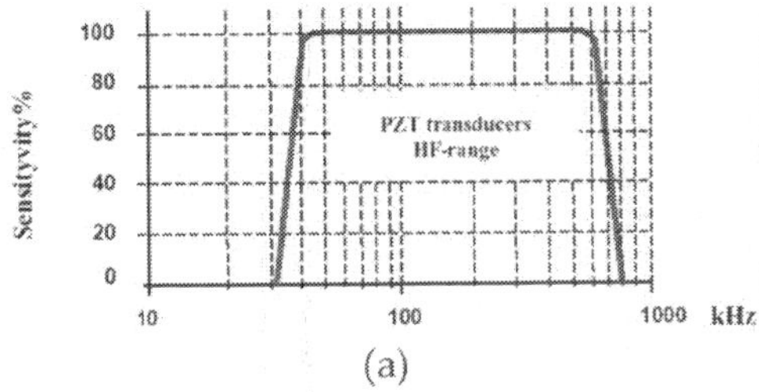

Figure 3. (a) AE sensor frequency bands and (b) acquisition system of AE system.

2.3.2. *Bridge model and material property*

A two-arch masonry bridge model was designed and the scale of the bridge model is 1:2 and it is 5.90 m long, 1.60 m wide, and 1.75 m high according to the theory of models. Bricks with a uniform size of 130×65×30 mm were made by hand. The mechanical properties of masonry brick and mortar were selected to better represent the real historical bridges. The bridge has two masonry abutments and a central pier. The abutments set on basements are built by concrete and anchored to the ground with special reinforcements, as shown in **Figure 4**. The details of the bridge can be found in our previous paper.

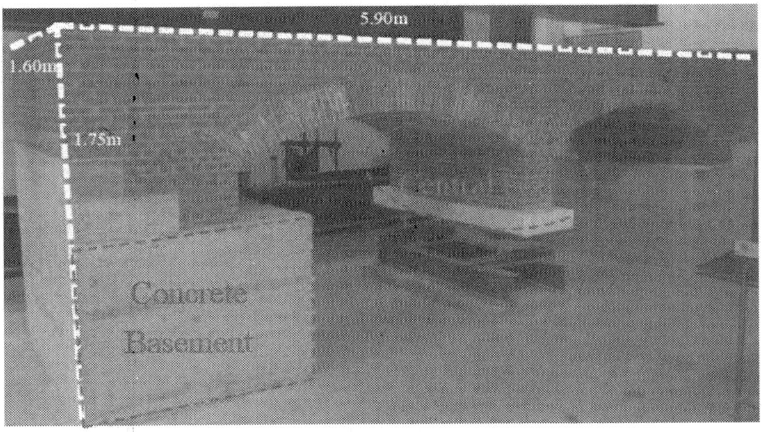

Figure 4. Bridge model and its dimension.

Before the tests, it is necessary to evaluate the mechanical properties of the bridge, such as the basic material (i.e. the masonry).

Figure 5. Mechanical tests for the masonry: (a) compression, (b) diagonal compression, (c) four-point bending test of arch, and (d) shear test.

As shown in **Figure 5**, different kinds of laboratory tests have been conducted to estimate the mechanical parameters of the bridge. Besides, tests about the mortar itself and on the concrete to support the abutments were also performed. The key mechanical parameters extracted are the Young's modulus E, tensile strength f_t, Poisson ratio v, tensile fracture energy G_F, and compressive strength f_c. The detailed information about the parameters related to our research can be found in **Table 1**.

Parameter	**γ [kg/m^3]**	**E [Pa]**	**ν**	**f_c [Pa]**	**f_t [Pa]**	**G_F [Nm]**
Value	1900	1.5×10^9	0.2	3×10^5	4.3×10^6	400

Table 1. Mechanical parameters of materials extracted from the tests.

2.3.3. *Tests of velocity field*

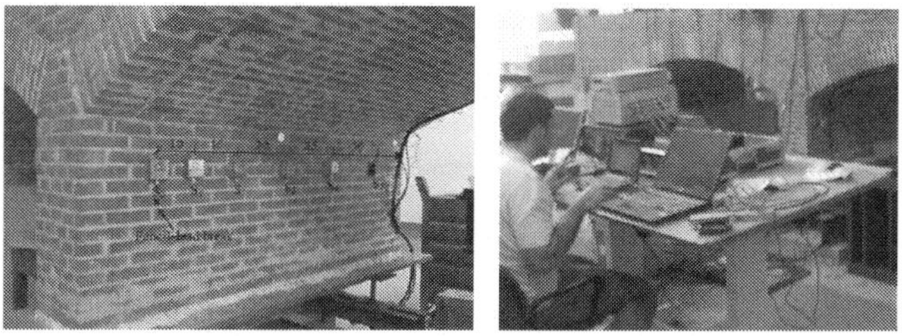

Figure 6. Wave velocity test: (left) the sketch of the pencil-lead break point and the sensor distribution and (right) scenario of the velocity test.

As shown in **Figure 6**, sensors (S_0-S_5) are adopted to monitor acoustic data, and the distance between two neighbored sensors increases from S_0-S_1 to S_4-S_5 with an increment value of 5 cm. During the test, the pencil was broken beneath the S_0 sensor, 5 cm away from the same surface, to analyze velocity propagation in the surface, as shown in **Figure 6**. At the same time, AE wave propagation inside the masonry bridge was also investigated through the same sensor array shown in **Figure 6**, and the pencil was broken on the opposite surface of the central pier but at the same corresponding position of sensor S_0.

The test results of the detected velocities are shown in Figures 7 and 8. The labeled velocity *V*-homogeneous is calculated based on the mechanical properties in Table 1 for comparison. The labeled velocity *V*-average is the average value of all the calculated velocities from the corresponding test.

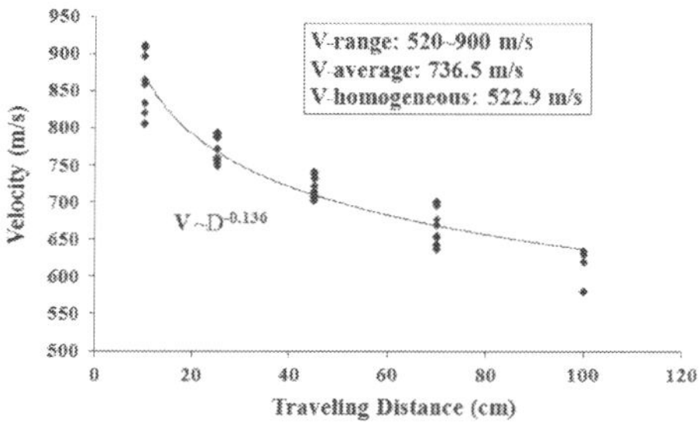

Figure 7. AE velocity on the masonry surface of the bridge model.

In **Figures 7** and **8**, the velocities, both on surface and inside the masonry, show that the velocity decreases clearly with the increasing propagating distance. The explanation is that the transient wave changes its mode from longitudinal to shear and/or to surface waves and vice versa due to the refraction and/or reflection during propagation in different phases. The longer the propagating distance is, the more mode change happens, which causes a bigger deviation of the actual wave propagation path Γ_i from the calculated one *ds*. This is just what we named the phenomenon of propagation delay in Equation (8).

From **Figures 7** and **8**, we can see that there is a great difference between the *V*-homogeneous and the *V*-average, so we cannot take the masonry as a homogeneous material in initial guess. In this case, the *V*-average from the field velocity test is used as the value in the initial guess to start the iteration of calculation for the source localization.

2.3.4. *Modified source localization method in masonry*

Based on the classical method, a modified method is proposed in this part. The basic idea of the source localization in masonry material is similar as that in concrete. However, modifications for propagation delay are implemented. We still take the geometry distance *ds* as the calculated path, as it is not possible to know the actual wave path Γ. However, time-delay modification can be made according to the velocity properties in **Figures 7** and **8** to reduce the effect of inhomogeneous property. In the proposed modified model, the classical model result in Equation (5) is modified into

$$\chi^2 = \sum_{i=2}^{n}(r_{i1})^2 = \sum_{i=2}^{n}\left[x_{10} - x_{i0} - (k_i t_i - t_1)v_1\right]^2, \tag{9}$$

where $k_i = (d_1 / d_i)^{\xi}$ is the modified factor used to modify the effects of propagation delay or the inhomogeneity. The parameter ξ, named as degree of the inhomogeneity, in k_i reflects the inhomogeneous degree of the material. The ξ is calculated from the pencil-lead break wave velocity filed test result, as shown in **Figure 7**. It shows the relation between the wave propagation distance and the calculated velocity. In ideal homogeneous materials, the value ξ is 0 because the wave velocity is a constant value with change of traveling distance. However, the value ξ will increase with the degree of the heterogeneity if the material is not homogeneous. In our research, the degree of the inhomogeneity ξ is 0.14 as calculated from the

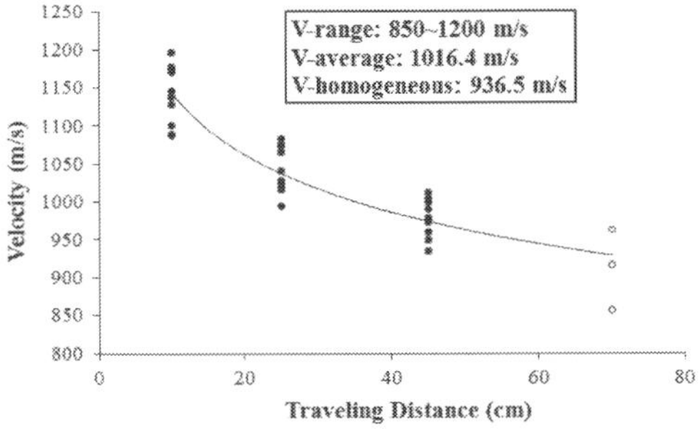

Figure 8. AE velocity inside the masonry model bridge.

relationship between the traveling distance and the velocity, as shown in **Figure 7**. The procedure to determine the crack source by the modified localization method is shown in **Figure 9**.

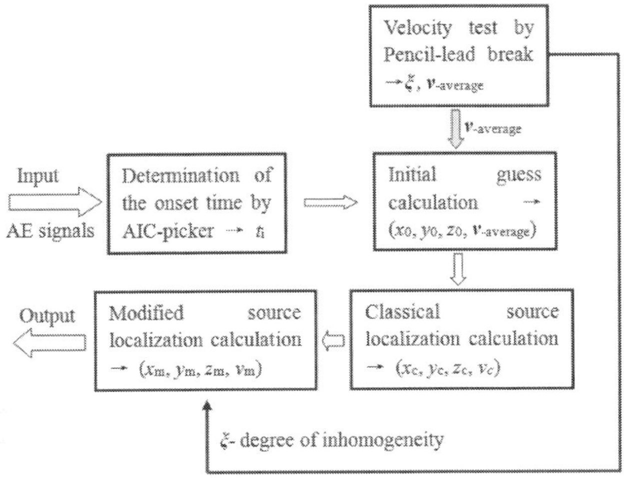

Figure 9. Flow chart of the proposed source localization method.

2.3.5. *Validation of the modified localization methods*

Pencil-lead break tests are conducted to produce AE source on the right-side surface of the masonry bridge. As shown in **Figure 10**, six sensors described in Section 2.3.1 are attached to the surfaces. Nineteen different points were selected to make a pencil-lead break (artificial

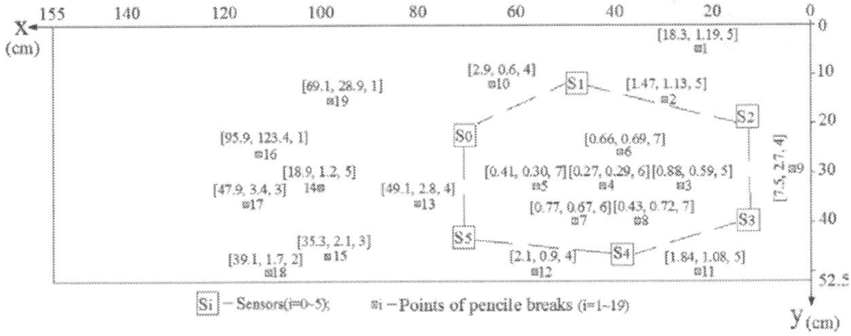

Figure 10. Crack source localization results from the traditional and the proposed modified methods. For [*a*, *b*, *c*], *a* is the average error from the traditional method, *b* is the average error from the modified method, and *c* is the source breaks used for each point.

source), and a pencil was broken eight times for each point, so totally 152 signal results are recorded. Nine hundred twelve AE events were obtained from the six sensors in this test and then all the 19 points were source localized following the procedure shown in **Figure 9**.

The localization results are shown in **Figure 10** for both the proposed modified and traditional methods. It is clearly shown that the location accuracy changes with the source break position. According to the accuracy, we divided the break points into three groups. As the dashed line shown in **Figure 10**, points 3 to 8 inside the central area of the sensor network have the best accuracy for localization for both methods. In the second group, points 2, 9, 10, 11, and 12, distributed on the nearby region of the sensor network, comprise the second group. Points in the third group are the rest of the points, far from the sensor network.

In the first group, the two methods both get the ideal result, the errors are all smaller than 6 mm, and most crack sources can be detected. In the second group, the proposed modified model gives better localized results than the traditional method. The errors value (*a* in **Figure 10**) in the traditional method is 15 to 75 mm, whereas the errors value (*b* in **Figure 10**) can be reduced to 6 to 27 mm in the proposed modified method. In this group, about half of the break sources can be monitored based on the value *c* in **Figure 10**. The result from the traditional method for the third group is hardly acceptable for its huge errors, but the proposed modified method can be still in good results. Although the errors are 3 or 4 cm, which are slightly large, the localized results are still acceptable if compared to the total size scale of the surface.

From the test results, compared to the traditional method, the proposed modified method can give better source localization results for the masonry structure in all the three group conditions.

3. Qualitative analysis

Qualitative analysis, also called the parameter-based AE technique, extracts only parameters but the signal itself. The parameters usually used are arrival time, amplitude, duration, number of oscillations, and so on. The typical methods employed in the parameter-based technique are ring-down counting and event counting, in which the event intensity is measured by the oscillation number NT, and the oscillation number NT increases with the signal amplitude.

To express the application of qualitative analysis, we applied the AE technique to monitoring one historical chapel, which is located in north Italy of Sacro Monte di Varallo. In this monitoring, our task is to investigate the stability of the painted plaster walls and injuries with the technical AE. The monitoring structure is the Chapel XVII of the Sacred Mountain of Varallo. The cumulative AE event, frequency distribution, time-scaling exponent, and *b*-value were adopted in this monitoring for the qualitative analysis.

3.1. Scenario of the monitored heritage chapel

The Sacred Mountain of Varallo is one of the oldest Italian constructions in its style. It was constructed in 1491 by San Francesco. The Sacro Monte was built on a rocky foundation, located on the slopes of Mount of Three Crosses, and River Sesia is on the left side. This natural formed terrace (600 m) soars up the historic center of Varallo (450 m) about 150 m.

The Sacred Mountain of Varallo consists of a basilica and 45 chapels, either isolated or inserted into the large monumental complexes, famous by more than 800 life-size painted statues, in wood and terracotta, which dramatically illustrate the life, passion, death, and resurrection of Christ. These interiors are vividly decorated with fresco paintings. **Figure 11** is a point view of Sacro Monte.

Figure 11. Overall view of the square of the courts.

Figure 12. Out view of the Chapel XVII.

The first monitored structure by AE technique was chosen as the Chapel XVII, which tells the story of the transfiguration of Christ on Mount Tabor, as shown in **Figure 12**. Chapel XVII was constructed starting from the foundations in 1572 but was finished only in the 1760s.

As shown in **Figure 13a**, the frescoes inside the Chapel XVII are the outcome of the brothers Montaldo, who also are the makers of the decoration of the dome of the Basilica of the Sacred Mountain. The statues attributed to Peter and John Francis Petera Varallo Camasco by Soldo were finished in the 1770s, as shown in **Figure 13b**.

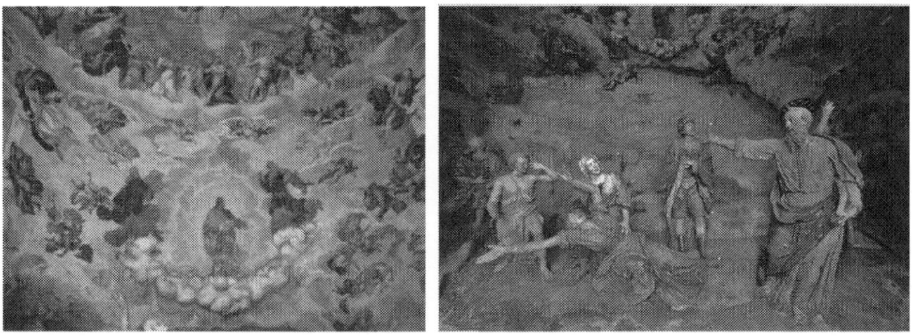

(a) (b)

Figure 13. Inside view of the Chapel XVII: (a) the frescoes and (b) the status of the Peter and John Francis Petera.

3.2. In-site monitoring by AE technique

The target is to set the sensors for the monitoring of AE signals from a lesion in the north wall of the chapel and a detachment of the fresco. On the other hand, some of the terracotta statues in the chapel were also analyzed (**Figure 14**).

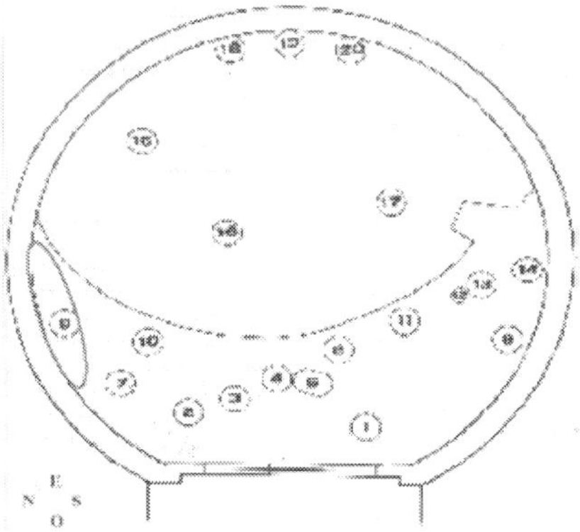

Figure 14. Top-to-down view of the Chapel XVII. The monitoring region is inside the red line.

Some necessary previous operations were carried out by professional restorers for attaching the AE sensors on the wall. First, areas were selected and prepared for monitoring with the films of Japanese paper, of which the surface was then stretched out a hand of light "Paraloid", as shown in **Figure 15a**. This is one kind of acrylic resin that is used in restoration as a consolidation at low concentrations (2–4%) or glue at higher concentrations. As shown in **Figure 15b**, the "Paraloid" here provides a good basis for the bonding of the AE sensors.

http://dx.doi.org/10.5772/63093

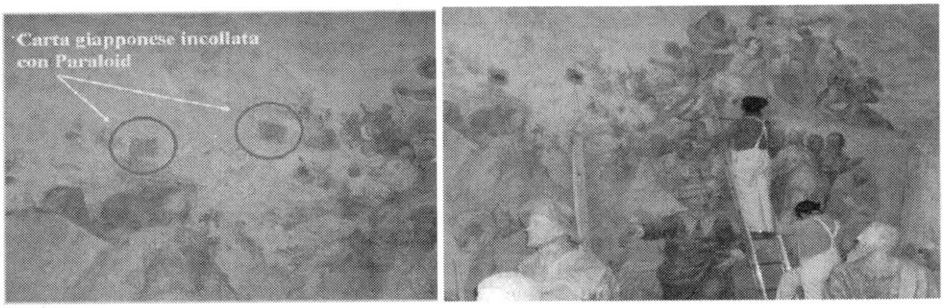

Figure 15. (a) "Paraloid" position with Japanese films and (b) bonding of the sensors.

The sensors have been arranged in positions useful to monitor simultaneously both the progress of the lesion and the detachment of the plaster. The detailed locations of the sensors are shown in **Figure 16**.

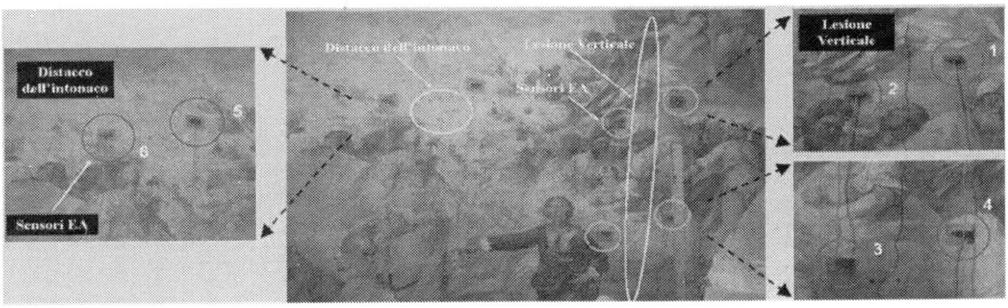

Figure 16. Detailed sensor positions.

The AE signals were picked up by a transducer and preamplified and transformed into electric voltage. Unwanted noisy signals were then eliminated by filtering frequencies, such as the vibration caused by the mechanical instrumentation, which is normally lower than 40 kHz. The signals were therefore analyzed by a measuring system that counts the signals that exceed a certain set voltage threshold measured in volts.

PZT sensors are set on a frequency range between 50 and 400 kHz. The data acquisition system consisted of six data storage provided trigger, six preamplified sensors, a central unit for the synchronization phase, and a threshold detector. From this monitoring system, microcrack localization is performed and the safe condition of the monitored specimen can be determined.

3.3. Results of the process based on the monitoring data

The monitoring work begun on 15:00 of April 28, 2011, and was planned to last for 6 months. In this part, only the monitoring data until 12:00 of June 4, 2011, were studied.

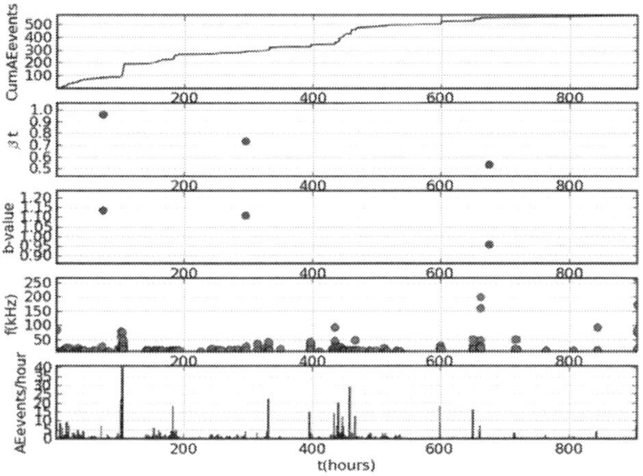

Figure 17. Results expression of sensors 1 to 4 near the lesion region.

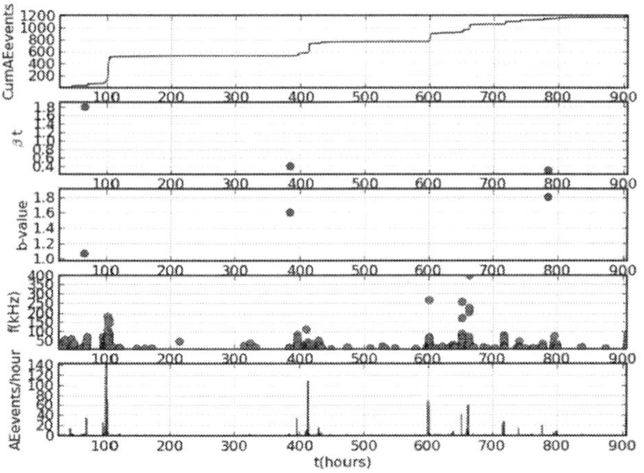

Figure 18. Results expression of sensors 5 and 6 near the detachment of the plaster.

Based on the monitoring data, the qualitative analysis of the data is mainly based on five techniques: the cumulative events analysis, the time dependence analysis, the *b*-value analysis, the amplitude of frequency analysis, and the event occurrence rate analysis. The results are shown in **Figures 17** and **18**. Considering the lesion condition in **Figure 17**, $0.5<b<1.0$ and $1.10<b<1.30$, both in average values of the three calculated points, means that the lesion region appears stable during the monitoring period and there are only some microcracks that happened. However, if we check the detachment condition of the plaster, $0.5<b<1.8$ and

1.10<*b*<1.80 show that a high-frequency event happens in this monitoring region and the detachment phenomena sometimes even occur during the monitoring period.

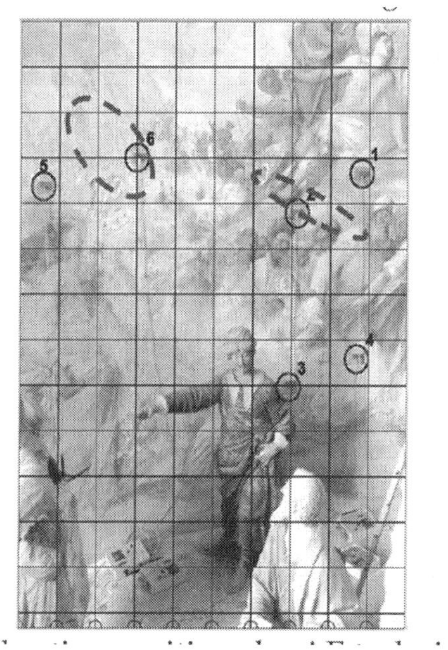

Figure 19. Monitored active positions by AE techniques are labeled by dashed lines.

According to the above qualitative analysis of the monitoring results, we know that the lesion region of the Chapel XVII appears to be stable during the monitoring period with some microcrack occurrence. As shown in **Figure 19**, the region around sensor 2 should be a more active region for lesion, and we should pay more attention on this region. On the contrary, the condition is not so good for the detachment area and the detachment phenomenon occurs with high frequency, especially for the region around sensor 6, as shown in **Figure 19**. Some maintenance measures should be implemented on this region.

4. Discussion and conclusions

Due to the increasing demand of structural retrofit and strengthening, monitoring techniques have received high attention. AE is one of the nondestructive monitoring techniques, which is widely employed for the cracking analysis in steel and concrete structures. However, the application of AE techniques to monitor masonry structures is complicated due to the fact that attenuation and wave propagation are highly dependent on the heterogeneity of the material.

Based on this problem, quantitative and qualitative analyses were proposed in this research to make the AE monitoring in masonry possible and reliable.

Crack localization is an important function in the quantitative analysis to offer the information of the active damage region. Considering the complicated material properties of masonry making the traditional localization method cannot be used directly, one modified method was proposed in this research. The proposed modified method of localization permits giving reasonable location results. In the proposed modified method, a modified factor k_i related to inhomogeneous or propagation delay is employed. The degree of heterogeneity ξ in k_i plays an important role to minimize the effect of the inhomogeneous of the material. The value ξ can be obtained by the pencil-lead break velocity field tests on the masonry structure and employed in the monitoring process. The ad hoc test results have shown that the results by the proposed modified method of localization are all located around the real crack positions. This proves that the proposed method can be a reliable and suitable one compared to the traditional one.

Besides the quantitative analysis, the qualitative analysis can also be used to monitor masonry structures. The application of the qualitative analysis was explained by monitoring one historical chapel. The cumulative events analysis, the time dependence analysis, the *b*-value analysis, the amplitude of frequency analysis, and the event occurrence rate analysis were adopted to analyze the stability of the monitoring areas.

In conclusion, quantitative and qualitative analyses proposed in this research can help to make good use of the AE technique to monitor masonry structures, and sometimes, better results can be driven if the two analysis methods can be used simultaneously.

Author details

Jie Xu*, Qinghua Han and Ying Xu

*Address all correspondence to: jxu@tju.edu.cn

School of Civil Engineering, Key Laboratory of Coast Civil Structure Safety, Tianjin University, Ministry of Education, 92 Weijin Rd, Nankai, Tianjin, China

References

[1] Dutta D, Sohn H, Harries KA, Rizzo P. A nonlinear acoustic technique for crack detection in metallic structures. Structural Health Monitoring. 2009;8(3):251–262.

[2] McCabe W, Koerner RM, Load AE. Acoustic emission behavior of concrete laboratory specimens. Journal of the American Concrete Institute. 1976;13:367–371.

[3] Köppel S, Vogel T. Localization and identification of cracking mechanisms in reinforced concrete using acoustic emission analysis. In: Proceedings of the Fourth International Conference on Bridge Management; Surrey. 2000.

[4] Carpinteri A, Lacidogna G, Niccolini G, Puzzi S. Critical defect size distributions in concrete structures detected by the acoustic emission technique. Meccanica. 2008;43:349–363.

[5] Holford KM, Pullin R, Lark RJ. Acoustic emission monitoring of concrete hinge joint models. In: Proceedings of the 26th European Conference on Acoustic Emission Testing; Berlin. 2004.

[6] Lovejoy SC. Development of acoustic emissions testing procedures applicable to conventionally reinforced concrete deck girder bridges subject tension cracking [thesis]. Oregon State University, Corvallis. 2006.

[7] Ohtsu M, Uchida M, Okamoto T, Yuyama S. Damage assessment of reinforced concrete beams qualified by acoustic emission. ACI Structural Journal. 2002;99(4):411–418.

[8] Carpinteri A, Lacidogna G. Damage diagnosis in concrete and masonry structures by acoustic emission technique. Facta Universitatis. 2003;3:755–764.

[9] Cullington DW, MacNeil D, Elliott J. Continuous acoustic monitoring of grouted post-tensioned concrete bridges. NDT & E International. 2001;34(2):95–105

[10] Giurgiutiu V, Zagrai, A, Bao JJ. Piezoelectric wafer embedded active sensors for aging aircraft structural health monitoring. Structural Health Monitoring. 2002;1(1):41–61.

[11] Golaski L, Gebski P, Ono K. Diagnostics of reinforced concrete bridges by acoustic emission. Journal of Acoustic Emission. 2002;2:83–98.

[12] Chang PC, Flatau A, Liu SC. Review paper: Health monitoring of civil infrastructure. Structural Health Monitoring. 2003;2(3):257–267.

[13] Carpinteri A, Invernizzi S, Lacidogna G. Structural assessment of a XVIIth century masonry vault with AE and numerical techniques. International Journal of Architectural Heritage. 2007;2:214–226.

[14] Melbourne C, Tomor AK. Application of acoustic emission for masonry arch bridges. Strain. 2006;42:165–172.

[15] Verstrynge E, Schueremans L, Gemert DV, Wevers M. Application of the acoustic emission technique to assess damage in masonry under increasing and sustained axial loading. In: NDTCE'09, Non-Destructive Testing in Civil Engineering; Nantes, France. 2009.

[16] Carpinteri A, Lacidogna G, Manuello A, Binda L. Monitoring the structures of the ancient temple of Athena incorporated into the cathedral of Syracuse. In: Proceedings of the 14th International Brick and Block Masonry Conference; Sydney, Australia. 2008.

[17] Tomor AK, Melbourne C. Monitoring masonry arch bridge response to traffic loading using acoustic emission techniques. In: Proceedings of the 5th International Conference on Arch Bridges; Madeira, Portugal. 2007.

[18] Ge M. Analysis of source location algorithms, part I: Overview and non-iterative methods. Journal of Acoustic Emission. 2003;21:14–28.

[19] Han QH, Xu J, Carpinteri A, Lacidogna G. Localization of acoustic emission sources in structural health monitoring of a masonry bridge. Structural Control and Health Monitoring. 2015;22(2):314–329.

Chapter 6

A NDT&E Methodology Based on Magnetic Representation for Surface Topography of Ferromagnetic Materials

Yanhua Sun and Shiwei Liu

Additional information is available at the end of the chapter

http://dx.doi.org/10.5772/62414

Abstract

Accurate evaluation is the final aim of nondestructive testing (NDT). However, the present electromagnetic NDT methods are commonly used to check the existence of defects, and all the tested targets only consist of concave defects (i.e., section-loss defects), such as holes, cracks, or corrosions, failing to evaluate the tested surface topography, which mainly consists of concave-shaped and bump-shaped features. At present, it is accepted that the commonly observed signals of the defects mainly manifest themselves in a single-/double-peak wave and their up/down directions of the peak wave can be easily changed just by changing the directions of either applied magnetization or pick-up units even for one defect. Unlike the present stylus and optical methods for surface topography inspections, a new electromagnetic NDT and evaluation (NDT&E) methodology is provided based on the accurate magnetic representation of surface topography, in which a concave-shaped feature produces "positive" magnetic flux leakages (MFLs) and therefore forms a "raised" signal wave but a bump-shaped feature generates "negative" magnetic fields and therefore leads to a "sunken" signal wave. In this case, the corresponding relationships between wave features and surface topography are presented and the relevant evaluation system for testing surface topography (concave, bumped, and flat features) is built. The provided methodology was analyzed and verified by finite element and experimental methods. Meanwhile, the different dimension parameters of height/depth and width of surface topography are further studied.

Keywords: Nondestructive testing and evaluation (NDT&E), surface topography, non-contact electromagnetic inspection, magnetic representation, dimension parameters, ferromagnetic materials

1. Introduction

Accurate evaluation is the final aim of NDT. However, present electromagnetic NDT methods (i.e., [1]) are commonly used to check the existence of defects, and all the tested targets only consist of concave defects (i.e., section-loss defects), such as holes, slots, break flaws, corrosion pits, notches, and cracks, failing to evaluate the tested surface topography which mainly consists of concave-shaped and bump-shaped features. Particularly, in electromagnetic NDT standards such as the American Society for Testing and Materials (ASTM) E570-09: 2009 (i.e., [2]) and British Standards (BS) EN 10246–4: 2000 (i.e., [3]), and models or specimens in large amounts of research and report. The reference discontinuities only consist of notch, hole, and hole and notch combination types in a standard specimen, as shown in **Figure 1**, as well as the models used in Li Xiao-meng's research on the stress-magnetism effect of ferromagnetic materials based on three-dimensional (3-D) magnetic flux leakage testing [4], Yong Zhang's adaptive method for channel equalization in MFL inspection [5], and Z.D. Wang's review of three magnetic NDT technologies [6], and so on, which are all concave defects. Thus, all the previous work focuses on the inspected objectives of concave defects, and all of the standard specimens for equipment calibration in engineering practice or for experiments or simulations in labs only consist of concave defects. At present, it is accepted that the commonly observed signals of the defects mainly manifest themselves in a single-/double-peak wave, and their up/down directions of the peak wave can be easily changed just by changing the directions of either applied magnetization or pick-up units even for one defect.

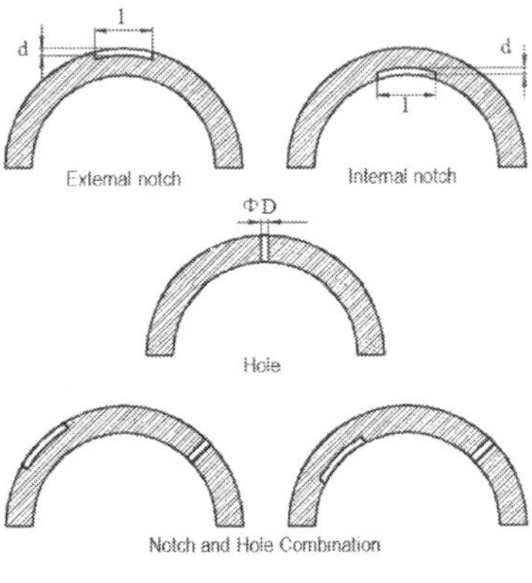

Figure 1. Standard specimens with reference defect consisting of concave-like defects in electromagnetic NDT.

Surface topography usually consists of concave (i.e., section loss), bumped (i.e., section gain), and flat features. At present, there are a number of useful techniques for testing surface topography [7], but three typical methods for detecting surface topography, namely, observation, stylus, and optical methods, are most commonly used in engineering practice [8–19]. The observation method can simply tell which surface is more rough. Although the stylus method may provide excellent vertical resolution of more than 1 nm, it must contact the test surface. The pressure formed by the contact between the stylus tip and the specimen surface is sufficient to cause permanent damage to a number of common engineering materials especially when used in high-precision roughness measurement, and a limitation for the radius of stylus also exists. The optical approaches would be much better if they are unaffected by some environmental factors such as surface cover or light intensity. In this manuscript, unlike the present stylus or optical methods for surface topography inspections of materials [19–27], a new electromagnetic NDT&E methodology is provided on the basis of the accurate magnetic representation of surface topography. Corresponding relationships between wave features and surface topography are presented and the relevant evaluation system for testing surface topography (concave, bumped, and flat features) is built. The provided methodology is analyzed and verified by finite element and experimental methods. Meanwhile, the different dimension parameters of height/depth and width of surface topography are further studied.

2. NDT&E methodology for surface topography

For a ferromagnetic material, a concave-shaped topography in its surface will produce "positive" magnetic flux leakages due to magnetic refractions, forming a "raised" signal wave. In contrast, a bump-shaped topography will generate "negative" magnetic fields region because of its self-magnetization regulation, leading to a "sunken" signal wave, as illustrated in **Figure 2**.

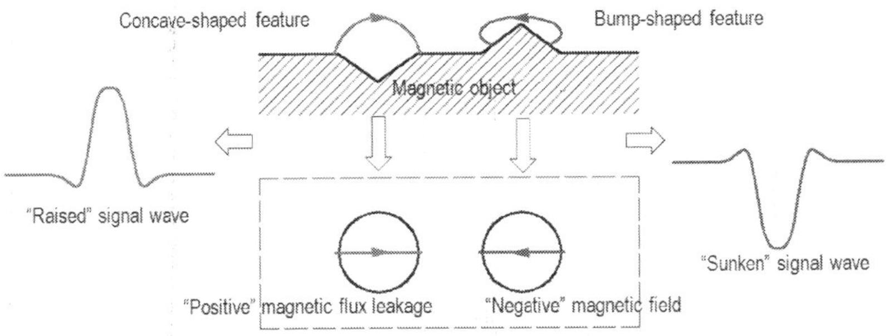

Figure 2. The magnetic representation of surface topography for ferromagnetic materials.

Accordingly, three models, namely triangular concave defects, bump defects, and their multifeature defects (three concave and two bump feature defects), were primarily built for simulations. By using finite element simulations and experimental methods, the magnetic field distribution in the vicinity of both types of defects and their signal waves was obtained, as shown in **Figure 3**, and magnetic scalar potential method was primarily used here. What can be drawn from **Figure 3a** is that the MFL produced by the concave-shaped defects such as cracks and corrosions is an increase, evidently being much stronger than the magnetic field in the region without any defects. In contrast, from **Figure 3c**, it can be seen that bump-shaped defects like protrusions produce an opposite/converse/negative magnetic field and thereafter forms a decrease vicinity of it as excepted. As a result, all picking-up (or measuring) points for magnetic field density near the concave defect are much stronger and finally a raised signal wave is formed. Conversely, those near the bump defect are much weaker and finally a sunken signal wave is formed, as shown in **Figure 3b** and **3d**, respectively. Finally, a continuous set of multifeature defects including concave and bump are modeled and simulated as shown in **Figure 3e**, and the signal of magnetic flux density above the surface of ferromagnetic material object is shown in **Figure 3f**, which constitutes a succession of raised and sunken features and directly reflects the defects.

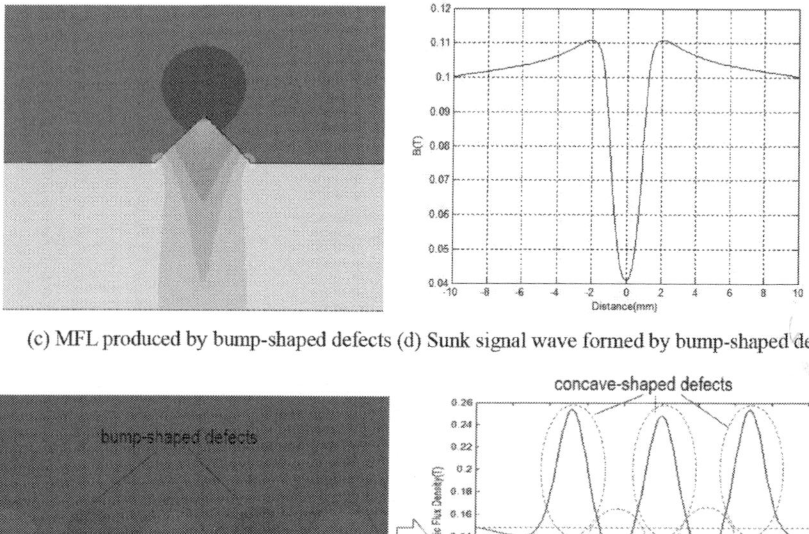

(c) MFL produced by bump-shaped defects (d) Sunk signal wave formed by bump-shaped defects

(e)MFL produced by multi-feature defects (f) continuous raised and sunk signals by multi-feature defects

Figure 3. Magnetic representation of concave-shaped and bump-shaped topography. (a) MFL produced by concave-shaped defects (b) Raised signal wave formed by concave-shaped defects (c) MFL produced by bump-shaped defects (d) Sunk signal wave formed by bump-shaped defects (e)MFL produced by multi-feature defects (f) continuous raised and sunk signals by multi-feature defects.

From the perspective of geometry, surface topography is composed by a series of micro concave-shaped defects/features and bump-shaped defects/features. Therefore, if these two types of features are inspected, surface topography can also be easily measured accordingly. In this case, by using magnetic sensors to directly induce the magnetic features with a certain liftoff distance above the measured surface, the corresponding surface topography can be inspected and therefore be evaluated [28–45]. Naturally, a non-contact magnetic inspection and evaluation methodology is provided here by using magnetic sensors such as Halls or magneto-resistive units with a certain liftoff distance to directly scan the corresponding magnetic characters caused by different surface topography, concave or bump, for instance, and then, the picking-up signals are processed through a data processing system where the surface topography is reflected ultimately, as shown in **Figure 4**.

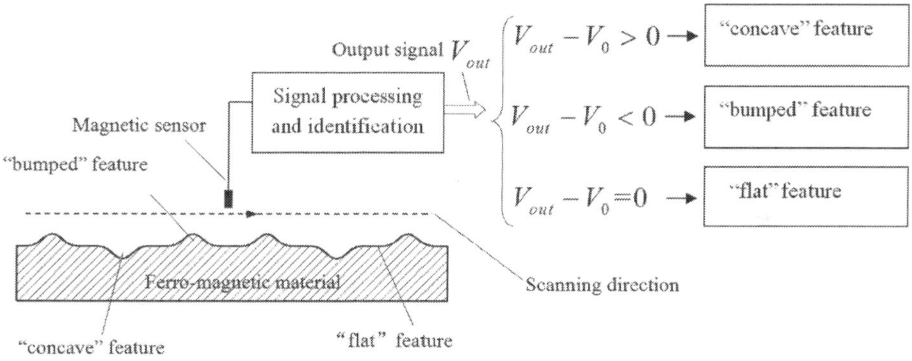

Figure 4. Non-contact magnetic inspection and evaluation methodology for surface topography.

The judgment of the output signals is completed by formulas (1)–(3), where V_{out} is the output voltage signal of magnetic sensor and V_0 represents the baseline voltage. When formula (1) is satisfied, we can definitely consider appearance of a concave feature. Similarly, bumped features and flat features can be indentified or inspected as formulas (2) and (3) are met, respectively. As usual, the residual magnetism in the inspected object is introduced by some manufacturing processes. Even if there is no residual magnetism, applying magnetization also can be easily done.

$$V_{out} - V_0 > 0 \tag{1}$$

$$V_{out} - V_0 < 0 \tag{2}$$

$$V_{out} - V_0 = 0 \tag{3}$$

Using the given inspection method, ferromagnetic material surface topography can be scanned and reflected by output signals; most importantly, the proposed inspection method is unaffected by other external elements such as surface perfection and light intensity. Here, the key issue is to confirm the different generating mechanisms of the two classified defects through their apparently different signal features.

3. Simulations

To implement the finite element simulation, a 2-D axisymmetric model is built in ANSYS. The simulation model consists of a pipe to show the signal characteristics of concave and convex, a coil used to magnetize the pipe, and a surrounding air region to conduct the magnetic field. All the models are meshed with PLANE 53 elements, which are commonly used in electromagnetic analyses. To precisely and quantitatively test the surface topography, the studies for the parameters of height/depth and width were also simultaneously and systematically included. Concaves and convex are meshed with elements of size 0.04 mm and other regions are meshed with elements of size 0.2 mm. Flux-parallel boundary condition is applied to the exterior nodes to confine the magnetic field inside the air region. Magnetizing current density of the coil is set to 1×10^7 A/m^2. After solving, a path is drawn above the pipe surface to extract the magnetic field distribution and show the MFL signal. To be specific, the dimension parameters of height/depth and width of surface topography were further studied, where the scalar magnetic potential method was primarily applied. According to the non-contact magnetic inspection method proposed here, three basic models were built, namely, triangular

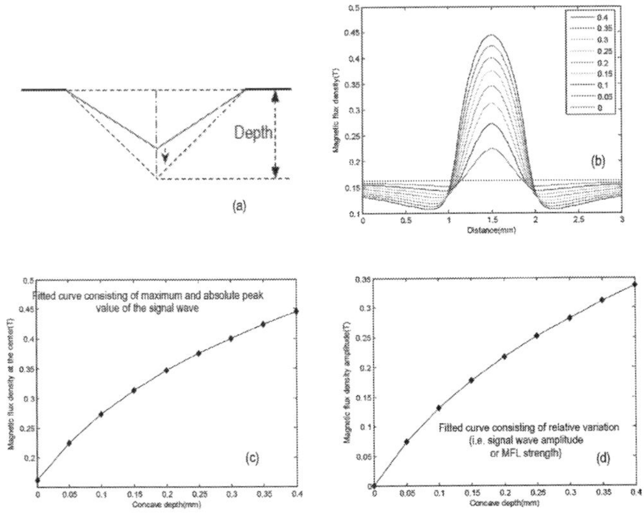

Figure 5. Detailed FEM results by separately varying the depth of the triangular "concave" defect. (a) schematic model of concave depth in simulation; (b) MFL signals in different "concave" depth; (c) Fitted curve consisting of maximum and absolute peak value of the signal wave(d) Fitted curve consisting of relative variation (i.e. signal wave amplitude or MFL strength).

"concave" defects (maximum width: 0.8 mm and maximum depth: 0.4 mm), triangular "bump" defects (maximum width: 0.8 mm and maximum height: 0.4 mm), and their multi-feature defects (three "concave" and two "bump" feature defects with the same maximum sizes), with the variable depth/height or width, as schematically illustrated in **Figures 5a–9a**, respectively. To summarise, the key characters of both the triangular "concave" defect and the "bump" defect are in accordance with the analyses mentioned previously. That is, the former generates "positive" MFL (i.e., an increase) but the latter forms "negative" magnetic field (i.e., a decrease), resulting in a raised test signal wave and a sunken wave, respectively.

Concretely, in **Figure 5a**, by separately varying the depth of the triangular "concave" defect with the decreasing unit of 0.05 mm per step (from 0 to 0.4 mm), a series of raised signal waves were generated as indicated in **Figure 5b**. The resulting fitted curve consisting of maximum and absolute magnetic flux density in the vicinity of the center of the defect (i.e., the maximum and absolute peak value of the signal wave) and the fitted curve consisting of relative variation (i.e., signal wave amplitude representing MFL strength) are displayed in **Figure 5c** and **5d**, respectively.

From **Figure 5**, it can be observed that the "positive" magnetic flux density in the vicinity of the "concave" defect is larger than that far away from the defect and increases with the increasing depth of the defect in approximately linear form. Briefly speaking, we can conclude that the signal wave amplitude (or the MFL strength) is approximately proportional to the depth linearly.

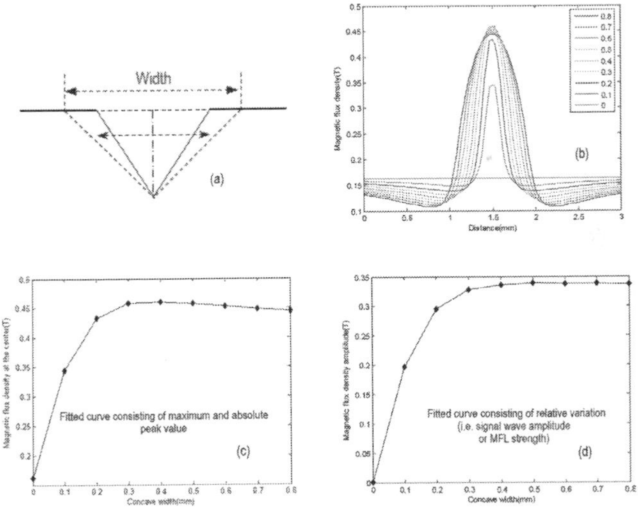

Figure 6. Detailed FEM results by separately varying the width of the triangular "concave" defect. (a) schematic model of concave width in simulation; (b) MFL signals in different "concave" width; (c) Fitted curve consisting of maximum and absolute peak value; (d) Fitted curve consisting of relative variation (i.e. signal wave amplitude or MFL strength).

On the other hand, in **Figure 6a**, by separately changing the width of the triangular "concave" defect with the decreasing unit of 0.1 mm per step (from 0 to 0.8 mm), the series of raised signal waves were also obtained as shown in **Figure 6b**. Likewise, the fitted curve consisting of maximum and absolute magnetic flux density in the vicinity of the center of the defects (i.e., the maximum and absolute peak value of the signal wave) and the fitted curve consisting of relative variation (i.e., the signal wave amplitude displaying the MFL strength) are given in **Figure 6c** and **6d**, respectively.

Which can be shown from **Figure 6** is that the amplitude of the "positive" magnetic flux density produced by the "concave" defect increases with the increasing depth in a nonlinear form, growing more rapidly at smaller value of width but decreasing gradually at higher value. That is, the signal wave amplitude (or the MFL strength) increases rapidly at first but then gradually decreases as the width grows. Most notably, the decrease of amplitude is limited and remains at a high level approaching the maximum value (i.e., peak value) during the whole change process.

Similarly, as demonstrated in **Figure 7a**, by separately varying the height of the triangular "bump" defect with the decreasing unit of 0.05 mm per step (from 0 to 0.4 mm), a series of sunken signal waves were obtained, as displayed in **Figure 7b**, and so did the fitted curve consisting of minimum and absolute magnetic flux density in the vicinity of the center of the defect (i.e., the minimum and absolute peak value of the signal wave) and the fitted curve consisting of relative variation (i.e., signal wave amplitude demonstrating the "negative" magnetic strength), as shown in **Figure 7c** and **7d**, respectively.

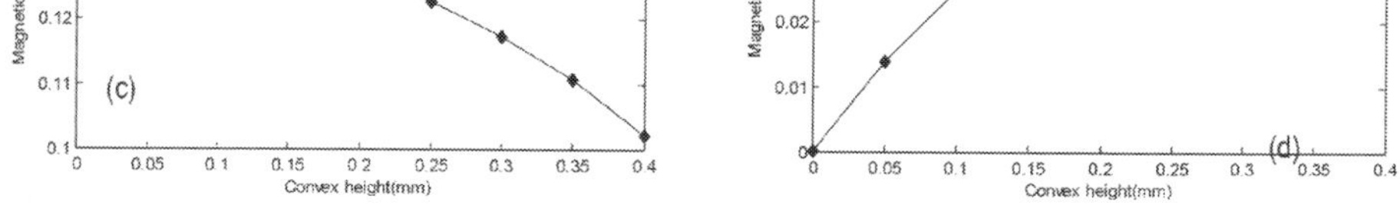

Figure 7. Detailed FEM results by separately varying the height of the triangular "bump" defect. (a) schematic model of bump height in simulation; (b) MFL signals in different "bump" height; (c) Fitted curve consisting of minimum and absolute peak value (d) Fitted curve consisting of relative variation (i.e. signal wave amplitude or "negative" magnetic strength).

Conversely, from **Figure 7**, it can be concluded that the sunken signal wave amplitude (i.e., peak-to-peak value indicating the "negative" magnetic strength) grows as the height of the triangular defect increases. In particular, **Figure 7c** intuitively demonstrates that the higher the "bump" defect is, the weaker the absolute magnetic field strength in the vicinity of the defect is, which also displays the existence of the "negative" magnetic fields.

Similarly to **Figure 6**, in **Figure 8a**, by separately varying the width of the triangular "bump" defect with the decreasing unit of 0.1 mm per step (from 0 to 0.8 mm), a series of concave-like signal waves were also obtained as presented in **Figure 8b**, and the fitted curve consisting of maximum and absolute magnetic flux density in the vicinity of the center of the defect (i.e., the minimum and absolute peak value of the signal wave) and the fitted curve consisting of relative variation (i.e., signal wave amplitude demonstrating the "negative" magnetic strength) are also obtained, intuitively displayed in **Figure 8c** and **8d**, respectively.

The results from **Figure 8** reveal the similar change trend from **Figure 7**, concerning the relationship between the amplitude of signal waves (indicating the "negative" magnetic strength) and the variable width of the defect. That is, the sunken signal wave amplitude (i.e., peak-to-peak value indicating the "negative" magnetic strength) increases with the increasing height of the triangular defect. However, compared with the results obtained from **Figure 6**

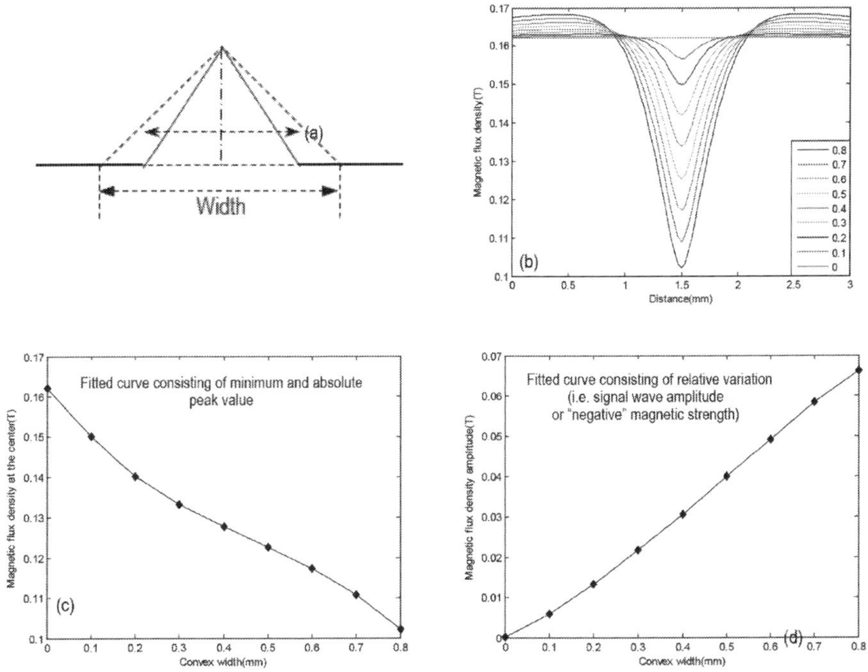

Figure 8. Detailed FEM results by separately varying the width of the triangular "bump" defect. (a) schematic model of bump width in simulation; (b) MFL signals in different "bump" width; (c) Fitted curve consisting of minimum and absolute peak value (d) Fitted curve consisting of relative variation (i.e. signal wave amplitude or "negative" magnetic strength).

illustrating the effect caused by the variable width, the linear relationship between the signal amplitude and the defect width is better in **Figure 8**.

Finally, in a like manner, the finite element modeling (FEM) results for the multifeature defects (three "concave" and two "bump" feature defects with the same maximum sizes) were obtained, as shown in **Figure 9**, from which the similar relationships between the signal amplitudes and the depth/height and width can be observed, and it can be concluded that the sensitivity of the identification can at least reach 0.02 mm in height and 0.02 mm in width of the defect. Actually and theoretically, the smaller features ("concave" or "bump") can be observed through the mechanisms. To be more concrete, **Figure 9a** and **9c** presents the variable height/depth and width in multifeature defects model, respectively, while **Figure 9b** and **Figure 9d** displays the MFL signals correspondingly.

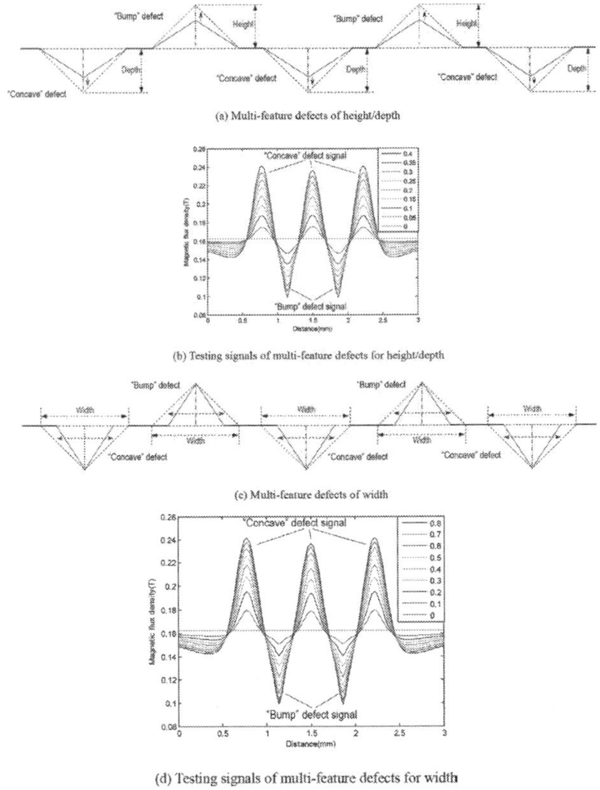

Figure 9. Detailed FEM results for the multifeature defects (three "concave" and two "bump" feature defects with the same maximum sizes). (a) Multi-feature defects of height/depth (b) Testing signals of multi-feature defects for height/depth (c) Multi-feature defects of width (d) Testing signals of multi-feature defects for width.

In addition, finite element modeling (FEM) simulations for multifeature defect (four "concave" and three "bump" features with the same sizes) distance or spacing are also applied, as illustrated in **Figure 10**. Eight kinds of models were simulated with different defect distances or spacing in the process of FEM, as shown in **Figure 10a**. Three "bump" defects and four "concave" defects are evenly spaced with the same distance value of D in every defect simulation model, and the defect distance is, respectively, 0.01, 0.02, 0.04,0.05, 0.065, 0.08, 0.10, and 0 mm, which refers to a smooth surface without any defects in different simulation models. The FEM results are distinctly shown in **Figure 10b**, where each line represents the corresponding MFL signals in the defect distance or spacing simulation.

From the MFL signals observed in **Figure 10b**, an interesting phenomenon can be seen in that magnetic flux density of multifeature defects with different distances are still able to be distinguished from each other, and based on which a conclusion can also be drawn that a defect distance or spacing resolution of 0.01 mm or higher can be reached using this proposed method of surface topography evaluation by electromagnetic methodology.

Equations (1–3) are the basic principal models for the judgment method of "concave," "bumped," and "flat" features in the proposed electromagnetic non-contact method. Certainly, the surface of the object to be tested is scraggly and has the features with different sizes and shapes. Based on the analyses of test signal amplitude displaying the features with different sizes in the manuscript and also as shown above, the different sizes or shapes affect the signal amplitude and thereafter represent surface roughness. Given a constant liftoff distance, the "bump," "concave," and "flat" test signals to be observed for presenting the surface status are

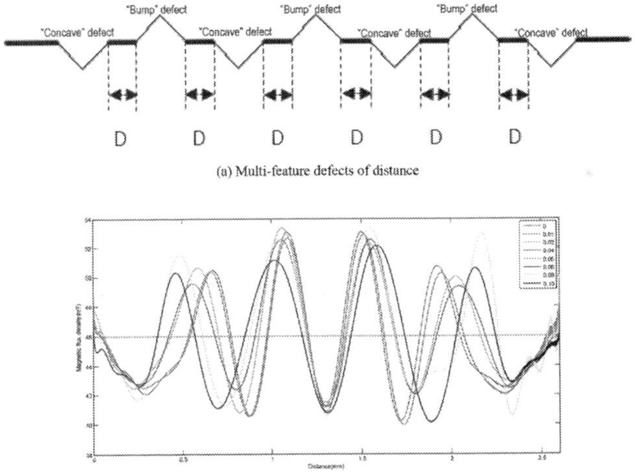

(a) Multi-feature defects of distance

(b) Testing signals of multi-feature defects for distance

Figure 10. Detailed FEM results for the multifeature defect distance (four "concave" and three "bump" features with same sizes). (a) Multi-feature defects of distance (b) Testing signals of multi-feature defects for distance.

exactly due to the various features such as different sizes and shapes. The proposed non-contact test principle is illustrated in **Figure 11**.

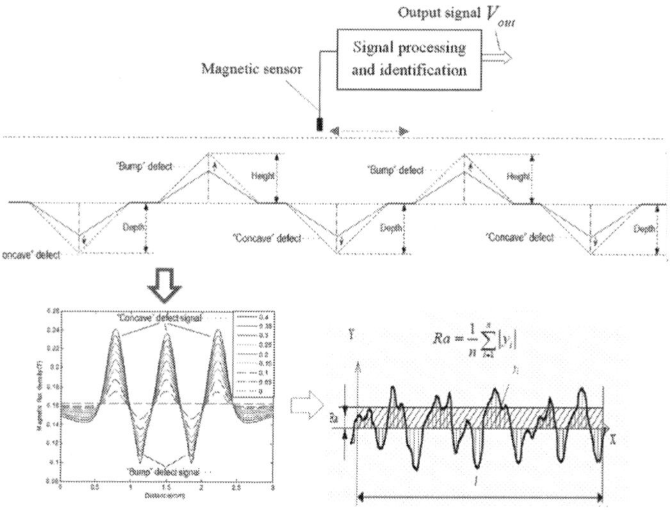

Figure 11. The explanation for non-contact test method at a given liftoff distance of magnetic sensor.

To further and better understand the principles of MFL generating mechanism for surface topography of ferromagnetic materials, three-dimensional FEM simulation was completed. The simulation model containing both "concave" and "bump" defects with magnetic field distribution in 3-D is shown in **Figure 12. Figure 12a** expressed simulation model of two typical kinds of defects with certain distances from each other, while **Figure 12b** represents local MFL signals of simulation model in **Figure 12a** and provides basic reference data for surface roughness evaluation, from which we can clearly see that the "concave" and "bump" feature inversely.

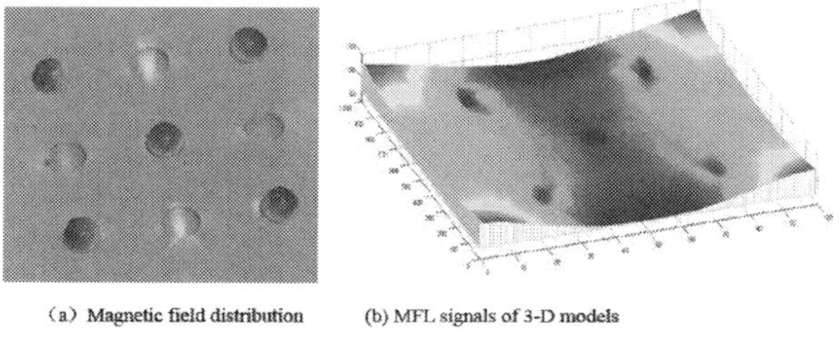

(a) Magnetic field distribution (b) MFL signals of 3-D models

Figure 12. Surface roughness inspection and its signals (3D).(a) Magnetic field distribution (b) MFL signals of 3-D models.

4. Experiments

To further confirm the validity of the non-contact magnetic inspection method,large amount of experiments were performed. According to the simulation models, four kinds of samples were applied in these experiments including triangular concave depth/width and triangular bump height/width. The experimental apparatus and process are briefly shown in **Figure 13** and its procedure and details will be described as follows. First, two groups of magnetizing coils connected by a Hall sensor are driven by electro magnetization and scanning drive modules, which provide magnetizing current and motor power and thus artificial defects on samples made of steel plate can be scanned. Then, defects signals sampled by acquisition card are sent to PC and finally processed and calculated where filtering and amplification are included.

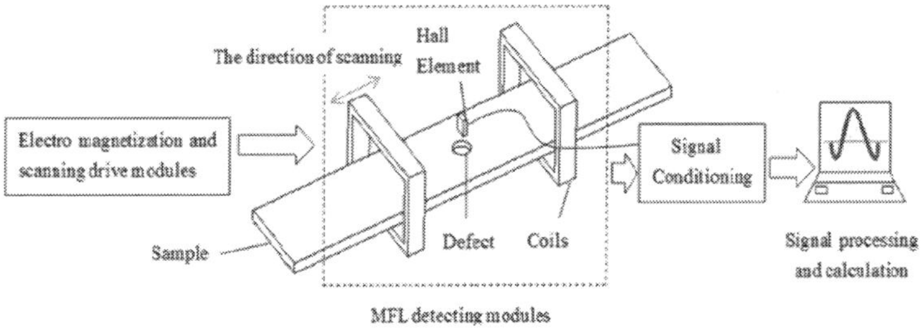

Figure 13. Experimental apparatus and process.

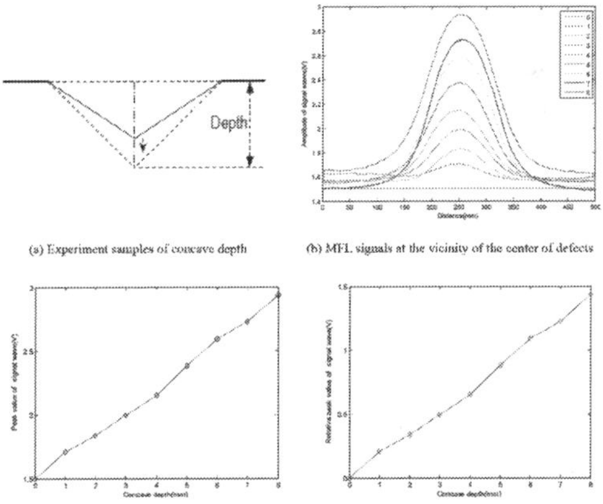

Figure 14. Experimental results for depth variation of triangular concave defects. (a) Experimental samples of concave depth (b) MFL signals in the vicinity of the center of defects. (c) Maximum or absolute peak value of MFL signal wave in (b). (d) Relative peak value of the signal wave in (b).

By scanning at the center in the vicinity of different depth of triangular defects as shown in **Figure 14a**, a series of raised MFL signals were generated as displayed in **Figure 14b** where the legend of 0–8 represents the defect depth of 0–0.4 mm with the increasing unit of 0.05 mm per step. Through calculating the maximum or absolute peak value and the relative peak value of the signal wave in **Figure 14b**, a battery of eigenvalues was acquired as displayed in **Figure 14c** and **14d**, respectively.

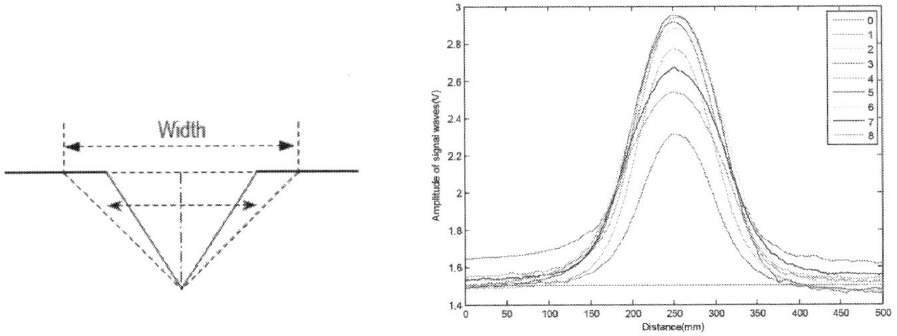

(a) Experiment samples of concave width (b) MFL signals at the vicinity of the center of defects

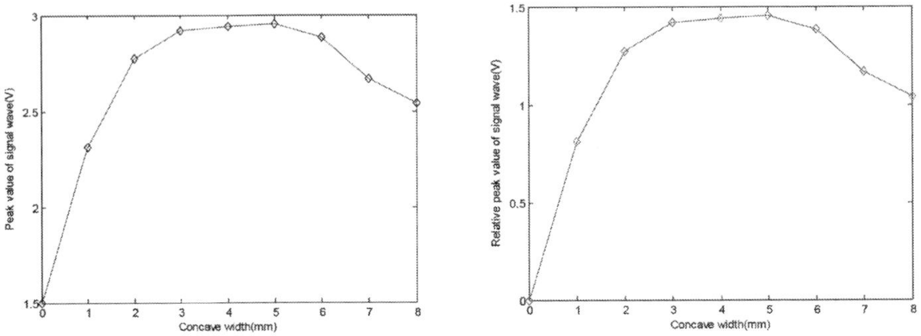

(c) Maximum or absolute peak value of MFL signal wave in (b) (d) Relative peak value of the signal wave in (b)

Figure 15. Experimental results for width variation of triangular concave defects. (a) Experimental samples of concave width. (b) MFL signals in the vicinity of the center of defects. (c) Maximum or absolute peak value of MFL signal wave in (b). (d) Relative peak value of the signal wave in (b).

Comparing the above experimental results with what has been done in the simulations part, it is not hard to find that the "positive" magnetic flux density in the vicinity of the center of concave defect is larger than that far away from the defect and manifests a similar change trend regarding relevant eigenvalues of maximum and relative peak values. By the consistency of simulations and experimental results, we can definitely conclude that the maximum or the relative peak value is approximately proportional to the concave depth linearly.

By the same token, another experiment was conducted on the samples of width of concave defects. After scanning above the surface of defects as shown in **Figure 15a**, a series of MFL signal waves were presented in **Figure 15b**, in which the legend of 0–8 presents the changing width of triangular concave defect from 0 to 0.8 mm with the increasing unit of 0.1 mm per step. After data processing, the maximum or absoLute Peak Value And The Relative Peak value of the signal wave in **Figure 15b** were acquired and displayed in **Figure 15c** and **15d**, respectively.

Observing from the changing trend of MFL signals of concave width defects and comparing them with the simulation results obtained previously, we can notice that the peak value of MFL signal waves increases rapidly with concave width at the beginning within a small range and remains at a high level as the width increases further. These changing rules are in well accordance with that of the simulations, which further demonstrates the correctness of the relationship between concave width and the peak value of signal waves and provides us the necessary information to predict the defect widths in turn.

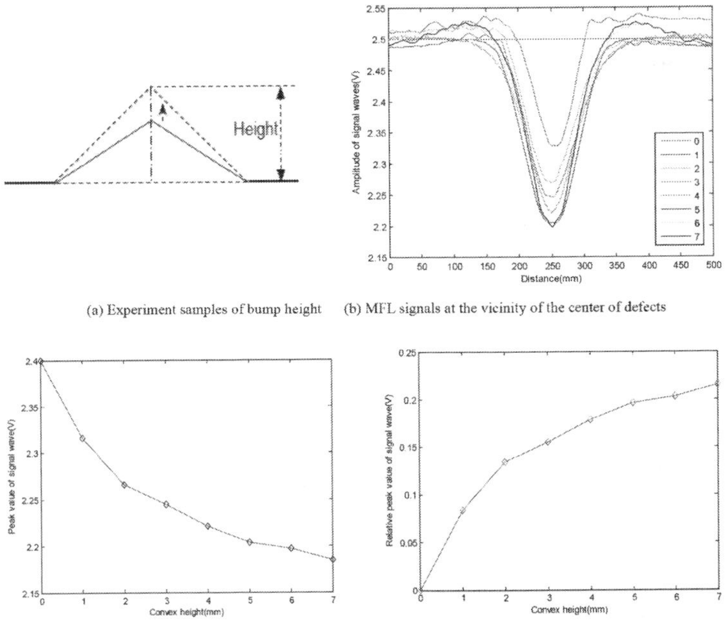

Figure 16. Experimental results for height variation of triangular bump defects. (a) Experimental samples of bump height (b) MFL signals in the vicinity of the center of defects (c) Maximum or absolute peak value of MFL signal wave in (b) (d) Relative peak value of the signal wave in (b).

Likewise, after scanning on the surface of the bump height samples shown in **Figure 16a** with a certain liftoff for the Hall sensor, MFL signal waves representing different values of bump height are expressed in **Figure 16b** where the legend of 0–7 reflects the bump heights of 0–0.35

mm with the increasing unit of 0.05 mm per step. After extracting eigenvalues for the signal waves in **Figure 16b**, the maximum or absolute peak value and the relative peak value of the signal wave in **Figure 16b** were acquired and displayed in **Figure 16c** and **16d**, respectively.

Reading from **Figure 16** and comparing it with simulations of **Figure 7** described earlier, it is obvious that the experimental results about the bump height are in good agreement with the simulations, all of which indicated the existence of the "negative" magnetic fields. Moreover, the parallel change trend also means the presence of a negative correlation between bump height and the peak value of MFL signal wave.

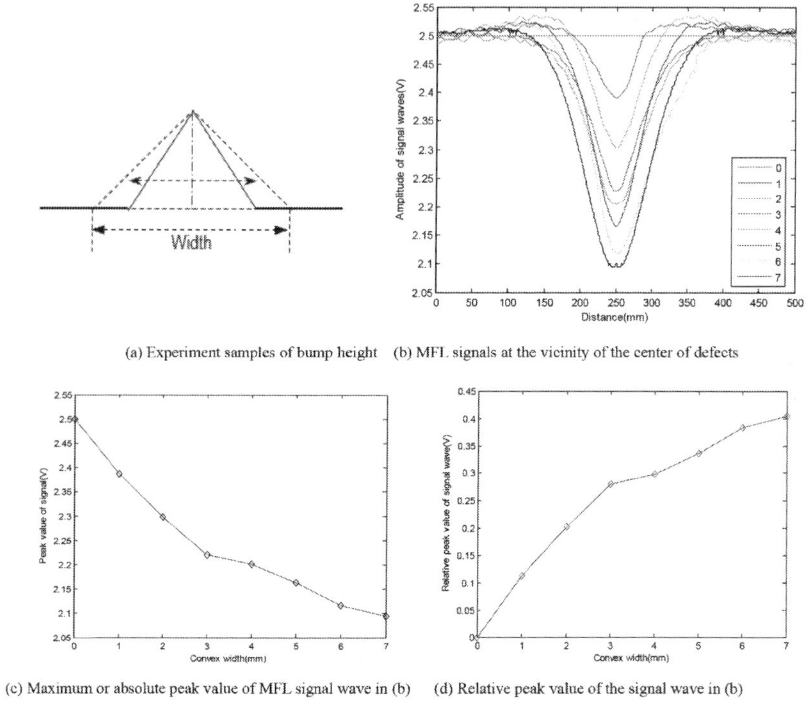

Figure 17. Experimental results for width variation of triangular bump defects. (a) Experimental samples of bump height (b) MFL signals in the vicinity of the center of defects (c) Maximum or absolute peak value of MFL signal wave in (b) (d) Relative peak value of the signal wave in (b).

Similarly, after scanning on the surface of the bump/convex width samples shown in **Figure 16a** with a certain liftoff by the Hall sensor, MFL signal waves representing different values of bump width are expressed in **Figure 16b** where the legend of 0–7 reflects the bump width of 0–0.7 mm with the increasing unit of 0.1 mm per step. After extracting eigenvalues for the signal waves in **Figure 16b**, the maximum or absolute peak value and the relative peak value of the signal wave in **Figure 16b** are acquired and displayed in **Figure 16c** and **16d**, respectively.

http://dx.doi.org/10.5772/62414

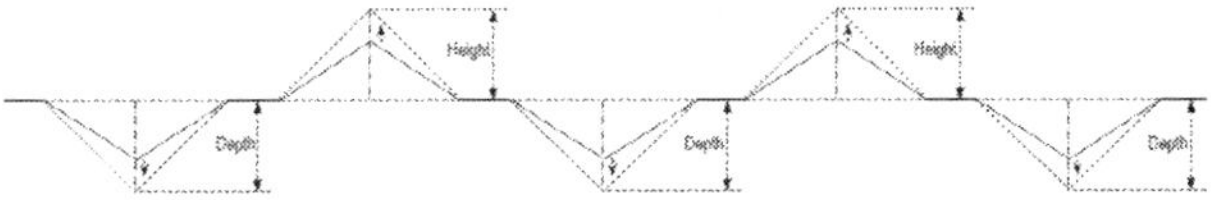

(a) Multi-feature defects of height/depth

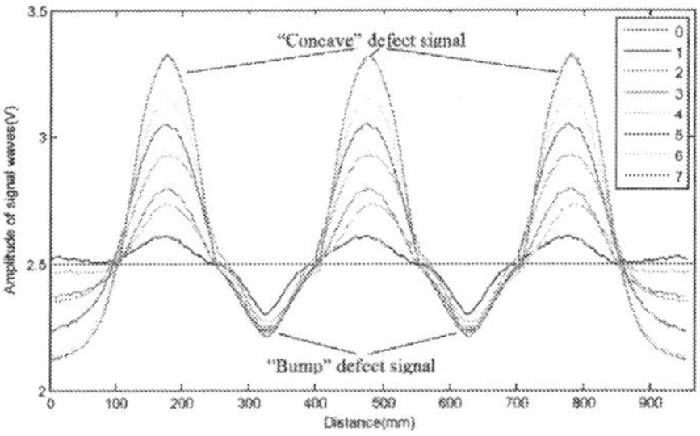

(b) Experiment results of multi-feature defects for height/depth

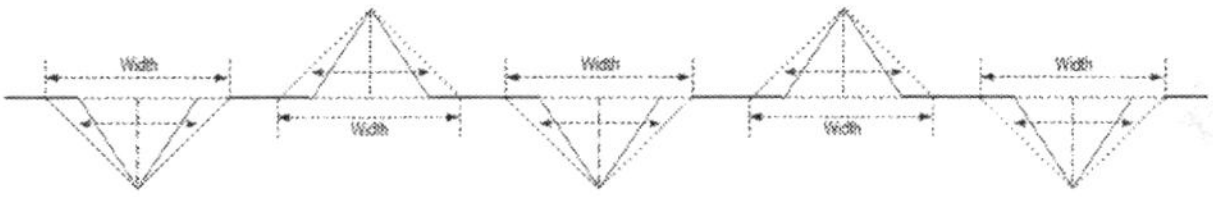

(c) Multi-feature defects of width

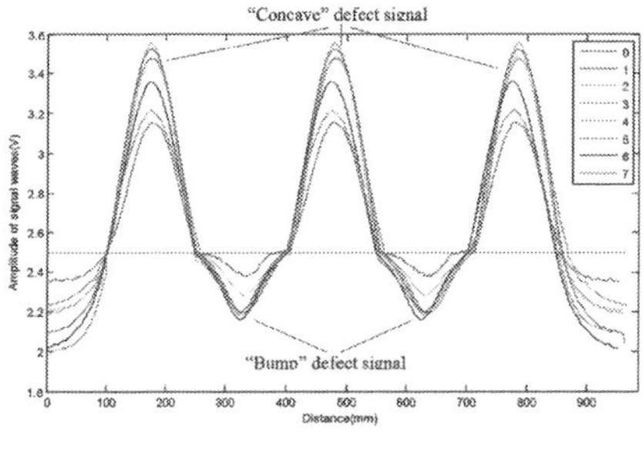

(d)Experiment results of multi-feature defects for width

Figure 18. Experiment results for the multi-feature defects (three "concave" and two "bump" feature defects with the same maximum sizes) (a) Multi-feature defects of height/depth (b) Experiment results of multi-feature defects for height/depth (c) Multi-feature defects of width (d)Experiment results of multi-feature defects for width.

Similarly to explanations for **Figure 16**, observing from **Figure 17(a)–(d)** and comparing it with simulation results of **Figure 8** described earlier, it is obvious that the experimental results about the bump width are in good agreement with the simulations; they indicate not only the existence of the "negative" magnetic fields but also the presence of a negative correctness relation between bump width and peak values of the MFL signal or relative peak values. To be specific, there exists a negative correlation between convex-shaped defect width and peak value of MFL signals, and vice versa for relative peak value of MFL signals. By the measured linear relationship between convex height and signal peak value or relative signal peak value, convex-shaped defect height can be attained if another variable of MFL signal is obtained in advance.

To verify the validity of the theory proposed in this paper and the reliability of the simulations done previously, experiments for multifeature defects, which include concave and bump defects simultaneously, were also conducted. One group of experimental samples are roughly drawn as shown in **Figure 18a**, where two same concave defect depth and three same bump defect heights vary from 0 to 0.7 mm with an increasing unit of 0.1 mm per step while other dimensions of these defects remain unchanged. Similarly, **Figure 18c** expresses another group of samples that double same concave-shaped defect widths vary with another three same bump defects from 0 to 0.7 mm while other size of these defects remain unchanged. The experimental results are separately shown in **Figure 18b** and **18d**, where the legend of 0–7 represents the variable value of 0–0.7 mm, which coincides well with simulation results in **Figure 9**, as well as further confirming the conclusion of the theory proposed here that a concave-shaped feature will produce "positive" magnetic flux leakages and form a "raised" signal wave while a bump-shaped feature will generate "negative" magnetic fields and lead to a "sunken" signal wave.

To summarise, surface topography or topography measurement can be divided into the measurement of concave and bump features. The combination of simulations and experiments simultaneously demonstrate the correctness and feasibility of the proposed method and the validity of the law that concave-shaped and bump-shaped features possess opposite magnetic field or magnetic flux leakage, and form raised and sunken signal waves, respectively. The change rules of depth/height and width further provided us detailed information about the relationship between surface topography and MFL signal eigenvalues.

5. Conclusions

The up/down directions of the peak wave of the commonly observed signals relate to not only the directions of either applied magnetization or pick-up units but also actually bear on the accurate magnetic representation of surface topography. The corresponding relationships between wave features of observed signals and surface topography features exist. That is, a concave-shaped feature produces "positive" magnetic flux leakages and therefore forms a "raised" signal wave but a bump-shaped feature generates "negative" magnetic fields and therefore leads to a "sunken" signal wave. The provided NDT&E methodology based on magnetic representation for surface topography of ferromagnetic materials is feasible (a raised signal wave representing a concave feature, a sunken signal wave representing a bumped feature, and a baseline wave representing a flat feature), and hopefully the further exploration of its applications can be further done. Surface topography can be accurately tested and evaluated by the provided methodology and the further relationship between signal wave amplitudes and surface parameters such as depth and width.

This study was funded by the National Natural Science Foundation of China (NNSFC) [Project No. 51475194] , the National Key Basic Research Program of China [Project No. 2014CB046706] and National Natural Science Foundation of China(NNSFC) [Project No. 51275193].

Author details

Yanhua Sun* and Shiwei Liu

*Address all correspondence to: yhsun@hust.edu.cn

Huazhong University of Science and Technology, Wuhan, China

References

[1] De Levie, R.: The influence of surface topography of solid electrodes on electrochemical measurements. Electrochimica Acta, 10(2): 113–130 (February 1965).

[2] ASTM E570-09, Standard Practice for Flux Leakage Examination of Ferromagnetic Steel Tubular Products, ASTM International, West Conshohocken, PA, 2009, DOI: 10.1520/E0570-09.www.astm.org.

[3] BS EN 10246–4.: Non-destructive testing of steel tubes – Part 4: Automatic full peripheral magnetic transducer/flux leakage testing of seamless ferromagnetic steel tubes for the detection of transverse imperfections. May (2007). http://www.freestd.us/soft/145444.htm.

[4] L. Xiao-meng, D. Hong-sheng, and B. Shi-wu, "Research on the stress-magnetism effect of ferromagnetic materials based on three-dimensional magnetic flux leakage testing," NDT & E International, vol. 62, pp. 50-54, 2014.

[5] Y. Zhang, Z. Ye, and X. Xu, "An adaptive method for channel equalization in MFL inspection," NDT & E International, vol. 40, pp. 127-139, 2007.

[6] Z. Wang, Y. Gu, and Y. Wang, "A review of three magnetic NDT technologies," Journal of Magnetism and Magnetic Materials, vol. 324, pp. 382-388, 2012.

[7] P. Benardos and G.-C. Vosniakos, "Predicting surface roughness in machining: a review," International journal of machine tools and manufacture, vol. 43, pp. 833-844, 2003.

[8] Poon, C.Y., Bhushan, B.: Comparison of surface topography measurements by stylus profiler, AFM and non-contact optical profiler. Wear, 190(1): 76–88 (November 1995).

[9] Vorburger, T.V., Teague, E.C.: Optical techniques for on-line measurement of surface topography. Precis Eng, 3(2): 61–83 (April 1981).

[10] Campbell, F. C.:Chapter1: Inspection methods—Overview and comparison. Inspection of metals: Understanding the basics. ASM International, Pages 1–20 (20) (April 01, 2013). http://www.asminternational.org/documents/10192/1849770/05372_Sample.pdf.

[11] Leach, R.: Chapter6-Surface topography measurement instrumentation. Fundamental principles of engineering nanometrology (second edition), Pages 133-204. http://www.sciencedirect.com/science/article/pii/B9780080964546100065.

[12] Pawlus, R., Śmieszek, M..: The influence of stylus flight on change of surface topography parameters. Precis Eng, 29(3): 272–280 (July 2005).

[13] P. Demircioglu, "Estimation of surface topography for dental implants using advanced metrological technology and digital image processing techniques," Measurement, vol. 48, pp. 43-53, 2014.

[14] T. Jeyapoovan and M. Murugan, "Surface roughness classification using image processing," Measurement, vol. 46, pp. 2065-2072, 2013.

[15] Wang, S.H., Jin, C.J., Tay, C.J., Quan, C., Shang, H.M.: Design of an optical probe for testing surface topography and micro-displacement. Precis Eng, 25(4): 258–265 (October 2001).

[16] Hocken, R.J., Chakraborty, N., Brown, C.: Optical metrology of surfaces. CIRP Ann Manu Technol, 54(2): 169–183 (2005).

[17] Brodmann, R., Gast, Th., Thurn, G., Wirtz, A., Optische Werke G.: Rodenstock. An optical instrument for measuring the surface topography in production control. CIRP Ann Manu Technol, 33(1): 403–406 (1984).

[18] Yim, D.Y., Kim, S.W.: Optical topography measurement of ground surfaces by light scattering. Int J Mach Tools Manu, 30(2): 283–289 (1990).

[19] Sherrington, I., Smith, E.H.: Modern measurement techniques in surface metrology: Part II; optical instruments. Wear, 125(3): 289–308. (1 August 1988).

[20] Whitehouse, D.J.: Surface metrology. Meas Sci Technol, 8(9): 955–972 (1997).

[21] Thomas, T.R.: Rough surfaces. London: Longman (1982).

[22] Ohlidal, M.: Comparison of 2-dimensional Fraunhofer approximation and 2-dimensional Fresnel approximation at analysis of surface-topography by angle speckle correlation. 2. Experimental results. J Mod Opt, 42(10): 2081–2094 (1995).

[23] Nadolny, K., et al.: Laser measurements of surface topography of abrasive tools using measurement system CLI 2000. Przegląd Elektrotechniczny (Elec Rev), 87(9a): 24–27 (2011).

[24] Łukianowicz, C., Karpiński, T.: Optical system for measurement of surface form and topography. Meas Sci Rev, 1(1): 151–154 (2001).

[25] Senchenko, E.S., Chugui, Yu.V.: Shadow inspection of 3D objects in partially coherent light. Meas Sci Rev, 11(4): 104–107 (2011).

[26] Kawate, E., Hain, M.: New scatterometer for spatial distribution measurements of light scattering from materials. Meas Sci Rev, 12(2): 56–61 (2012).

[27] Sherrington, I., Smith, E.H.: Modern measurement techniques in surface metrology: Part I; Stylus instruments, electron microscopy and non-optical comparators. Wear, 125(3): 271–288 (1 August 1988).

[28] K. Tsukada, M. Yoshioka, Y. Kawasaki, and T. Kiwa, "Detection of back-side pit on a ferrous plate by magnetic flux leakage method with analyzing magnetic field vector," NDT & E International, vol. 43, pp. 323-328, 2010.

[29] M. Le, J. Lee, J. Jun, J. Kim, S. Moh, and K. Shin, "Hall sensor array based validation of estimation of crack size in metals using magnetic dipole models," NDT & E International, vol. 53, pp. 18-25, 2013.

[30] S. Mukhopadhyay and G. Srivastava, "Characterisation of metal loss defects from magnetic flux leakage signals with discrete wavelet transform," Ndt & E International, vol. 33, pp. 57-65, 2000.

[31] Dutta, S.M., Ghorbel, F.H., Stanley, R.K.: Simulation and analysis of 3-D magnetic flux leakage. IEEE Trans Magn, 45(4): 1966–1972 (2009).

[32] Krause, T.W., Donaldson, R.M., Barnes, R., Atherton, D.L.: Variation of the stress dependent magnetic flux leakage signal with defect depth and flux density. NDT E Int, 29(2): 79–86 (1996).

[33] G. Kopp and H. Willems, "Sizing limits of metal loss anomalies using tri-axial MFL measurements: A model study," NDT & E International, vol. 55, pp. 75-81, 2013.

[34] Altschuler, E., Pignotti, A.: Nonlinear model of flaw detection in steel pipes by magnetic flux leakage. NDT E Int, 28(1): 35–40 (1995).

[35] Goktepe, M.: Non-destructive crack detection by capturing local flux leakage field. Sens Actuat A, 91(1): 70–72 (2001).

[36] Y. Zhang, Z. Ye, and C. Wang, "A fast method for rectangular crack sizes reconstruction in magnetic flux leakage testing," Ndt & E International, vol. 42, pp. 369-375, 2009.

[37] S. Saha, S. Mukhopadhyay, U. Mahapatra, S. Bhattacharya, and G. Srivastava, "Empirical structure for characterizing metal loss defects from radial magnetic flux leakage signal," Ndt & E International, vol. 43, pp. 507-512, 2010.

[38] M. Le, J. Lee, J. Jun, and J. Kim, "Estimation of sizes of cracks on pipes in nuclear power plants using dipole moment and finite element methods," NDT & E International, vol. 58, pp. 56-63, 2013.

[39] Uetake, I., Saito, T.: Magnetic flux leakage by adjacent parallel surface slots. NDT E Int, 30(6): 371–376 (1997).

[40] C. Coughlin, L. Clapham, and D. Atherton, "Effects of stress on MFL responses from elongated corrosion pits in pipeline steel," NDT & E International, vol. 33, pp. 181-188, 2000.

[41] Amineh, R.K., Koziel, S., Nikolova, N.K., Bandler, J.W., Reilly, J.P.: A space mapping methodology for defect characterization from magnetic flux leakage measurements. IEEE Trans Magn, 44(8): 2058–2065 (2008).

[42] Ravan, M., Amineh, R.K., Koziel, S., etc.: Sizing of 3-D arbitrary defects using magnetic flux leakage measurements. IEEE Trans Magn, 46(4): 1024–1033 (2010).

[43] D. Mukherjee, S. Saha, and S. Mukhopadhyay, "Inverse mapping of magnetic flux leakage signal for defect characterization," NDT & E International, vol. 54, pp. 198-208, 2013.

[44] Chen, Z., Preda, G., Mihalache, O., Miya, K.: Reconstruction of crack shapes from the MFLT signals by using a rapid forward solver and an optimization approach. IEEE Trans Magn, 38(2): 1025–1028 (2002).

[45] N. Gloria, M. Areiza, I. Miranda, and J. Rebello, "Development of a magnetic sensor for detection and sizing of internal pipeline corrosion defects," NDT & E International, vol. 42, pp. 669-677, 2009.

Chapter 7

Application of Non-destructive Testing for Measurement of Partial Discharges in Oil Insulation Systems

Tomasz Boczar, Andrzej Cichoń, Daria Wotzka,
Paweł Frącz, Michał Kozioł and Michał Kunicki

Additional information is available at the end of the chapter

http://dx.doi.org/10.5772/62409

Abstract

The subject area regards to metrology and measurement methods applied for non-destructive investigation of electrical discharges occurring in oil insulation systems of high-voltage devices. The main aim of performed research studies is a detailed and multivariate analysis of physical phenomena associated with generation of electrical partial discharges (PD), which occur in oil insulation of electrical equipment. An important cognitive component was the verification whether the form of PD has an effect on the energy contribution of the physical phenomena associated with their generation. For investigating the physical processes associated with generation of PD, a system for modelling, the study and analysis of physical phenomena associated with their generation in insulating oil were designed and implemented. In particular, the PD were simulated in three setups: (1) a surface system, (2) needle-needle system in insulating oil and (3) needle-needle system in insulating oil with gas bubbles. In these experimental setups, optical signals (IR, UV and visible), ultra–high frequency electromagnetic and high-energy X-ray radiation, acoustic emission and thermal images were registered. Recorded signals were subjected for multi-variant investigation and analyses in the time and frequency domains. The contribution of particular physical phenomena was determined.

Keywords: partial discharge, non-destructive testing, signal processing and analysis, paper-oil insulation, power transformer, diagnosis, non-conventional diagnosis of PD

1. Introduction

The generation and development of partial discharges (PD) in solid, liquid and gas dielectrics associate numerous physical phenomena. The most important are occurrence of a current pulse, emission of electromagnetic waves of ultra–high frequency (UHF), generation of optical radiation and high-energy radiation (X-/gamma rays), chemical transformations in the isolation, an elastic shock deformation and the associated acoustic wave generation. Based on these phenomena, diagnostic methods have been developed for detection, measurement and localization of PD, which may occur in insulating systems of power devices [1–8]. Currently, the assessment of electrical discharges shall apply the following non-destructive methods: electric [9–13], gas chromatography and acoustic emission (AE) [14–20] as well as the optical method [21–23]. Measurements are also performed as approximation of the amount of emitted heat, light and pressure changes in the PD generation area. It should be noted that each method has its advantages and limitations in application. In the normal operation of power devices, PD measurement by the electrical method is due to high levels of electromagnetic interference practically impossible. In the method of gas chromatography, serious problem is regarding the possibility of falsifying the test results while getting and transporting the insulating oil samples. Whereas, in the EA method there are problems associated with the localization of PD generation area location. At the same time, the information obtained through the application of these methods is a direct reason for their intensive development and improvement. Works carried out so far for the estimation of their applicability in the diagnosis of insulation systems of power devices have focused mainly on issues related to the selection and improvement of the measuring apparatus. They also concerned improvement of the recording methods, the analysis and interpretation of measurement results obtained mainly within modern signal processing algorithms and specialized computer software applied in this field. High-power transformer diagnostics that is based on PD measurement issues a problem of selection of indicators that would enable for correlation of measurements results with particular defects types, which are the source of discharges. These indicators should properly characterize the phenomena of formation and development of PD in insulating systems. Therefore, it appeared necessary to execute experimental studies aimed in understanding the physical processes associated with the generation of PD and thus enabling the evaluation of the energy contribution of individual phenomena using signal processing methods. In consequence, the results obtained with different testing methods enabled for linking and identification of the dominant processes whose measurement is essential for a proper evaluation of the measured isolation. Implementation of such analyses should help to reduce the risk of errors in the diagnosed isolation evaluation, which in turn can have a direct impact on economic and financial dimension.

The main cognitive objective of the research works was a detailed and multi-variant study of the physical phenomena associated with formation and development of electrical PD generated in liquid dielectrics. PD were measured with the following methods: acoustic, electric, gas chromatography, optical spectroscopy. The following research hypothesis was set up in the research: it is possible to experimentally investigate the physical processes associated with generation of the fundamental forms of PD that may occur in insulating oil and their reciprocal relation. This applies particularly to the development of a relationship and contribution of

http://dx.doi.org/10.5772/62409

various forms of radiated energy. Consequently, it allowed identifying those physical phenomena, associated with the formation of PD, which are dominant in terms of energy and thus have the greatest impact on the development of aging processes in the studied isolation. An important cognitive element was estimation of whether the form of a PD affects the energy contribution of the physical phenomena associated with their generation.

The results obtained may be applied for monitoring and comprehensive evaluation of the considered isolation. Inclusion during the tests performed on-line, all of the results obtained by different methods may increase the confidence of inference about the technical condition of the insulation equipment, and thus it may contribute to increase their work reliability, extend their operation time and avoid unexpected failures, which in turn can bring measurable economic benefits. This applies in particular to large power transformers, in which investment cost in relation to the total value of equipment used for transmission, distribution and distribution of electricity represents about 20%. Moreover, the expected cost of transformer failure in extreme cases can exceed up to five times its purchase price, while the total costs of the non-invasive diagnostic measurements of transformer insulation systems do not exceed 10% of its repairs cost. The above-mentioned findings justify the broad diagnostic procedures conducted within the research, the scope of which is correlated with the technical and economic importance of the measured electric power equipment.

2. The experimental part

To investigate the physical phenomena associated with generation of PD, a research station allowing for simulation and studying the phenomena of PD generation in insulating oil was designed and practically implemented. Also, the spark gaps for modelling of PD in surface (SURF), needle-needle (NN) and needle-needle with gas (air) bubbles (NNB) systems were designed and manufactured. The multi-variate experimental studies, performed under laboratory conditions, consisted of simultaneous recording of acoustic emission, electrical and optical signals, electromagnetic radiation of ultra–high frequency and high-energy X-, alpha, beta and gamma rays, which were generated by three considered fundamental forms of PD. The mentioned methods were applied during all planned trails. The trails included PD generation at different supply voltage levels, starting from 0 V and finishing by the breakdown voltage, which was different in the sub-sequential trials.

A professional test apparatus which enables measurement and analysis of acoustic emission (AE) signals was applied: PULSE DynXI system type 3050-B-A4 from Bruel&Kjael, numerical measuring instrument type 945A from SVAN, instrumentation transducers 4514-B-001 100mV/g and 4513-B-002 500mV/g, amplifiers with the power suppliers type Nexus from Bruel&Kjael, instrumentation cards Acquitec CH-3160, instrumentation computer Dell E6400 ATG, instrumentation computer Atlas ATXB-150. The AE signals were recorded for each type of PD modelling setup and for various levels of the supply voltage. Results of studies related to the AE are depicted in Section 3.3.

In order to implement the proposed research works connected with measurement and analysis of optical radiation generated by the modelled basic forms of PD, a spectrometer, HR4000 from Ocean Optics, equipped with multi-grating which allows for analysis of optical radiation spectra in the range from at least 270 to 1700 nm with a resolution of about 0.5 nm, was applied. The recorded signal is supplied to the grid by optical fibre. The sensor can be immersed in the insulating oil, which enabled measurements in the very vicinity of the PD generation location. The studies included measurement trails conducted at various distances of the sensor to the PD generation area. Results of studies related to the spectrophotometry are depicted in Section 3.2.

Also, a professional UVollé camera from OFIL, which enables registration of PD in the ultraviolet spectrum (in the range from 250 to 280 nm) and visible region, was applied. This camera can not be immersed in oil, thus UV measurements were performed at larger distances to the PD generation area, outside of the tank, where the electrodes were mounted. The analysis of results depicted no UV radiation outside of the tank; thus, this analysis is not included in this chapter.

Furthermore, the equipment included a PDS 100 device from Double that allows for measurements of the electromagnetic disturbances levels in the range from 50 MHz to 1 GHz. Additional measurement and research apparatus include: a multi-channel PD measurement and detection system with the electrical method, equipped with PD detectors, an external module for interferences elimination, UHF probe and software to analyze and evaluate the measuring results, according to the requirements of IEC 60270–2000 standard. Results of studies related to the electrical and UHF methods are depicted in Section 3.1.

Another instrument, which was partly built and partly purchased, is a setup for study of high-energy radiation comprising X-, gamma, alpha and beta ray sensors and equipment for acquisition of pulses representing the measured value of radiation (counters). Similar as it was the case of UV camera, the high-energy radiation sensor was mounted near the oil surface, at a larger distance to the PD generation area. The analysis of results depicted no high-energy radiation outside of the tank, thus this analysis is not included in this chapter. If interested, the reader is referred to [24,25], where the authors present results of PD high-energy radiation measurements in a gas-insulated system.

During an 8-hour period of time, the emitted infrared radiations were registered using an infrared camera, type ThermoGear G100EX, from NEC that allows for registration of electromagnetic radiation in the range from 8 to 13 microns, which is in the mid-infrared (MIR) frequency band (typically 5–30 microns). The camera is equipped with NS9500Lite software that is used to analyze the thermal images. Results of studies related to the infrared technology are depicted in Section 3.4.

The measuring setup along with the applied instrumentation and chosen preliminary results of the considered research were described in detail in [26], thus its presentation was omitted in this chapter. For archiving and processing of the measured data, numerical procedures implemented in the Matlab programming environment were applied.

The physical and chemical properties of applied insulation oil were investigated by means of a chromatographic analysis of collected oil samples, which was conducted by a specialist chemical laboratory. Results of studies related to chromatography are depicted in Section 3.5.

3. Analysis of measurement results

All gathered signals were subjected to a comprehensive processing and analysis. In the following sections, the achieved results are given. The particular measurement methods are juxtaposed for the three measuring setups for better comparison.

3.1. Results of analysis obtained using the electrical and UHF methods

The first step in the analysis of data gathered with the electrical and UHF method was to consider the background environment. **Figure 1** depicts the spectrum of UHF sweep, at which there were no significant interferences detected. In **Figures 2–4** the UHF amplitude spectra for PD measured in the needle-needle system with air bubbles, the needle-needle system and the surface system are depicted, respectively. Based on these spectra, three frequencies for each setup were selected, at which the UHF measurements were performed:

- For the NN system: 350, 480, 580 MHz
- For the NNB system: 350, 460, 560 MHz
- For the SURF system: 390, 485, 570 MHz

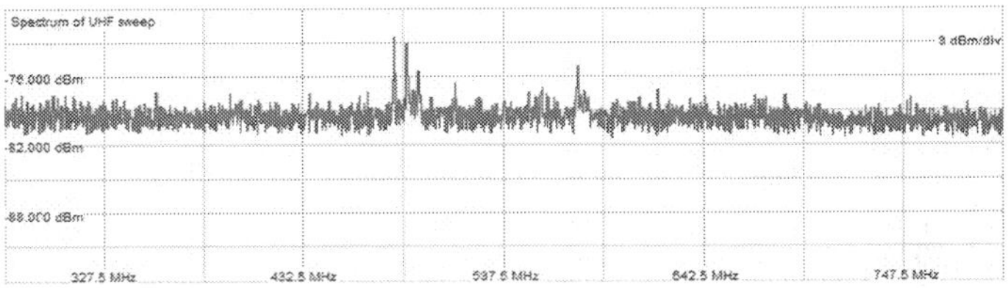

Figure 1. The UHF amplitude spectrum for background noise.

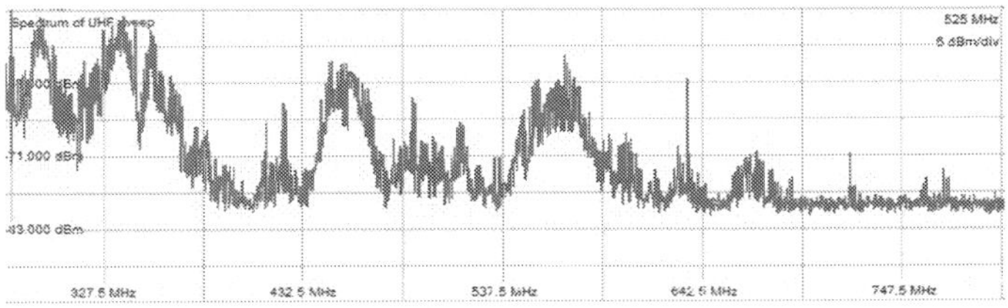

Figure 2. The UHF amplitude spectrum for PD measured in the needle-needle system with air bubbles.

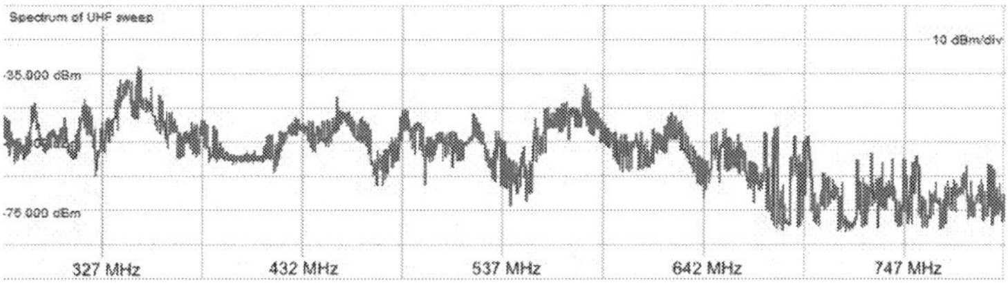

Figure 3. The UHF amplitude spectrum for PD measured in the needle-needle system.

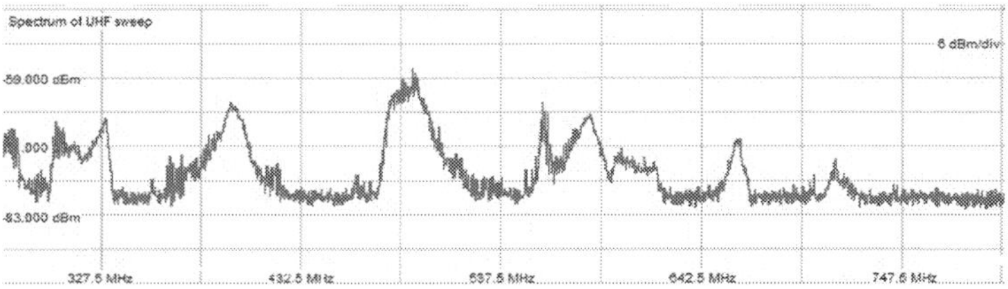

Figure 4. The UHF amplitude spectrum for PD measured in the surface system.

The gathered charges, measured with the electrical method and voltages, measured with the UHF method were subjected to further analysis. The collected data were depicted in the form of histograms, illustrating the relation of the apparent charge and voltage values, in the function of time that was registered during 1 min for each trail. An example is depicted in **Figure 5**.

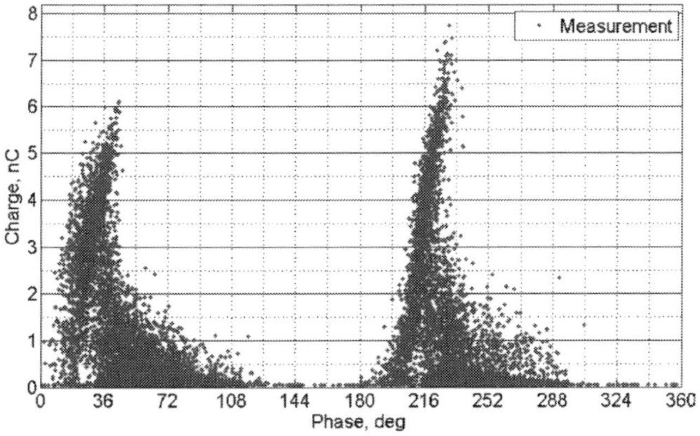

Figure 5. Charge values measured with the electrical method at the frequency f = 480 MHz, during the generation of PD at supply voltage U = 29 kV in the NN system.

For each phase of the supply voltage, the maximal intensity was chosen. The resulted curve was subjected for an interpolation using local regression and weighted linear least squares with a second-degree polynomial model. An example is depicted in **Figure 6**.

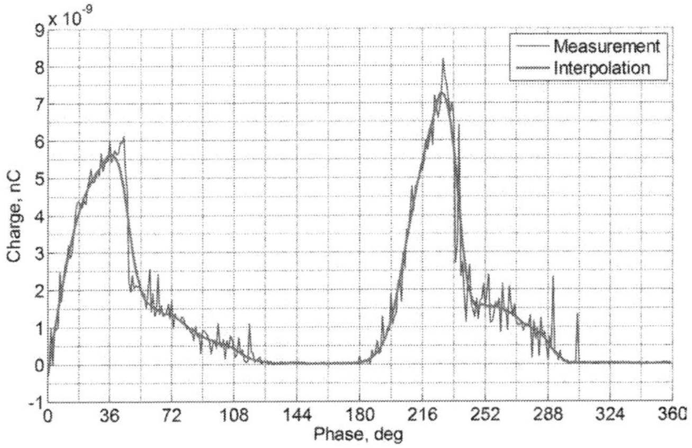

Figure 6. Envelope over the maximal charge intensities and its interpolation as the function of phase. Data measured for $f = 480$ MHz, $U = 29$ kV in the NN system.

The mentioned procedure was conducted for all measured signals. For better comparison for each of the considered PD type and generation voltage, the particular interpolates were plotted in one chart. An example is depicted in **Figure 7**.

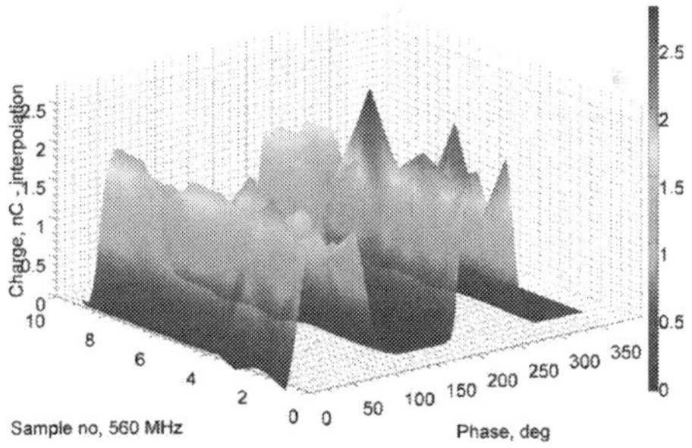

Figure 7. Interpolates over 10 measurements performed with the electrical method for $f = 560$ MHz, $U = 56$ kV in the NNB system.

The gathered interpolates for same PD type and generation voltage were averaged. The achieved arithmetical means were plotted into one chart to observe the influence of the PD generation voltage on the registered data. In **Figures 8–13**, the dependencies for the charge values and voltage values measured in the NN system are presented.

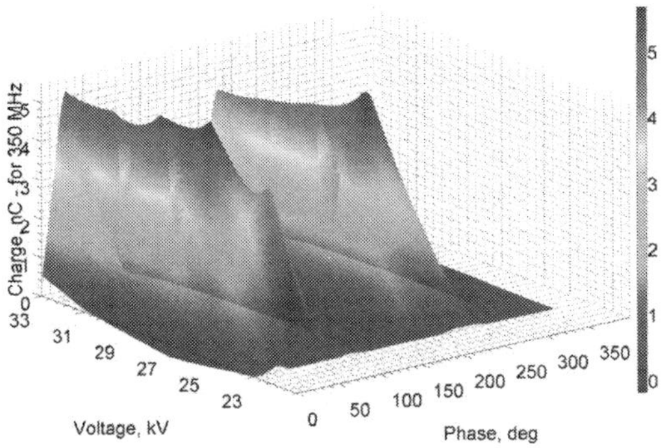

Figure 8. Charges measured at $f = 350$ MHz.

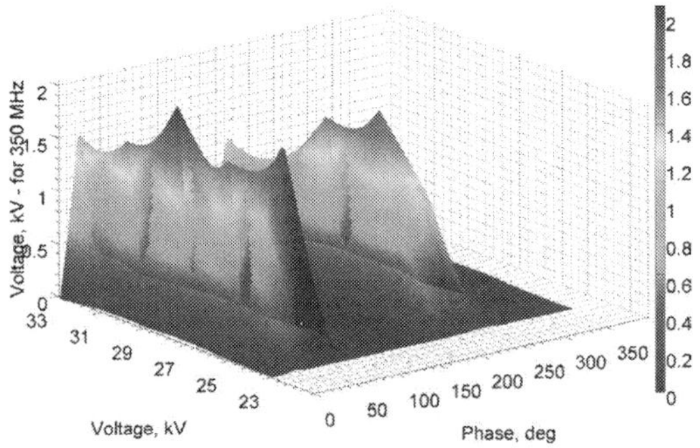

Figure 9. Voltages measured at $f = 350$ MHz.

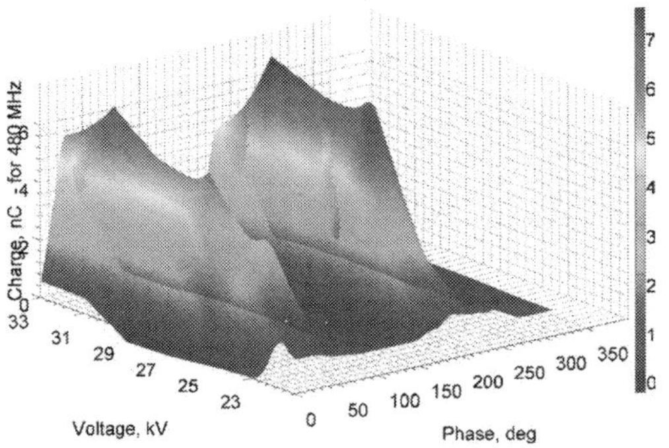

Figure 10. Charges measured at $f = 480$ MHz.

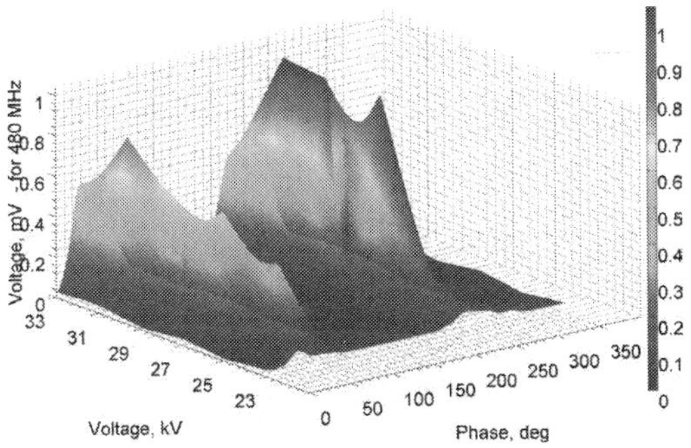

Figure 11. Voltages measured at $f = 480$ MHz.

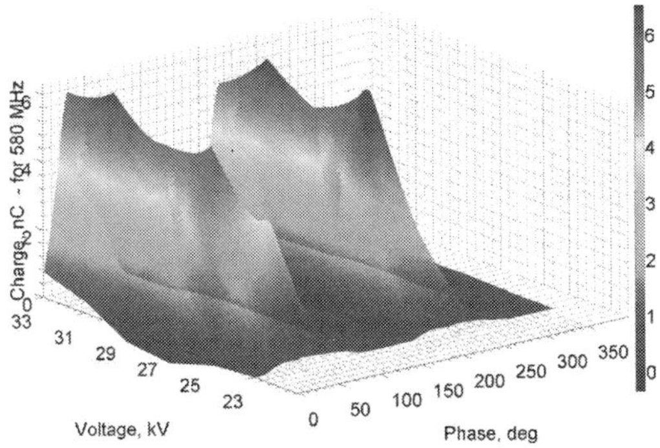

Figure 12. Charges measured at $f = 580$ MHz.

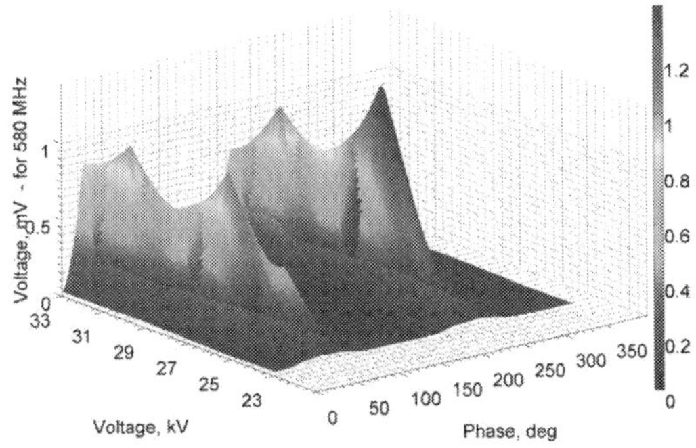

Figure 13. Voltages measured at $f = 580$ MHz.

In **Figures 14–19**, the arithmetical means over the interpolates in the function of phase and PD generation voltage, measured in the SURF system, are presented.

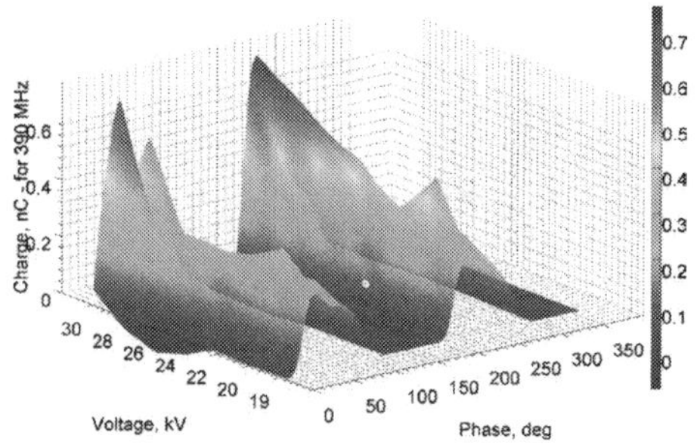

Figure 14. Charges measured at $f = 390$ MHz.

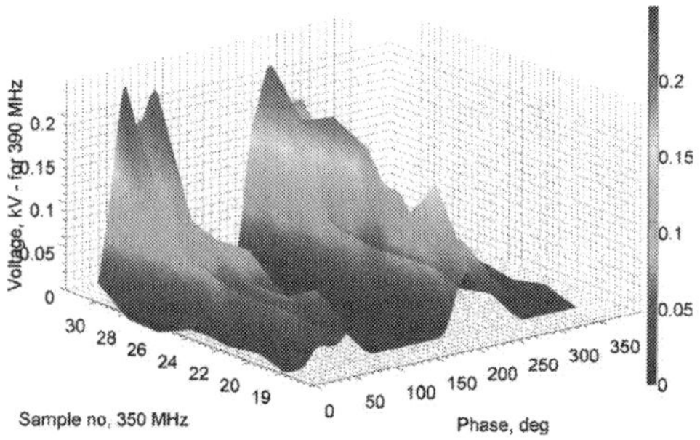

Figure 15. Voltages measured at $f = 390$ MHz.

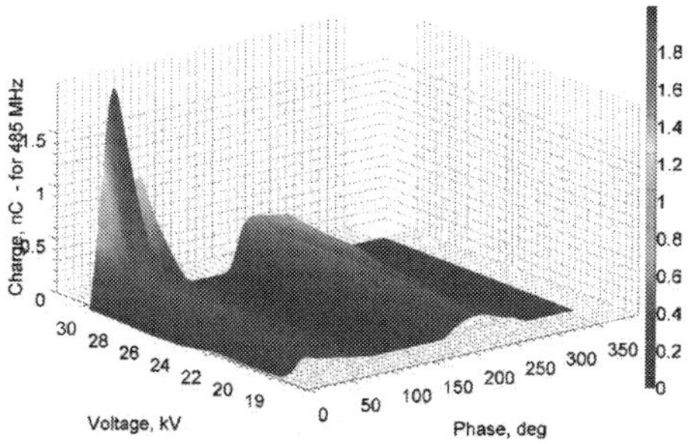

Figure 16. Charges measured at $f = 485$ MHz.

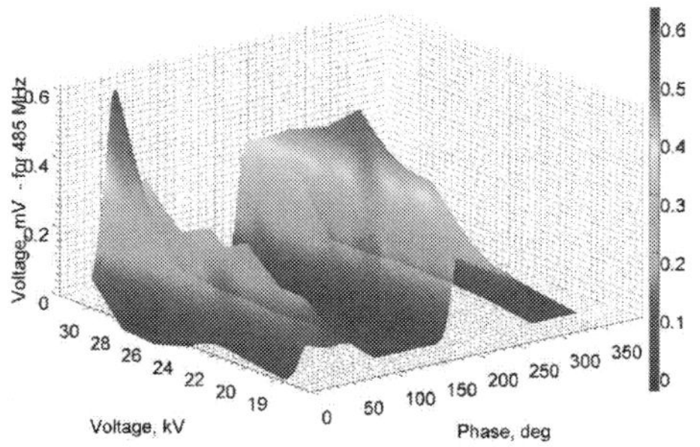

Figure 17. Voltages measured at $f = 485$ MHz.

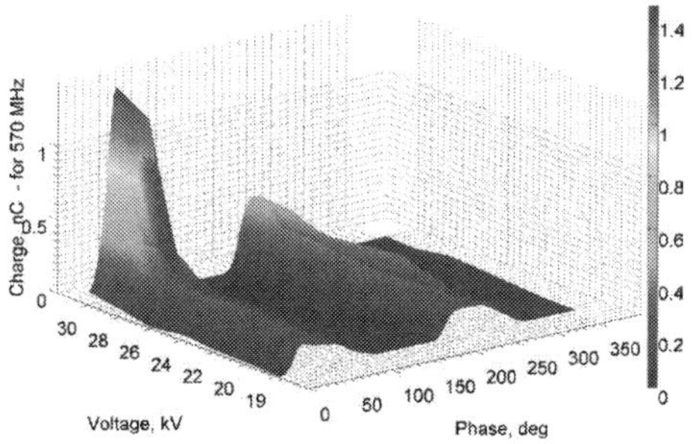

Figure 18. Charges measured at $f = 570$ MHz.

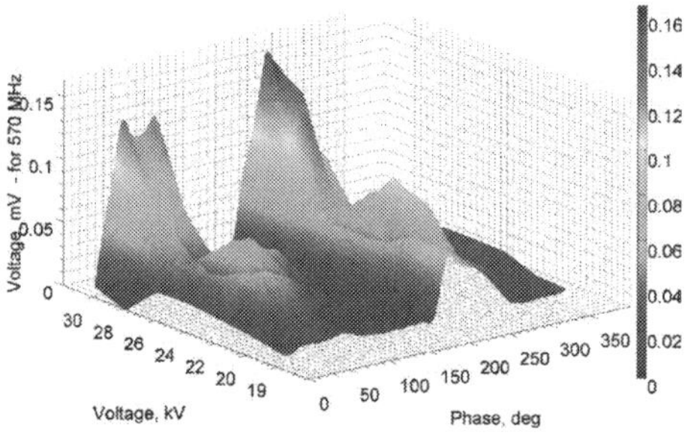

Figure 19. Voltages measured at $f = 570$ MHz.

The next type of analysis considered calculation of the empirical distribution (histograms) of the registered data for each of the phase of the PD generation voltage. In **Figures 20–25**, the results for charges and voltages registered at the selected frequencies, while generating PD at $U = 30$kV in the SURF system, are presented. In the case of charges, registered with the electrical method, there are no significant differences between the three considered frequency ranges to recognize. In the case of voltages, registered with the UHF method, the densities are different for the particular frequency ranges.

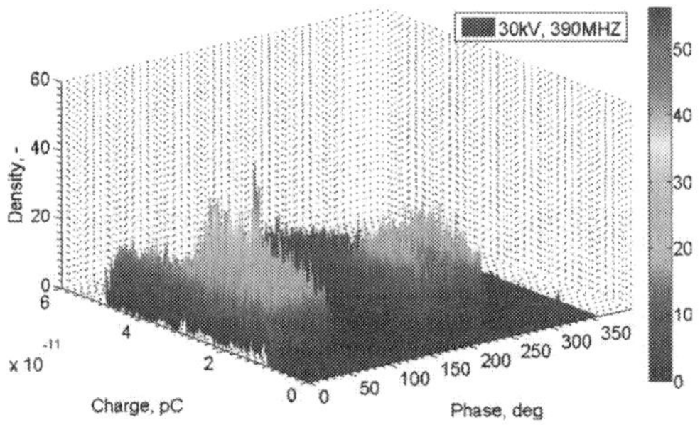

Figure 20. Charges measured at f = 390 MHz.

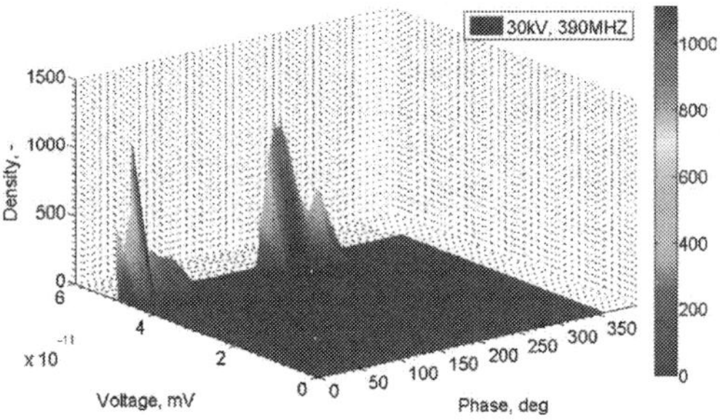

Figure 21. Voltages measured at f = 390 MHz.

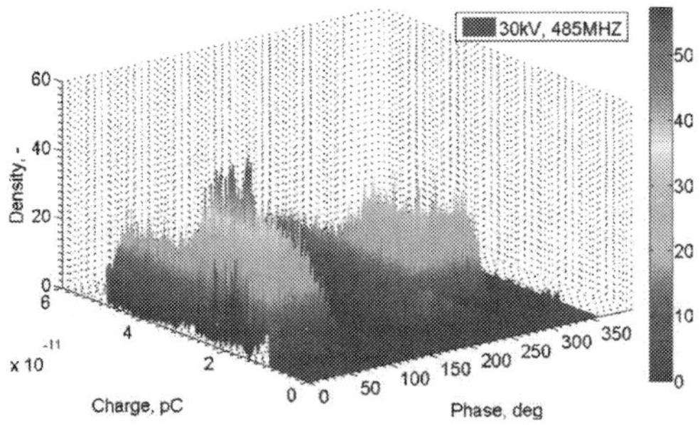

Figure 22. Charges measured at f = 485 MHz.

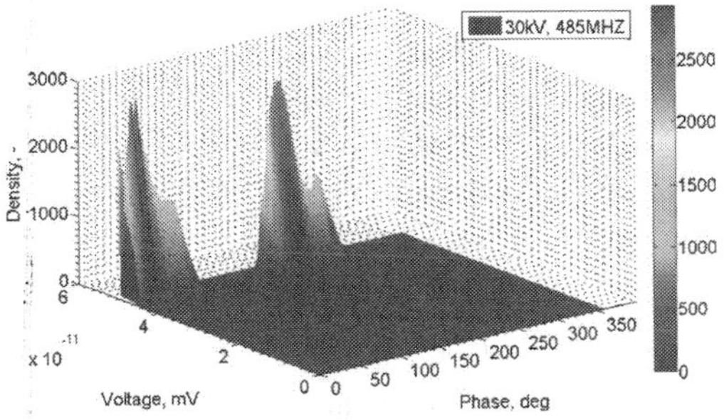

Figure 23. Voltages measured at $f = 485$ MHz.

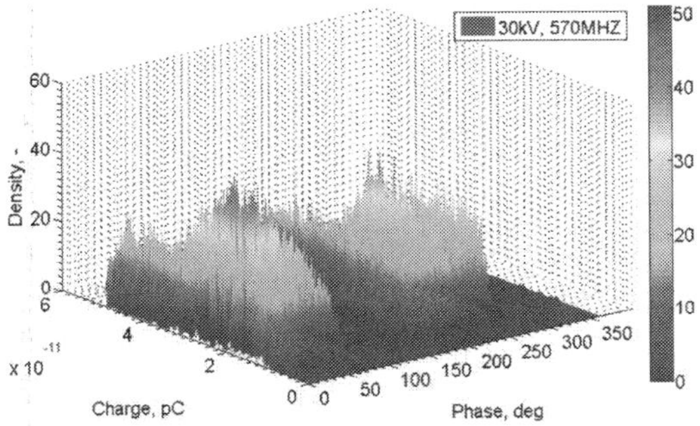

Figure 24. Charges measured at $f = 570.$MHz.

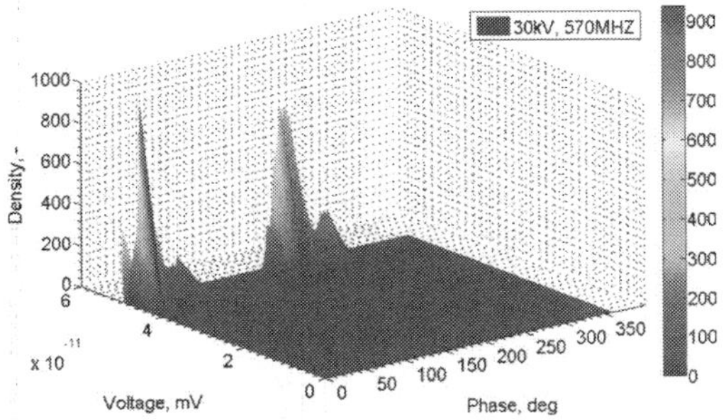

Figure 25. Voltages measured at f = 570 MHz.

In **Figures 26–31**, the results of histograms calculated for the supply voltage phases over charges registered at a chosen frequency $f = 570$ MHz, while generating PD at various voltage levels in the SURF system, are presented.

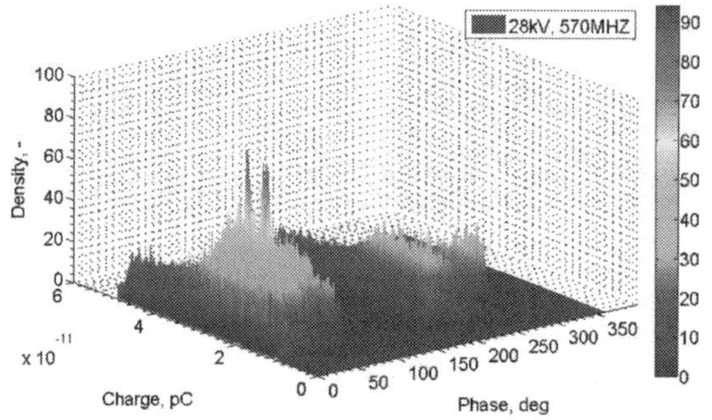

Figure 26. $U = 28$ kV.

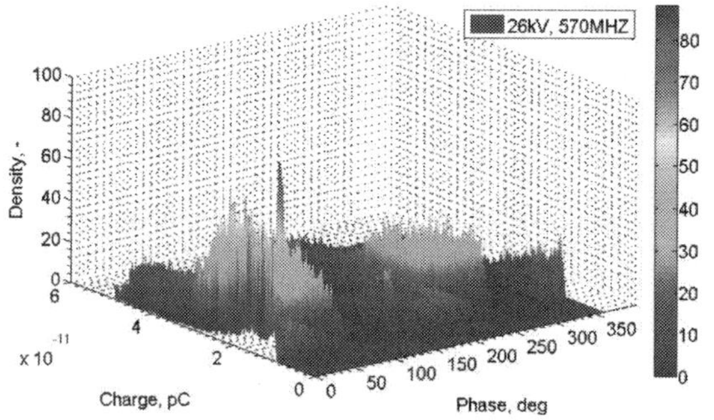

Figure 27. $U = 26$ kV.

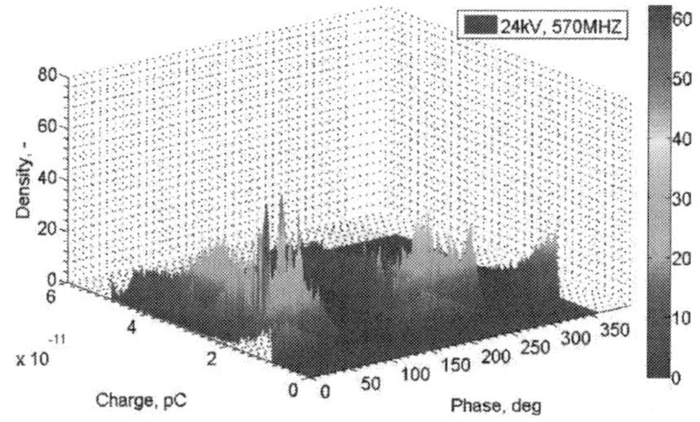

Figure 28. $U = 24$ kV.

http://dx.doi.org/10.5772/62409

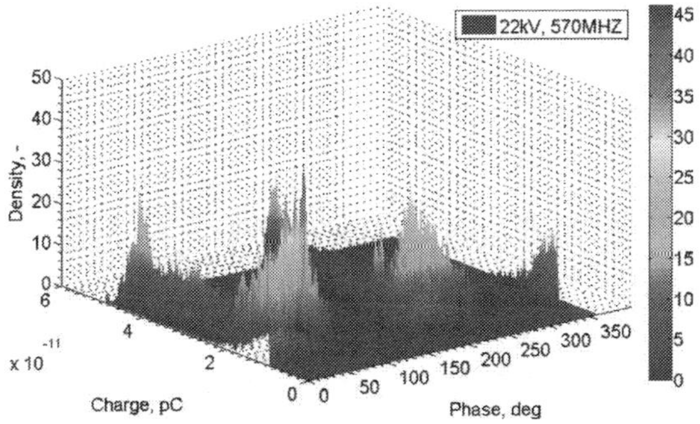

Figure 29. $U = 22$ kV.

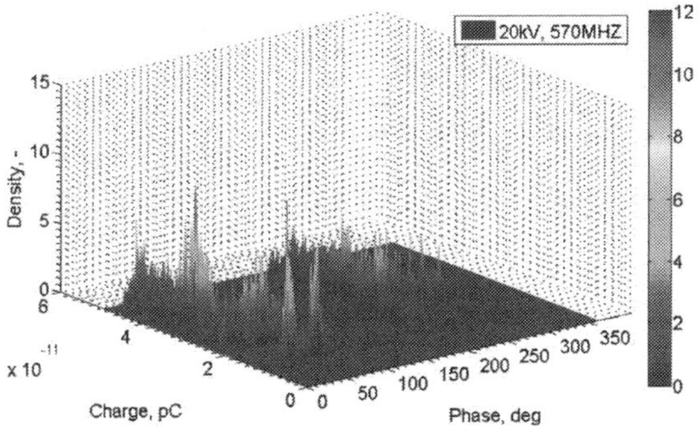

Figure 30. $U = 20$ kV.

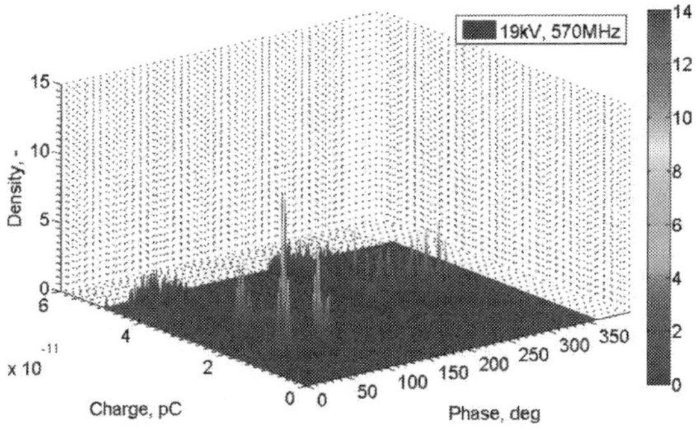

Figure 31. $U = 19$ kV.

In **Figures 32–34**, the results for histograms calculated for the supply voltage phases over charges registered while generating PD at U = 56 kV for various frequencies f in the NNB system are presented. In this electrode system, it was observed that all charts are normally distributed independently of the supply voltage phase.

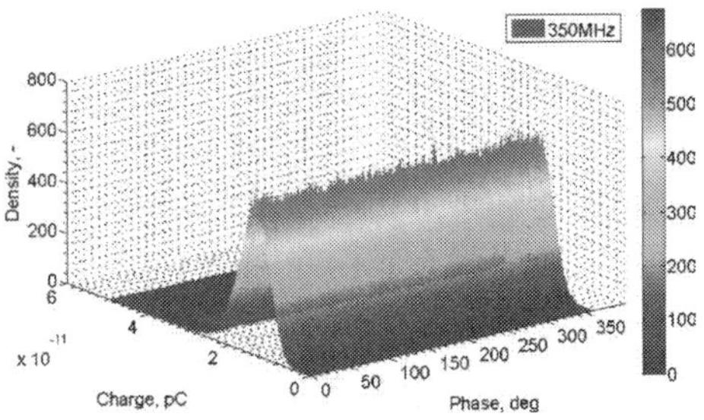

Figure 32. f = 350 MHz.

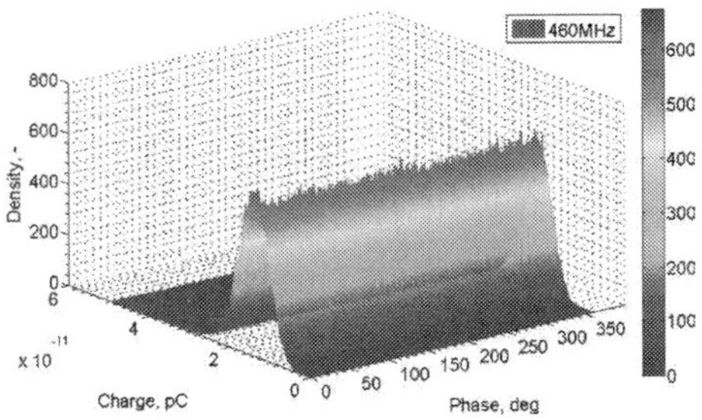

Figure 33. f = 460 MHz.

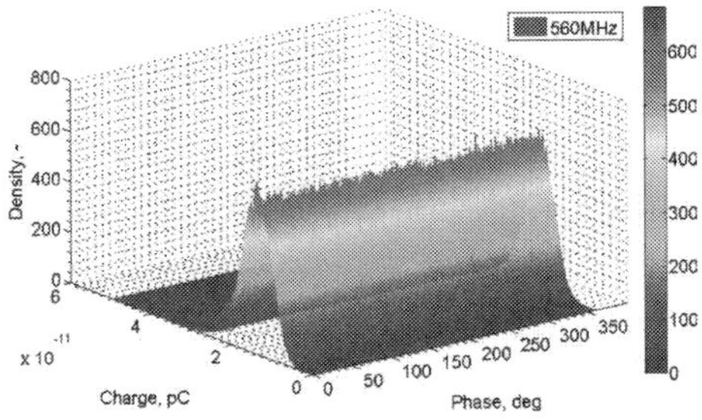

Figure 34. f = 560 MHz.

3.2. Results of analysis obtained using the spectrophotometry method

3.2.1. Results related to the needle-needle system

Measurements using the spectrophotometer were performed for different voltage values and at several distances between the measuring sensor and the PD generation area. In **Figures 35–37**, signals registered in the needle-needle system for distances 6, 7.5 and 8 mm are presented, respectively. Analysis of the charts resulted in the statement that there is no significant influence of the voltage value at the registered intensities at the three considered distances. The shape of the spectra remains similar when registered at the particular distances.

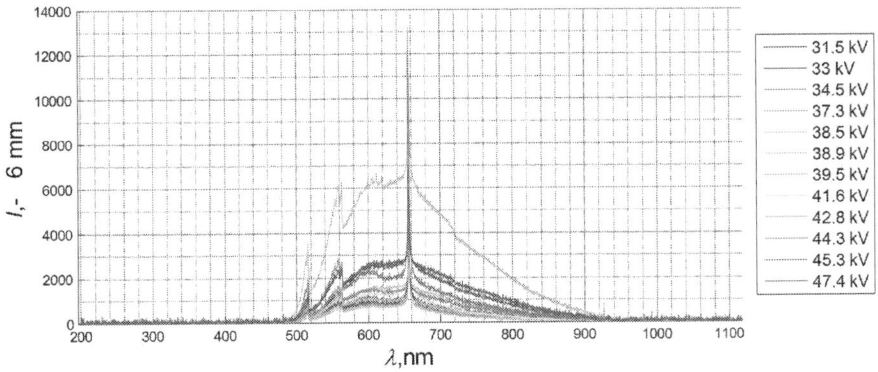

Figure 35. Spectra of the optical radiation emitted by PD generated at various supply voltage levels. The measuring sensor was placed at a distance equal to 6 mm from the HV electrode.

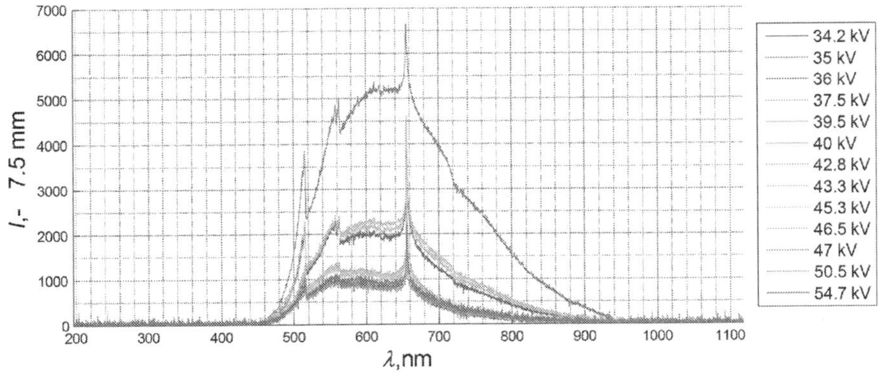

Figure 36. Spectra of the optical radiation emitted by PD generated at various supply voltage levels. The measuring sensor was placed at a distance equal to 7.5 mm from the HV electrode.

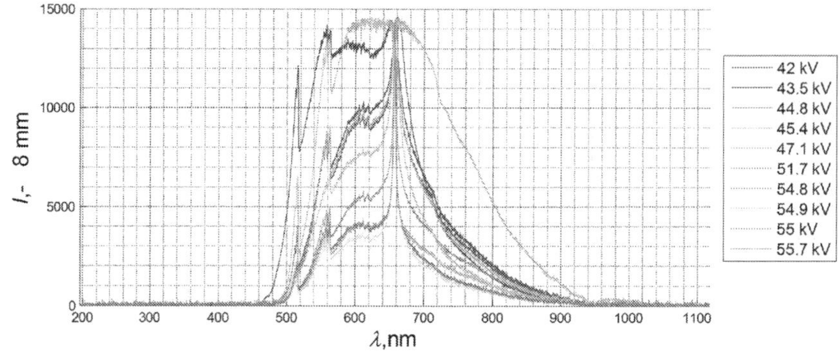

Figure 37. Spectra of the optical radiation emitted by PD generated at various supply voltage levels. The measuring sensor was placed at a distance equal to 8 mm from the HV electrode.

To confirm the above-mentioned visual analysis of voltage dependence, the maximal intensities in the registered spectra were selected for each of the trails. This type of analysis resulted in the presentation of the direct dependence of the intensity on the voltage, which is presented in **Figures 38–40** for the considered distances, respectively. In charts also, a linear approximation is presented together with the determination coefficient R^2, which was calculated from the Pearson equation.

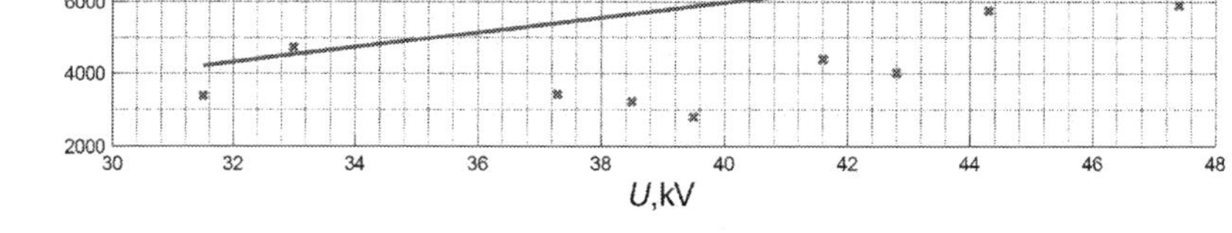

Figure 38. Maximal intensity registered at the distance 6 mm, in the function of PD generation voltage.

Analyzing the curve depicted in **Figure 38**, one may consider an increasing dependency of the voltage on intensity. Although the very low R^2 coefficient implies that the linear curve is not well-approximated. Reciprocal dependency was observed for data gathered at the distance of 7.5 (**Figure 39**) and 8 mm (**Figure 40**). In these cases, the R^2 values also imply low approximation quality. Based on the above, the lacking dependence of the voltage on the intensity was confirmed.

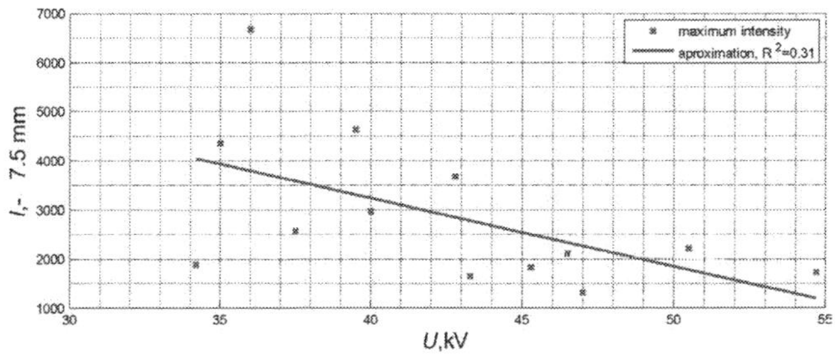

Figure 39. Maximal intensity registered at the distance 7.5 mm, in the function of PD generation voltage.

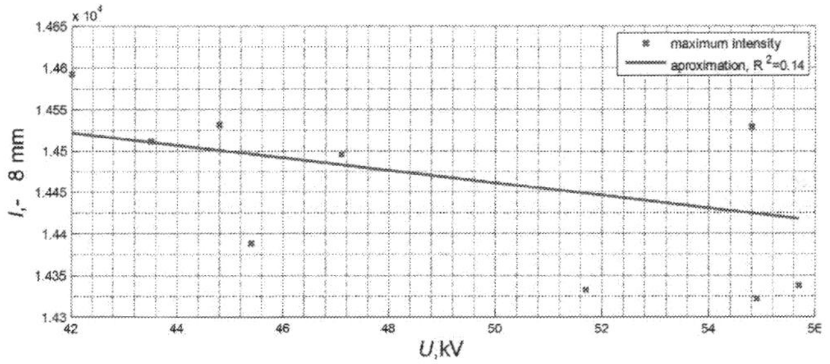

Figure 40. Maximal intensity registered at the distance 8 mm, in the function of PD generation voltage.

In the next step, the energy contained in the registered spectra was determined based on defined integrals calculated within the considered frequency range. In **Figures 41–43**, the estimated integrals in the function of supply voltage value are presented for the considered distances, respectively.

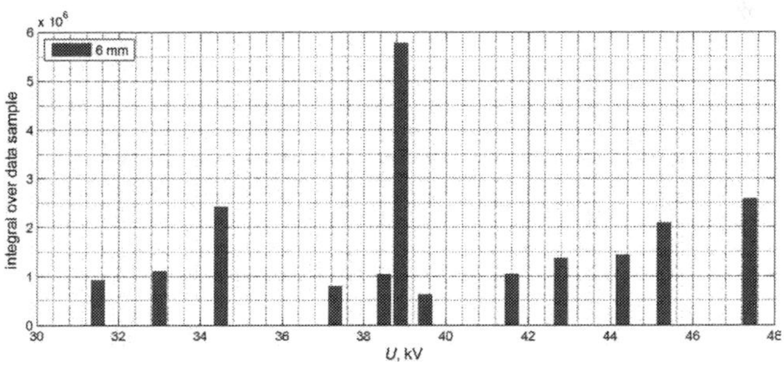

Figure 41. Integrals calculated for intensity spectra in the function of the PD generation voltage, while the sensor was installed at a distance equal to 6 mm.

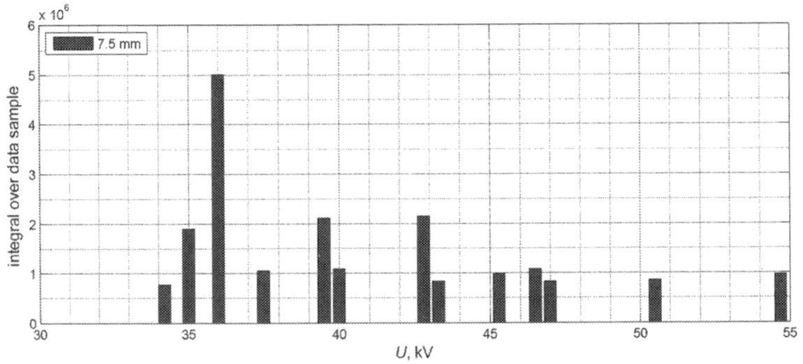

Figure 42. Integrals calculated for intensity spectra in the function of the PD generation voltage, while the sensor was installed at a distance equal to 7.5 mm.

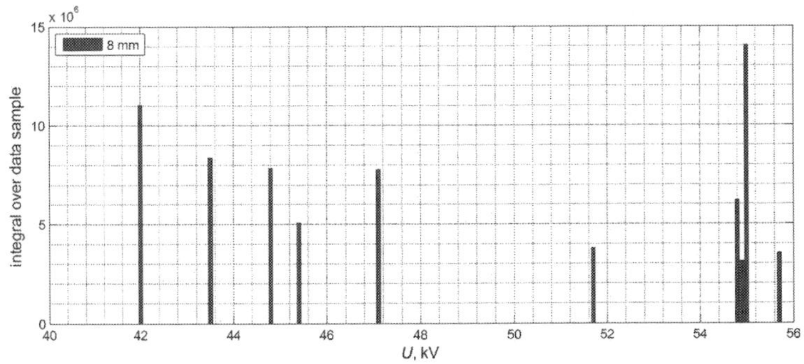

Figure 43. Integrals calculated for intensity spectra in the function of the PD generation voltage, while the sensor was installed at a distance equal to 8 mm.

At the distance of 6 mm, the highest energy was estimated for spectra registered at a supply voltage U = 39 kV. At the distance of 7.5 mm, the highest energy was estimated for spectra registered at a supply voltage U = 36 kV. At the distance of 8 mm, the highest energy was estimated for spectra registered at a supply voltage U = 55 kV. No significant dependency of the voltage value on the energy was recognized, contained in the spectra regardless of the distance at which the measuring sensor was installed.

3.2.2. Results related to the needle-needle system with air bubbles

In **Figures 44–47**, signals registered in the needle-needle system with air bubbles for distances 5, 6, 7.5 and 8 mm are presented, respectively. Analyses of the charts resulted with the same statement, as it was the case of NN system, namely, that there is no significant influence of the voltage value on the registered intensities at the considered distances. The shape of the spectra remains similar when registered at the particular distances and is also very similar as it was the case of NN system.

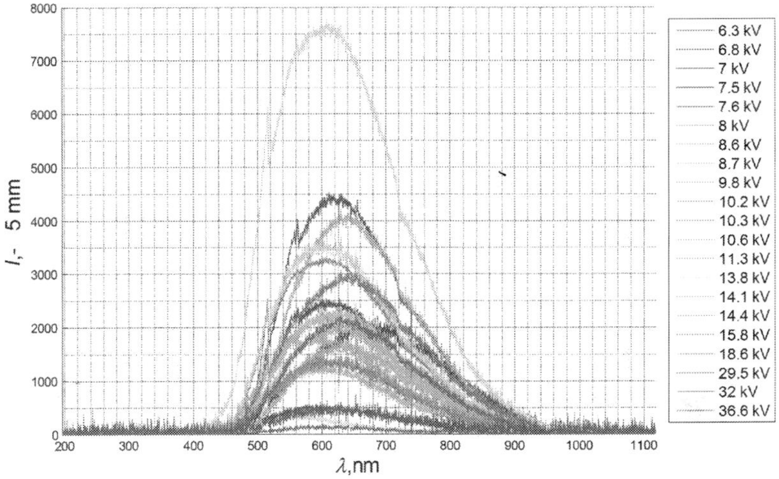

Figure 44. Spectra of the optical radiation emitted by PD generated at various supply voltage levels. The measuring sensor was placed at a distance equal to 5 mm from the HV electrode.

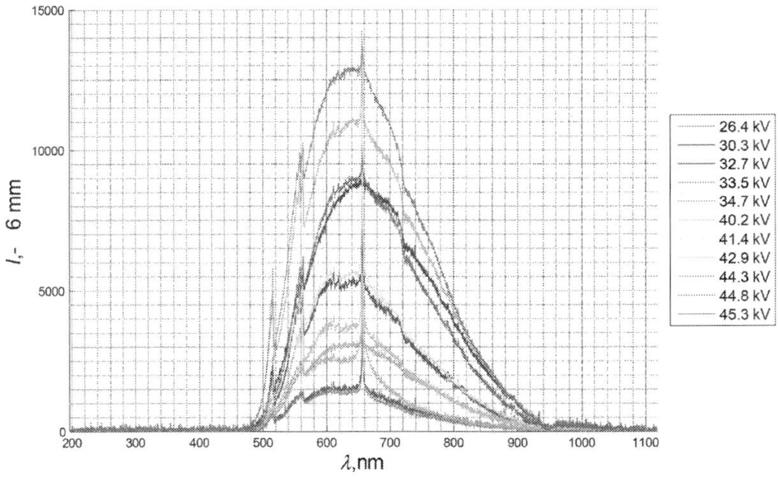

Figure 45. Spectra of the optical radiation emitted by PD generated at various supply voltage levels. The measuring sensor was placed at a distance equal to 6 mm from the HV electrode.

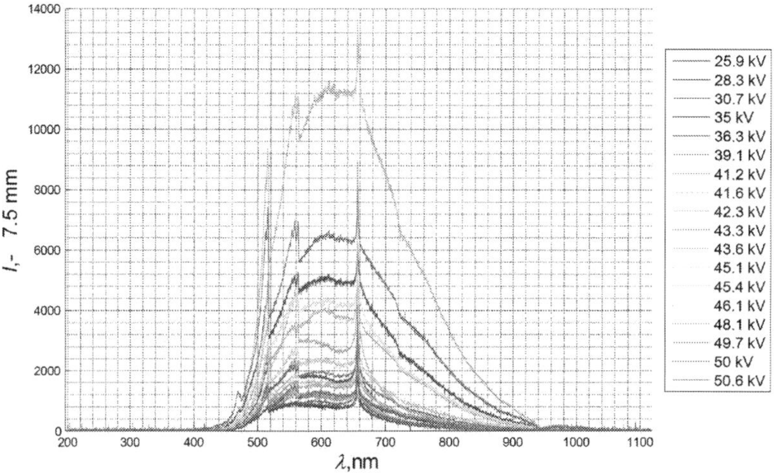

Figure 46. Spectra of the optical radiation emitted by PD generated at various supply voltage levels. The measuring sensor was placed at a distance equal to 7.5 mm from the HV electrode.

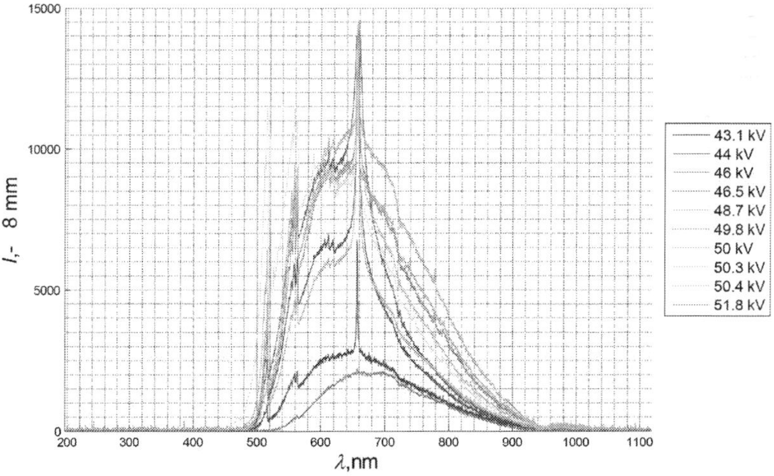

Figure 47. Spectra of the optical radiation emitted by PD generated at various supply voltage levels. The measuring sensor was placed at a distance equal to 8 mm from the HV electrode.

Based on the relationships, depicted in **Figures 48–51**, the lacking dependency of the voltage on the registered intensities for the considered distances was confirmed.

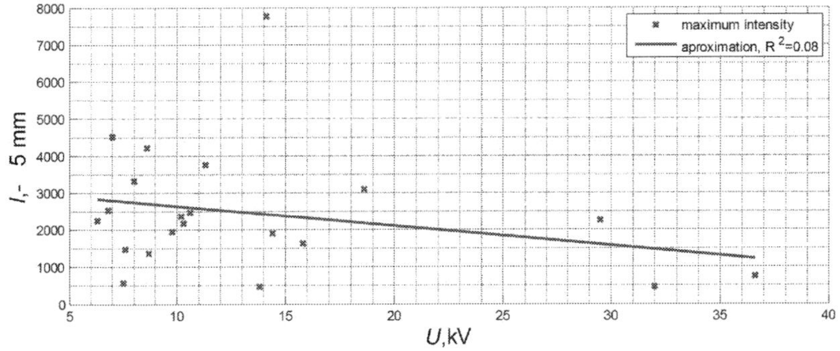

Figure 48. Maximal intensity registered at the distance 5 mm, in the function of PD generation voltage.

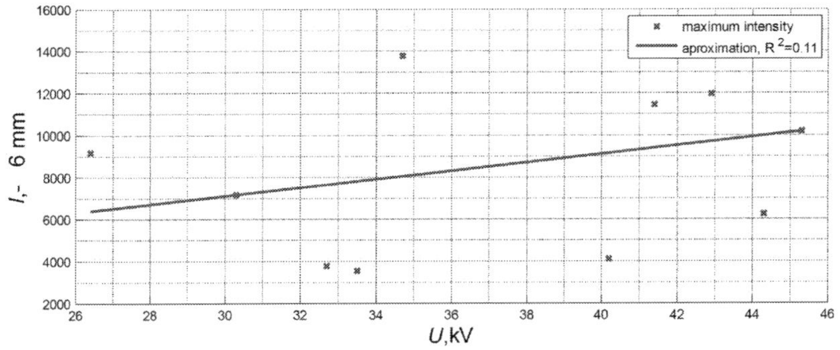

Figure 49. Maximal intensity registered at the distance 6 mm, in the function of PD generation voltage.

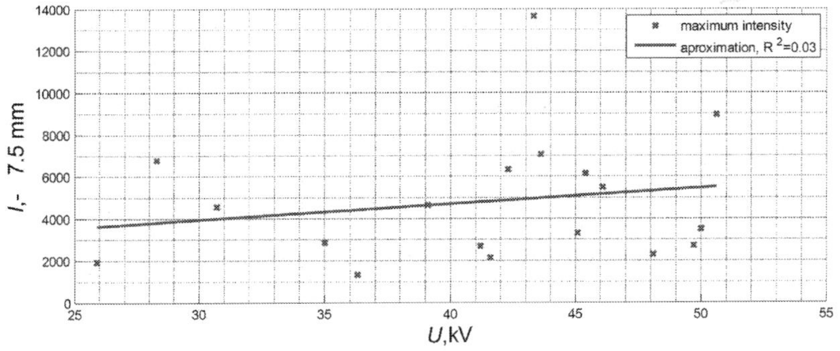

Figure 50. Maximal intensity registered at the distance 7.5 mm, in the function of PD generation voltage.

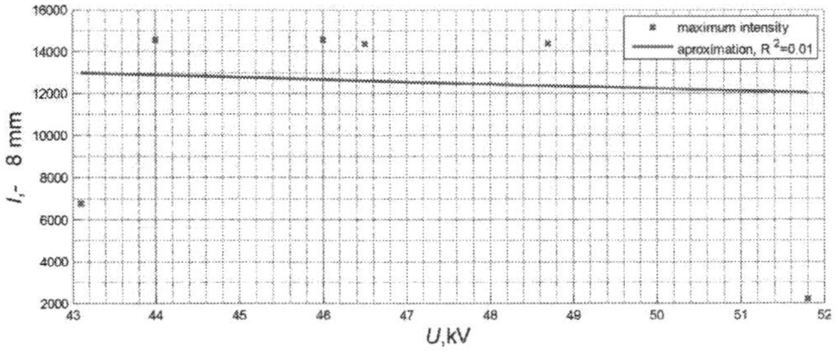

Figure 51. Maximal intensity registered at the distance 8 mm, in the function of PD generation voltage.

The energy values contained in the registered spectra were determined for the analyzed supply voltage levels. The calculation results are presented in **Figures 52–55** for the considered distances, respectively.

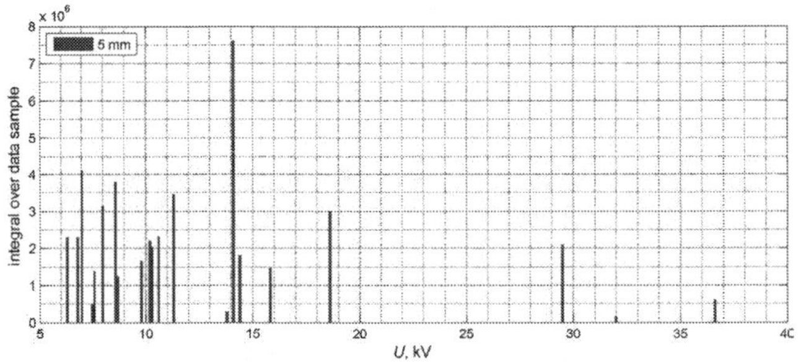

Figure 52. Integrals calculated for intensity spectra in the function of the PD generation voltage, while the sensor was installed at a distance equal to 5 mm.

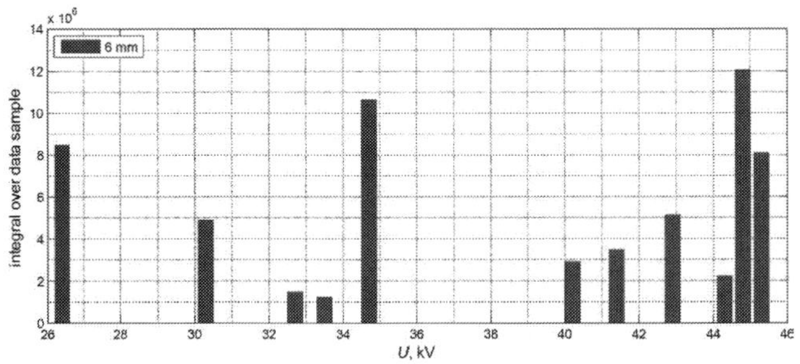

Figure 53. Integrals calculated for intensity spectra in the function of the PD generation voltage, while the sensor was installed at a distance equal to 6 mm.

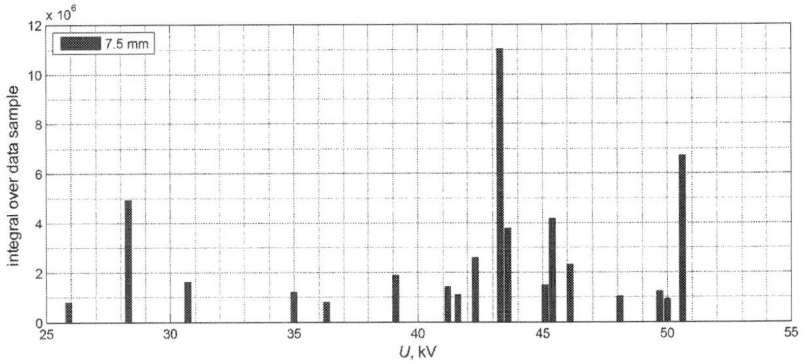

Figure 54. Integrals calculated for intensity spectra in the function of the PD generation voltage, while the sensor was installed at a distance equal to 7.5 mm.

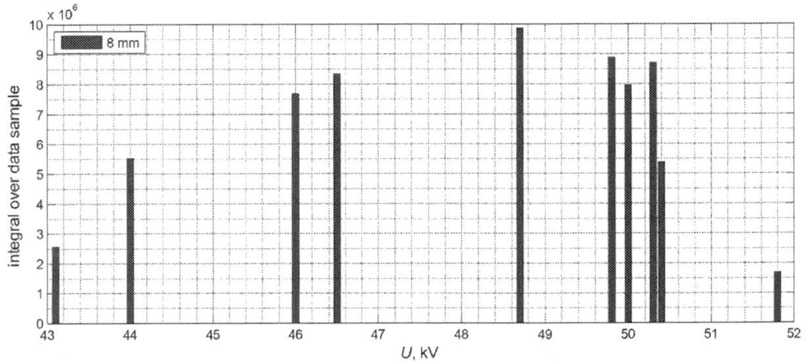

Figure 55. Integrals calculated for intensity spectra in the function of the PD generation voltage, while the sensor was installed at a distance equal to 8 mm.

At the distance of 8 mm, high energies were estimated for spectra registered at the supply voltage in the range from 46 to 51 kV. At the distance of 7.5 mm, the highest energy was estimated for spectra registered at a supply voltage $U = 33$ kV. At the distance of 6 mm, the highest energy was estimated for spectra registered at a supply voltage $U = 45$ kV. At the distance 5 mm, the highest energy was estimated for spectra registered at a supply voltage $U = 14$ kV. No significant dependency of the voltage value on the energy was recognized, contained in the spectra regardless of the distance at which the measuring sensor was installed.

3.2.3. Results related to the surface system

In contrast to the previously considered NN and NNB systems, in the SURF system a clear dependency of voltage on the registered intensity spectra was observed. In **Figures 56–58**, intensities registered in the surface system for distances 4, 6 and 8 mm are presented, respectively. The shape of the spectra remains similar when registered at the particular distances.

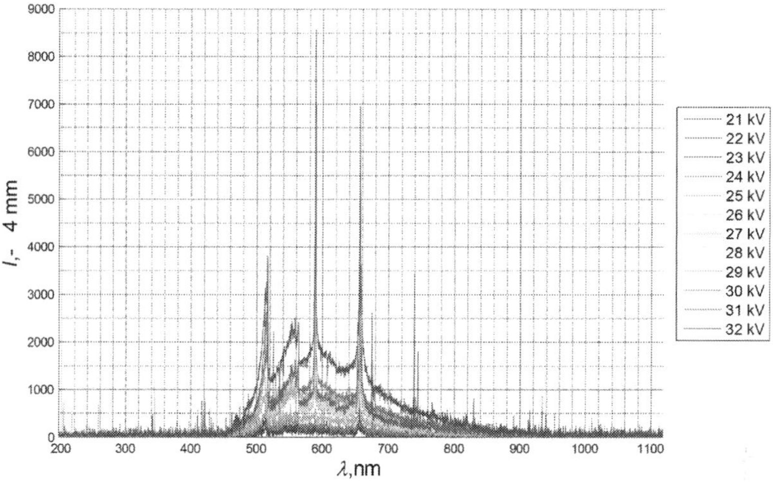

Figure 56. Spectra of the optical radiation emitted by PD generated at various supply voltage levels. The measuring sensor was placed at a distance equal to 4 mm from the HV electrode.

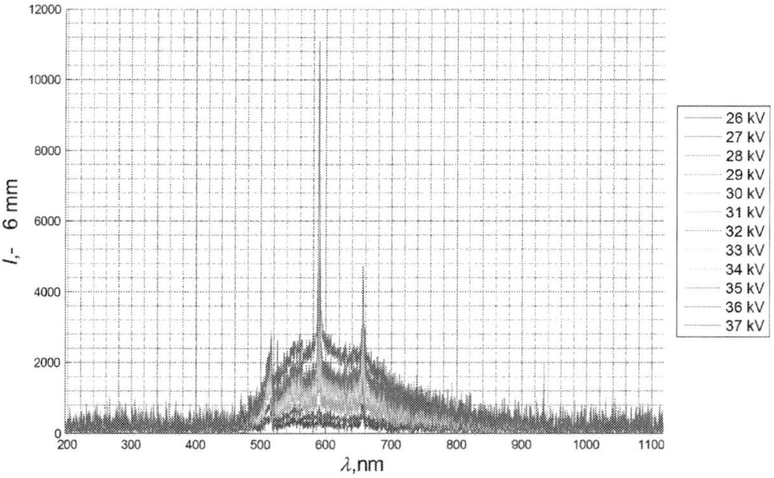

Figure 57. Spectra of the optical radiation emitted by PD generated at various supply voltage levels. The measuring sensor was placed at a distance equal to 6 mm from the HV electrode.

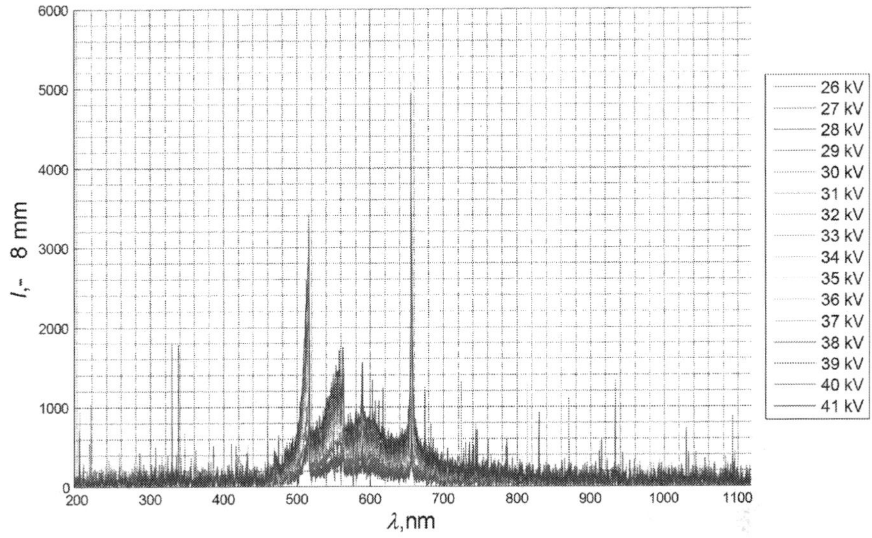

Figure 58. Spectra of the optical radiation emitted by PD generated at various supply voltage levels. The measuring sensor was placed at a distance equal to 8 mm from the HV electrode.

The above-mentioned clear observed dependency was confirmed by the relationships depicted in **Figures 59–61**. Also the linear approximation, which indicates higher R^2 values, confirms the given statement.

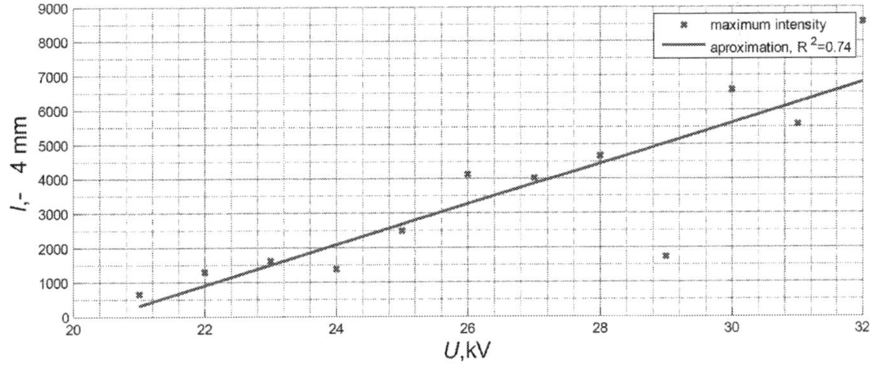

Figure 59. Maximal intensity registered at the distance 4 mm, in the function of PD generation voltage.

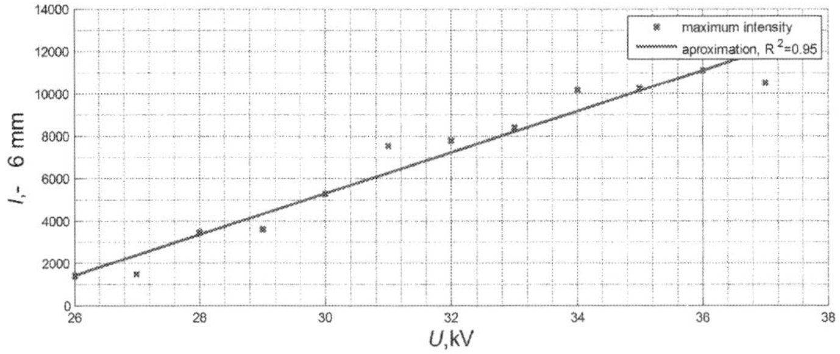

Figure 60. Maximal intensity registered at the distance 6 mm, in the function of PD generation voltage.

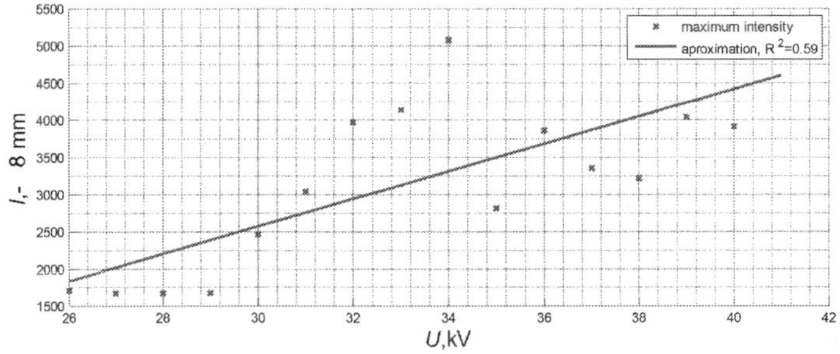

Figure 61. Maximal intensity registered at the distance 8 mm, in the function of PD generation voltage.

The energy values contained in the registered spectra were determined for the analyzed supply voltage levels. The calculation results are presented in **Figures 62–64** for the considered distances, respectively.

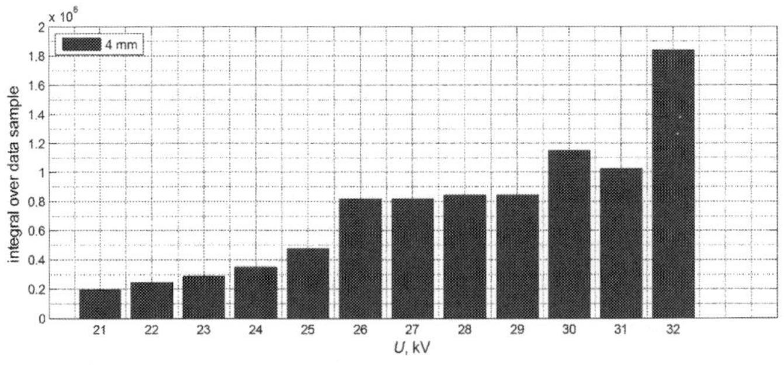

Figure 62. Integrals calculated for intensity spectra in the function of the PD generation voltage, while the sensor was installed at a distance equal to 4 mm.

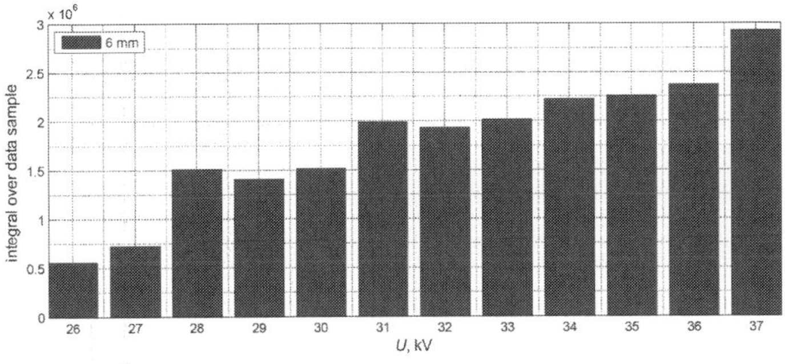

Figure 63. Integrals calculated for intensity spectra in the function of the PD generation voltage, while the sensor was installed at a distance equal to 6 mm.

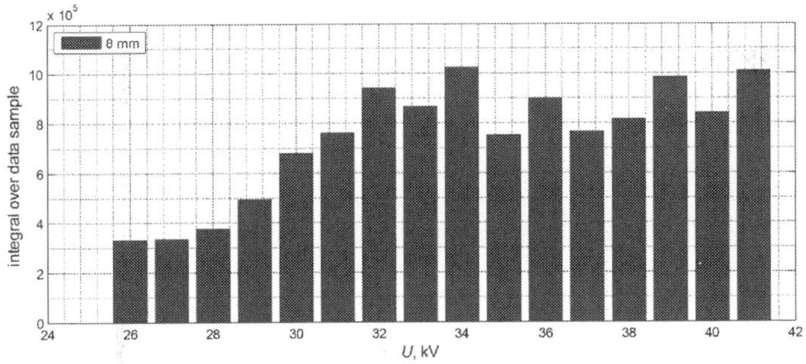

Figure 64. Integrals calculated for intensity spectra in the function of the PD generation voltage, while the sensor was installed at a distance equal to 8 mm.

Analysis of the energies calculated for the spectra registered at various distances enabled for the statement that these energies increase with the increasing value of PD generation voltage. Further, it was observed that at smaller distances (4 and 6 mm) the relationship is monotonic, despite to signals registered at 8 mm, where the energy remains almost constant, after the supply voltage achieves the value of 32 kV.

3.3 Results of analysis obtained using the acoustic emission method

For each of the regarded PD modelling systems and for various supply voltage values, over one hundred trails were recorded using the AE method. In the NN and SURF systems, a broad range of voltage values was analyzed, but for the NNB system only the 25 kV was involved. The results consider time runs and power spectral densities. In **Figures 65, 67,** and **69**, time runs of AE signals registered while generating PD at selected voltage levels in the SURF, NN and NNB systems are presented, respectively. In **Figures 66, 68,** and **70**, the corresponding power spectral densities (PSD) are presented. For the PSD spectra, the local maxima (extreme values) were calculated and were marked with red crosses on the charts. These values were subjected for further analysis in the frequency domain.

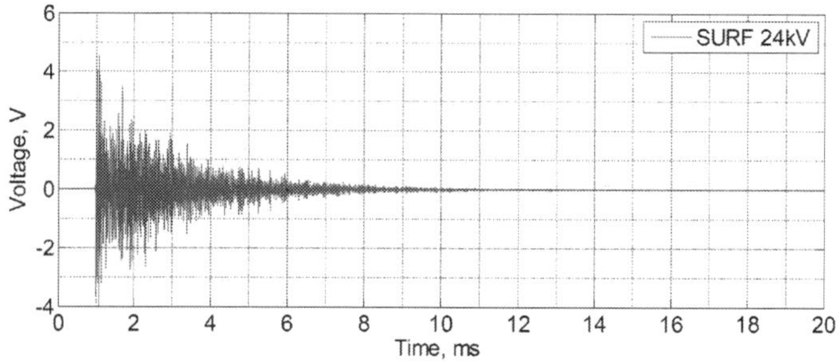

Figure 65. Time run of AE in the SURF system.

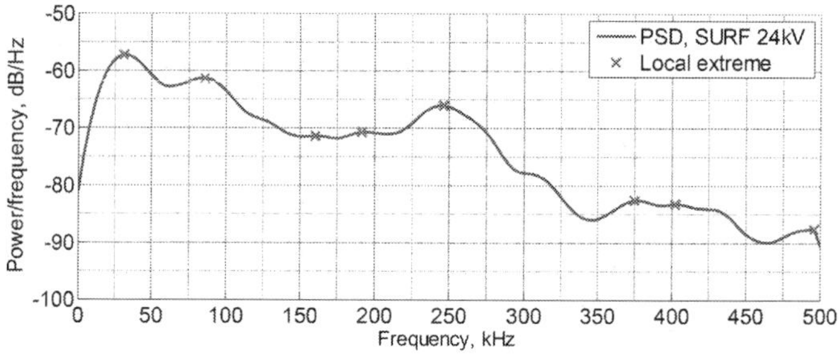

Figure 66. The corresponding PSD spectra.

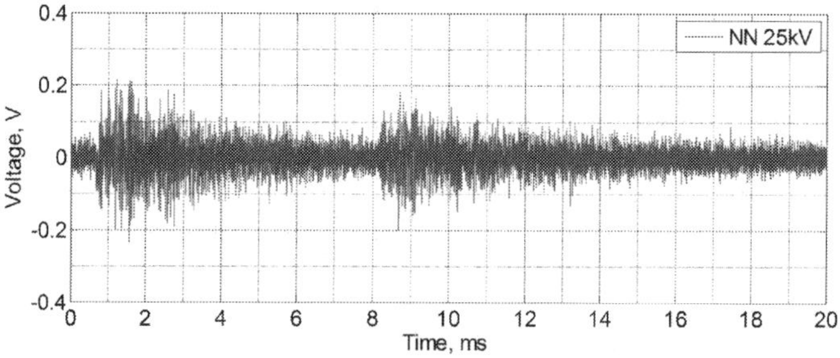

Figure 67. Time run of AE in the NN system.

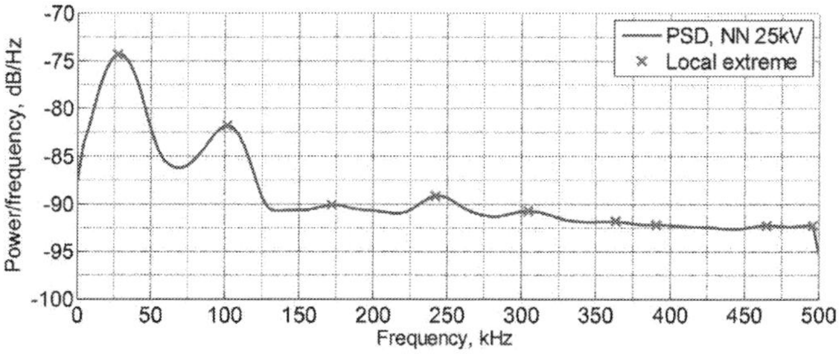

Figure 68. The corresponding PSD spectra.

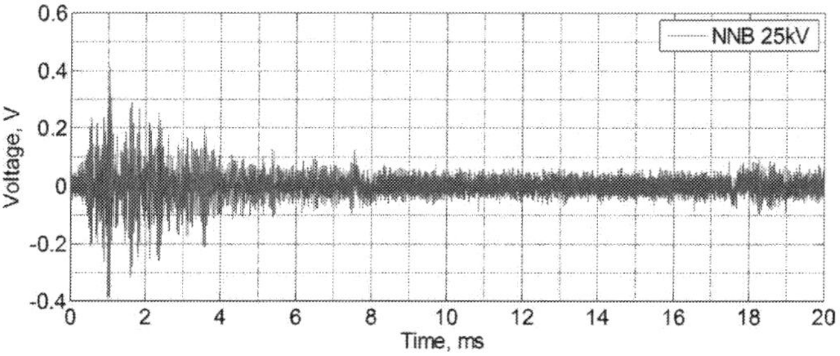

Figure 69. Time run of AE in the NNB system.

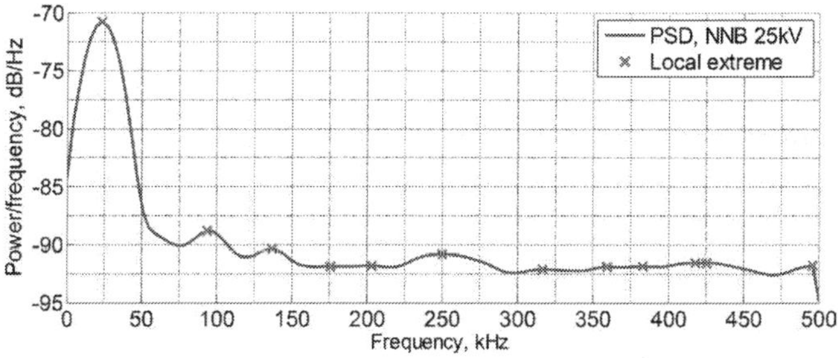

Figure 70. The corresponding PSD spectra.

The above-mentioned local extremes were subjected for empirical distribution analysis (histograms) with respect to the particular frequency values contained in the PSD spectra. In **Figures 71–73**, example results are presented for the three electrodes arrangements: NN (**Figure 71**), SURF (**Figure 72**), and NNB(**Figure 73**). The charts depict the distributions of local extremes in the PSD spectra, calculated for all measured data samples.

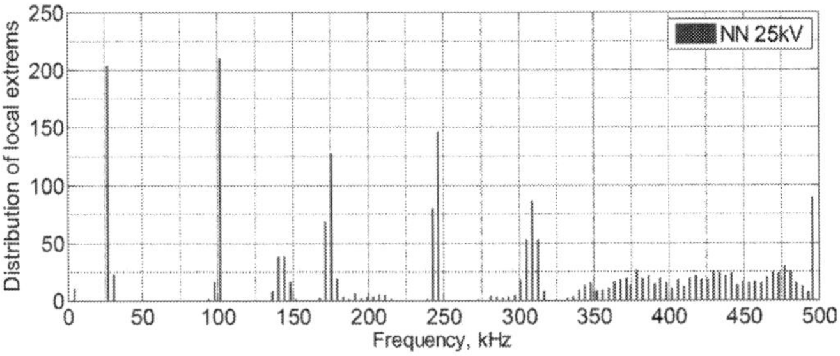

Figure 71. Empirical distributions (histograms) of local extreme values determined for all PSD spectra with respect to frequency for the NN system.

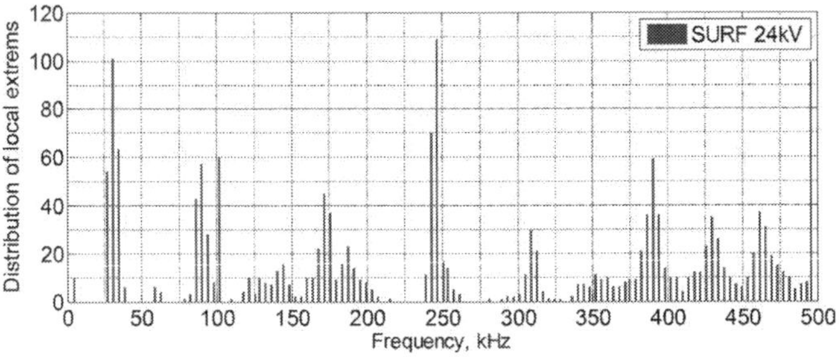

Figure 72. Empirical distributions (histograms) of local extreme values determined for all PSD spectra with respect to frequency for the SURF system.

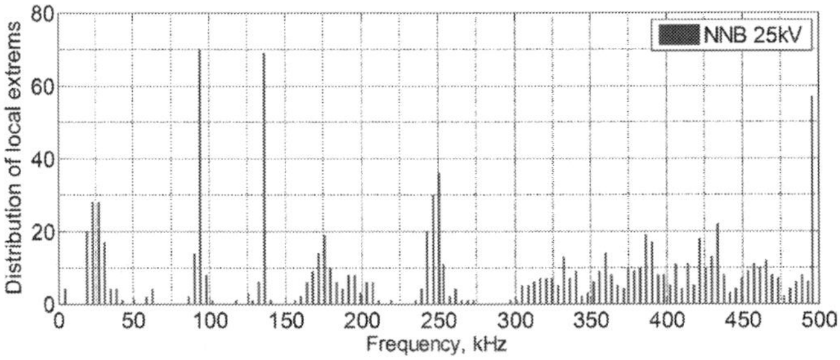

Figure 73. Empirical distributions (histograms) of local extreme values determined for all PSD spectra with respect to frequency for the NNB system.

For better comparison purposes, the particular charts were combined into a single figure, what has made possible to observe if any dependency of the PD generation voltage on the achieved results exists. In **Figures 74–75**, the mentioned relationships for the NN (**Figure 74**) and SURF

(**Figure 45**) are depicted. From the charts presented, it may be implied that there is no significant influence of PD generation voltage on the frequency components included in the PSD spectra, which were previously calculated for the AE signals registered in the NN and SURF system.

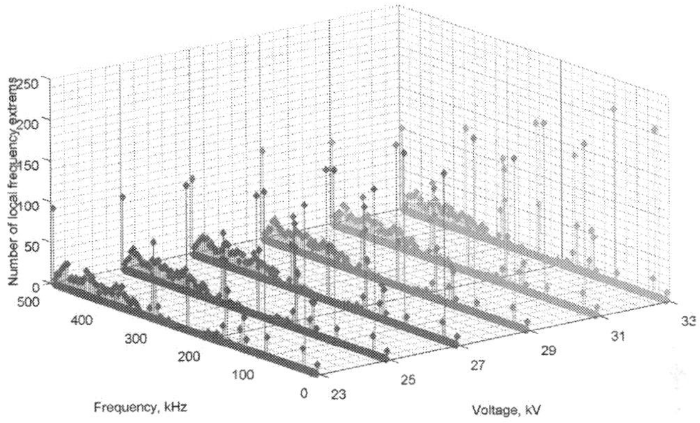

Figure 74. Relationship between PD generation voltage and the empirical distribution of local extreme values determined from the PSD spectra of AE signals registered in the NN system.

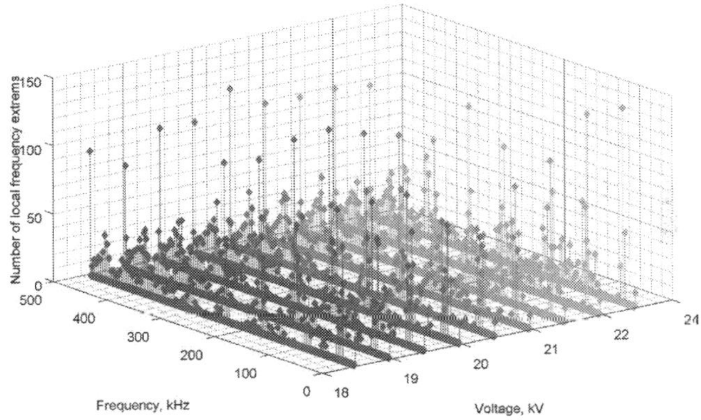

Figure 75. Relationship between PD generation voltage and the empirical distribution of local extreme values determined from the PSD spectra of AE signals registered in the SURF system.

3.4 Results of analysis obtained using the thermal imaging

The thermal imaging method was applied for all the earlier described measuring setups. Based on the results achieved, it was stated that there is no significant increase of temperature after such a long time as 8 hours. Since the applied camera was not able to be immersed in the insulation oil, the eventual temperature variations, which are supposed to occur in the vicinity of the PD generation area, were not possible to detect during the performed research. In **Figures 76–77**, the example result of thermal image and the corresponding photo made in the visible frequency range are presented.

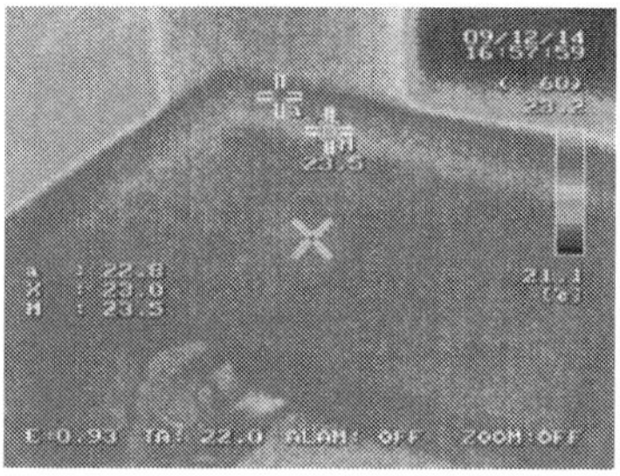

Figure 76. Example image made with the thermal camera.

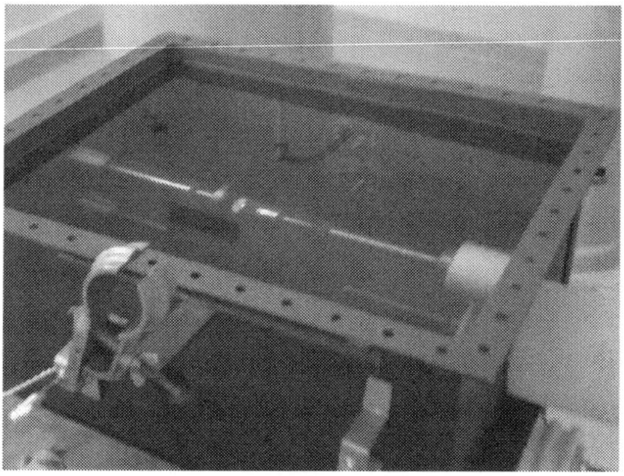

Figure 77. The corresponding photo of the investigated object.

3.5 Results of analysis obtained using the chromatography method

In order to analyze the chemical parameters of the insulating oil applied in the research, six oil samples were collected: first sample — at the beginning of the study, second sample — after 3 hours of PD generation, third, fourth, fifth and sixth samples — each after subsequent hour of PD generation. In **Figures 78–91**, the dendencies are presented showing the influence of PD generation time on selected parameters determined using the chromatography method.

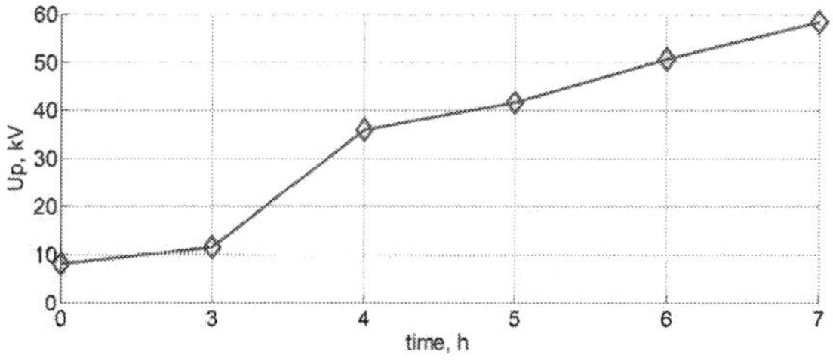

Figure 78. Up.

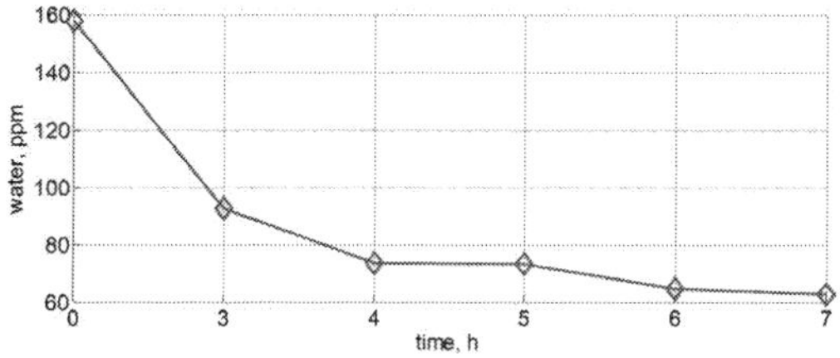

Figure 79. Water.

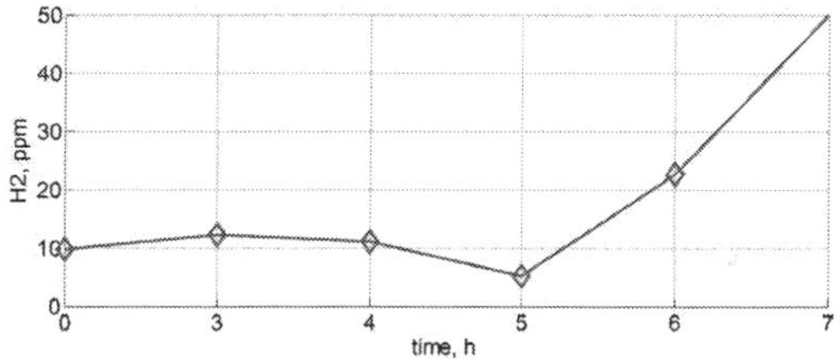

Figure 80. H2.

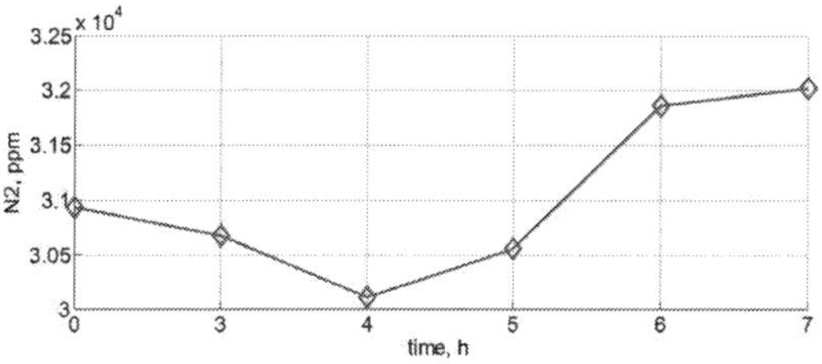

Figure 81. N2.

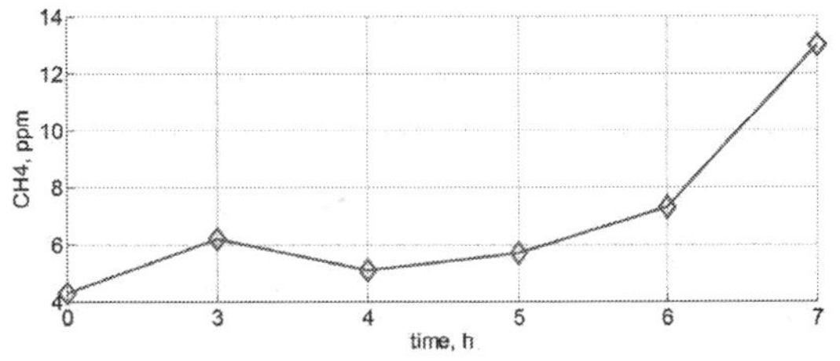

Figure 82. CH4.

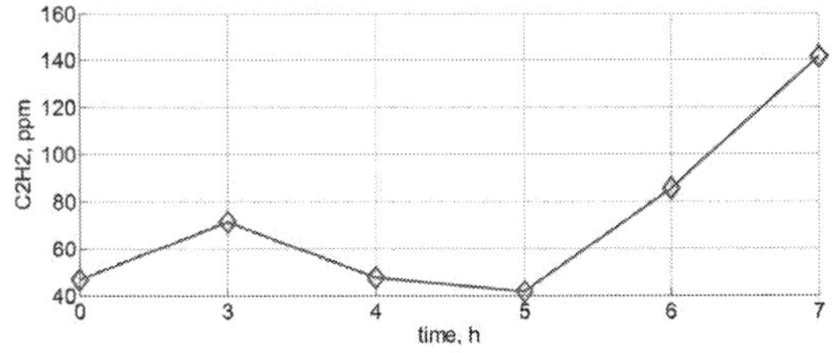

Figure 83. C2H2.

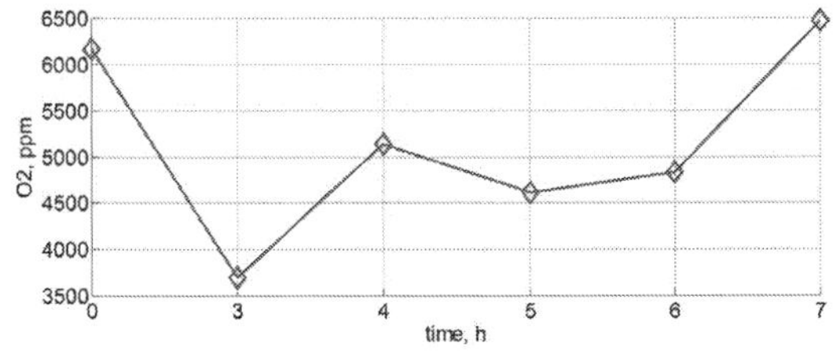

Figure 84. O2.

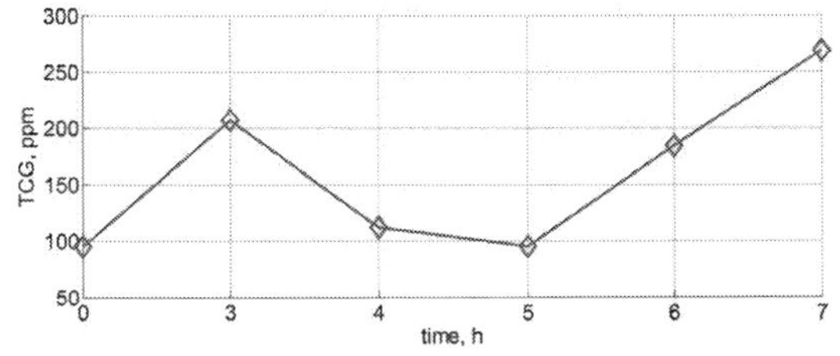

Figure 85. TCG.

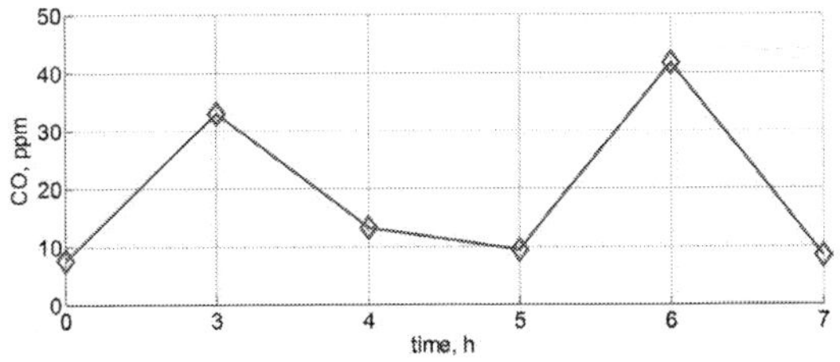

Figure 86. CO.

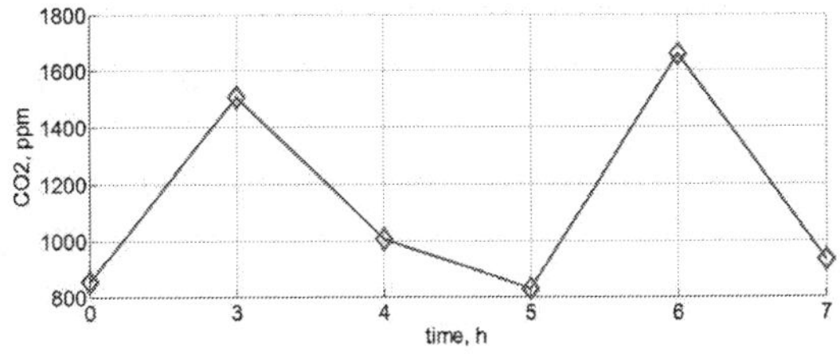

Figure 87. CO2.

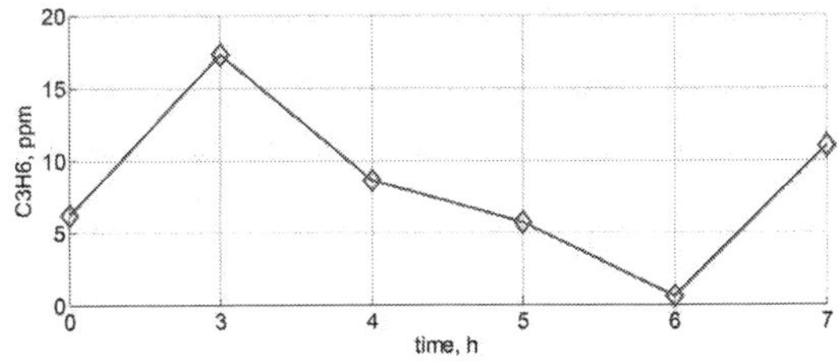

Figure 88. C3H6.

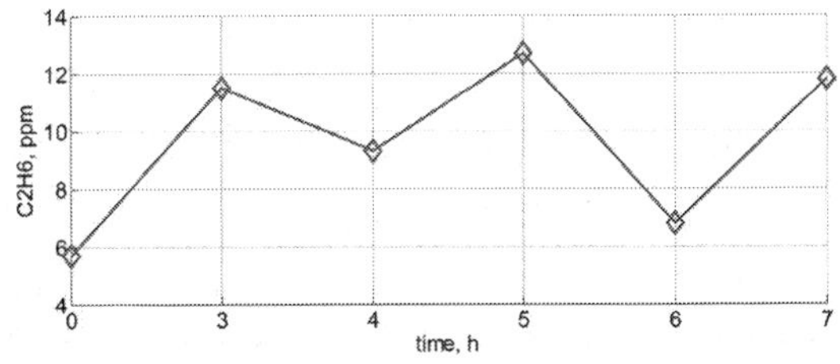

Figure 89. C2H6.

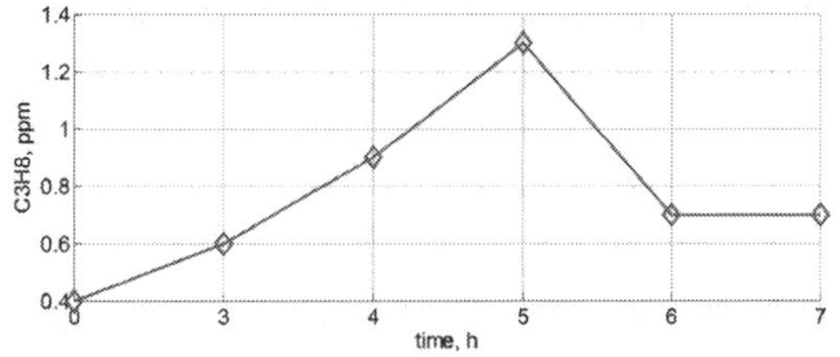

Figure 90. C3H8.

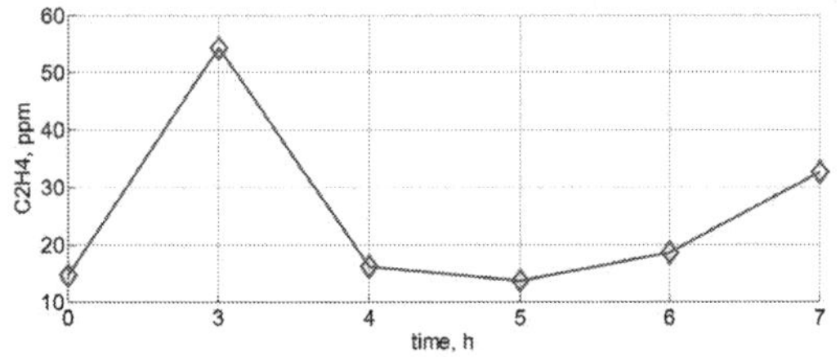

Figure 91. C2H4.

The breakdown voltage "Up" increases with time of PD generation. Similar relationship was observed for the following parameters: H2, N2, CH4 and C2H2. The water content decreases with the PD generation time. There is no significant influence of the PD generation time on the following parameters: O2, TCG, CO, CO2, C3H6, C3H8, C2H6, C2H4.

4. Conclusions

When considering the electrical and UHF methods, one can state the following conclusions:

- The most significant frequencies which were selected based on the UHF amplitude spectra are similar for all three electrodes arrangements.
- For data gathered in the NN and SURF systems, the empirical distribution of charges, registered with the electrical method, depicted no significant differences between the three considered frequency values. In the case of voltages, registered with the UHF method, the densities are different for the particular frequency ranges.
- For data gathered in the NNB system, it was observed that all charts, depicting empirical distributions of data gathered with the electric and UHF methods, are normally distributed independently of the phase of the supply voltage.

When considering the spectrophotometry method, the following conclusions can be stated:

- Analysis of the optical spectra, registered during PD generation at different voltage levels and distances in the NN and NNB systems, resulted with the statement that there is no significant influence of the voltage value at the registered intensities. This was confirmed by presentation of the maximal intensities, determined from the registered spectra, in the function of PD generation voltage. The lacking dependency was also confirmed by determining the energy of the registered optical spectra. For both NN and NNB systems, there was recognized no significant dependency of the voltage value on the energy contained in the spectra regardless of the distance at which the measuring sensor was installed.
- Analysis of the optical spectra, registered during PD generation at different voltage levels and distances in the SURF system, resulted with the statement that there is a clear dependency of voltage on the registered intensity spectra, which was farther depicted by presentation of the maximal intensities and energy values in the function of PD generation voltage.

When considering the acoustic emission, the conclusion was stated that there is no significant influence of PD generation voltage on the frequency components included in the PSD spectra, which were previously calculated for the AE signals registered in the NN and SURF system.

Analysis of UV and MIR images resulted with the statement that the is no such radiation apparent outside the tank. The same regards to high energy signals.

The most important conclusions related to the chromatography method are the following: the breakdown voltage Up and the H2, N2, CH4 and C2H2 values increase with the time of PD generation. Only the water content decreases with the PD generation time. The other parameters are not influenced by the PD generation time.

5. Summary

The main cognitive-scientific aim of the project is a detailed and multi-variate research study of physical phenomena associated with the generation of basic types of PD, which occur in the electro-insulation oil.

Based on the current knowledge, implied on literature reviews carried out by the authors as part of previous works and technical expertise, regarding the considered subject matter, it was stated that until now systematic research aimed to understanding of the phenomenon of PD generation in insulation oil has not been performed, which would include analyses of all physical phenomena simultaneously, i.e., under identical measurement conditions. In addition, no assessment was made of their interrelationships.

The experimental works performed within the research tasks involved measurement of acoustic emission and electrical signals, high-frequency electromagnetic and high-energy (X-rays/gamma) radiation, optical and thermal images, which are the common phenomena generated by PD. Three systems were designed and built for modelling of different types of PD: a surface system, a needle-needle system and a needle-needle system with gas bubbles. In the final effect, the most effective methods of PD measurement and evaluation were selected.

The research hypothesis that it is possible to investigate experimentally the physical processes related to generation of basic PD forms which can occur in insulation oil, was confirmed.

It should be emphasized that the objective and multivariate investigation of the PD phenomenon is an extremely important issue, not only from the cognitive point of view but also has an utilitarian value. This relates to application of the obtained dependencies for correlation of achieved results and by this means for improvement of the diagnostic methods applied for evaluation of high-voltage insulating systems of power transformers, which are essential in the system of arranging the distribution and transmission of electricity. Thus, the results obtained may have a measurable trade and industrial importance, having a direct impact on economic and financial aspects, and indirectly to the development of civilization, and a very strong social aspect.

Acknowledgements

The work was co-financed from funds of the National Science Centre (NCS) as part of the OPUS programme, project no.: 2013/09/B/ST8/01736.

The work was co-financed by the European Regional Development Fund "Increase of Scientific Research and Innovation for Enterprises in Terms of Sustainable Development Through the Creation of a Modern Diagnostics Laboratory of Surge Voltage" at the Opole University of Technology – Part I (2010–2011) and Part II (2011–2013), Project no RPO. 01.03.0101-16-007/10-00 and WND-RPOP.01.03.01-16-007/10.

Author details

Tomasz Boczar, Andrzej Cichoń, Daria Wotzka*, Paweł Frącz, Michał Kozioł and Michał Kunicki

*Address all correspondence to: daria@wotzka.eu

Opole University of Technology, Opole, Poland

References

[1] Jacob N.D. ; McDermid W.M. ; Kordi B.: On-line monitoring of partial discharges in a HVDC station environment, IEEE Transactions on Dielectrics and Electrical Insulation, Vol. 19, no 3, 2012, pp. 925 – 935

[2] Cavallini A. ; Montanari G.C. ; Tozzi M. ; Chen, Xiaolin: Diagnostic of HVDC systems using partial discharges, IEEE Transactions on Dielectrics and Electrical Insulation, Vol. 18, no 1, 2011, pp. 275 – 284

[3] Pompili M. ; Bartnikas R.: On partial discharge measurement in dielectric liquids, IEEE Transactions on Dielectrics and Electrical Insulation, Vol. 19, no 5, 2012, pp. 1476 – 1481

[4] Lindell E. ; Bengtsson T. ; Blennow J. ; Gubanski S.M.: Evaluation of a PD measuring system for repetitive steep voltage waveforms, IEEE Transactions on Dielectrics and Electrical Insulation, Vol. 18, no 1, 2011, pp. 246 – 255

[5] El-Hag A.H. ; Saker Y.A. ; Shurrab I.Y.: Online Oil Condition Monitoring Using a Partial- Discharge Signal, IEEE Transactions on Power Delivery, Vol. 26, no 2, 2011, pp. 1288 – 1289

[6] Lopatkiewicz R. ; Nadolny Z. ; Przybylek P.: The Influence of Water Content on Thermal Conductivity of Paper Used as Transformer Windings Insulation, 10th IEEE International Conference on the Properties and Applications of Dielectric Materials (ICPADM), JUL 24–28, 2012, India.

[7] Montanari G.C. ; Cavallini A.: Partial discharge diagnostics: from apparatus monitoring to smart grid assessment, IEEE Electrical Insulation Magazine, Vol. 29, no 3, 2013, pp. 8 – 17

[8] Junhao L. ; Quanwei H. ; Xuefeng Z. ; Xiu Y. ; Yongfen Y. ; Yanming Y.: Partial-Discharge Characteristics of Free Spherical Conducting Particles Under AC Condition in Transformer Oils, IEEE Transactions on Power Delivery, Vol. 26, no 2, 2011, pp. 538 – 546

[9] Yoshida M. ; Kojima H. ; Hayakawa N. ; Endo F. ; Okubo H.: Evaluation of UHF method for partial discharge measurement by simultaneous observation of UHF signal and current pulse waveforms, IEEE Transactions on Dielectrics and Electrical Insulation, Vol. 18, no 2, 2011, pp. 425 – 431

[10] Rodrigo A. ; Llovera P. ; Fuster V. ; Quijano A.: Influence of high frequency current transformers bandwidth on charge evaluation in partial discharge measurements, IEEE Transactions on Dielectrics and Electrical Insulation, Vol. 18, no 5, 2011, pp. 1798 – 1802

[11] Reid A.J. ; Judd M.D. ; Fouracre R.A. ; Stewart B.G. ; Hepburn D.M.: Simultaneous measurement of partial discharges using IEC60270 and radio-frequency techniques, IEEE Transactions on Dielectrics and Electrical Insulation, Vol. 18, no 2, 2011, pp. 444 – 455

[12] Jian L. ; Tianyan J. ; Harrison R.F. ; Grzybowski S.: Recognition of ultra high frequency partial discharge signals using multi-scale features, IEEE Transactions on Dielectrics and Electrical Insulation, Vol. 19, no 4, 2012, pp. 1412 – 1420

[13] Robles G. ; Sanchez-Fernandez M. ; Albarracin Sanchez R. ; Rojas-Moreno M.V. ; Rajo-Iglesias E. ; Martinez-Tarifa J.M.: Antenna Parametrization for the Detection of Partial Discharges, IEEE Transactions on Instrumentation and Measurement, Vol. 62, 932 – 941

[14] Boczar T. ; Zmarzly D.: The application of correlation analysis to acoustic emission pulses generated by partial discharges, Materials Evaluation, Vol. 62, pp. 2004, 935 – 942, 2004

[15] Wotzka D. ; CichońA. Boczar T.: Modeling and Experimental Verification of Ultrasound Transmission in Electro Insulation Oil, Archives of Acoustics, Vol. 37, no. 1, 2012, pp. 19 – 22

[16] Szmechta M. ; Boczar T. ; Frącz P.: Frequency and Time-Frequency Analysis of Acoustic Cavitation Noise in Insulating Oils, Acta Physicsa Polonica A, Vol. 120, 2011, pp. 744 – 747

[17] Borucki S. ; CichońA.: The influence of the power transformer load on vibroacoustic signal analysis results, Przegląd Elektrotechniczny Vol. 86, 2010, pp. 45 – 47

[18] Borucki S. ; Boczar T. ; Cichon A.: Investigation of the acoustic pressure distribution occurring around an aerial substation adjacent to apartment buildings, Archives of Acoustics, Vol.32, No. 4, 2007, pp. 291 – 297

[19] CichońA. ; Borucki S. ; Wotzka D. ; Szmajda M.: Characteristic of acoustic emission signals generated by the contacts of the selector, Acta Physica Polonica A, Vol. 122, Issue 5, 2012, pp. 804 – 807

[20] Wotzka D. ; Zmarzly D. ; Boczar T.: Numerical simulation of acoustic wave propagating in a spherical object filled with insulating oil, Acta Physica Polonica A, Vol. 118, Issue 6, 2010, pp. 1272 – 1275

[21] Fracz P. ; Boczar T. ; Zmarzly D. ; et al.: Analysis of optical radiation generated by electrical discharges on support insulator, Acta Physica Polonica A, Vol. 124, Issue 3, 2013, pp. 413 – 416

[22] Fracz P. ; Boczar T. ; Borucki S. ; CichońA. ; Zmarzly D.: Results of optical signals analyses emitted by electrical discharges recorded with UV camera, Acta Physica Polonica A, Vol. 122, Issue 5, 2012, pp. 814 – 817

[23] Fracz P. ; Zmarzly D. ; Boczar T.: Characteristic of surface partial discharges measured with ultraviolet camera, Acta Physica Polonica A, Vol. 127, Issue 3, 2015, pp. 715 – 718

[24] Zmarzły D. ; Nagi L. ; Borucki S. ; Boczar T.: Analysis of ionizing radiation generated by partial discharges, Acta Physica Polonica A, Vol. 125, Issue 6, 2014, pp. 1377 – 1379

[25] Zmarzły D. ; Nagi L. ; Frącz P. ; Boczar T.: Detection of high-energy ionizing radiation generated by electrical discharges in oil, IEEE Transactions on DEIS, 2016, accepted for publication

[26] Boczar T. ; Cichoń S. Kunicki M. ; Kozioł M. Wotzka D.: Indicator Analysis of Partial Discharges Measured Using Various Methods in Paper-Oil Insulation, IEEE Transactions on DEIS, 2016, under review.

Chapter 8

Non-Destructive Techniques Applied to Monumental Stone Conservation

Beatriz Menéndez

Additional information is available at the end of the chapter

http://dx.doi.org/10.5772/62408

Abstract

Non-destructive techniques have always been used in the study of built cultural heritage because of the high cultural value of the concerned objects and the need to preserve them as intact as possible. In this chapter, different non-destructive techniques applied to the conservation of historical building are presented. The selected techniques concern the measurement of some physical properties of the building materials measured at the surface: water absorption, permeability, water content, cohesion, hardness and so on; the actual conditions of the building: stress state, deformation, crack growth and so on; and in-depth physical properties: mechanical properties, inner structure of walls, damp location and salt content. Some of these techniques are used for inspection of the building at a given time, whereas others can be applied for long periods of time to investigate the evolution of the building or of one of its parts (e.g., crack propagation) with time.

After presenting the physical background of each method, the main objective of this review is to focus on the applications, especially to discuss which information can be supplied and to present published results in each case. Some techniques are very simple and require very inexpensive equipment but others, which are mainly adaptations of field geophysical techniques, use more sophisticated technology and require post-acquisition treatments based on more complex physical principles.

Finally, some examples of combinations of different techniques are presented because a unique method cannot provide all the information needed to understand the weathering processes taking place in a building. The final goal of the studies is to contribute to the preservation of the cultural heritage. The choice of the conservation strategies and methods should be based on a deep knowledge of the building, and non-destructive testing is usually the only way to get it.

Keywords: Non-destructive techniques, Stone conservation, Built cultural heritage, "in situ" studies, Monumental stone

1. Introduction

Conservation of built cultural heritage requires a deep knowledge of the building itself, its construction and restoration history, its structure, the nature and properties of the employed construction materials, their degradation degree, etc. but also of the external/internal conditions affecting the building: stress state, damp localization and humidity degree, climatic conditions, traffic, etc. Only a very superficial part of the building can be directly reached and studied. Due to their cultural value, sampling in order to determine the composition and properties of building materials is not always possible or desirable: this implies that non-destructive "in situ" inspection becomes the only way to study the building. In some cases, samples can be taken and studied in the laboratory but there are other features that can only be determined on site in the monument: cracks opening, stress and strain evolution or even the inner structure of walls. Some properties are measured not because their variation constitutes a risk for the integrity of the monument but because its change will affect the aesthetics of the building, as colour changes due to the application of some restoration products or procedures (consolidation, hydrofugation and cleaning).

According to Fitzner a study of a cultural heritage building in order to conserve and preserve it will follow three steps: anamnesis, diagnosis and therapeutical steps [1, 2]. The anamnesis includes the monument identification, location, history, environment and so on. The diagnosis step determines the building materials, their properties and state of deterioration to decide the need (or not) of preservation measures. When necessary, the final step consists of the conception, tests, application of therapeutical measures and of a long-term survey and maintenance of the building. For each of these phases, non-destructive techniques are nowadays necessary.

"In situ" non-destructive testing (NDT) have been employed during long time on conservation studies of monuments. For example, Karsten tube has been used since the middle of 19th century to measure water absorption of building materials [3]. This method, consisting in a continuous water supply on the surface of the materials, can be considered as an example of techniques that can be non-destructive in some cases, that is, when done on stone or brick surface, and destructive when applied on water-sensitive surfaces as plaster renders or frescoes. At the same period, geophysical techniques started to be used for archaeological campaigns. More recently, some geophysical techniques started to be applied for laboratory characterization of porous materials and then to "in situ" study of buildings and particularly to historic building conservation. Most of the NDT applied to cultural heritage have been implemented and tested in laboratory or field geology before being applied "in situ" on historical buildings. Cosentino et al. [4] use the term "micro-geophysics" to discuss about these techniques because the equipment has to be "miniaturized" to be used on buildings. An important difference between cultural heritage studies compared to geological ones is the fact that more than one face could be inspected and then better results than in classical geophysical investigations can be obtained. One major technical problem is the size of the probes.

NDT techniques can be applied "in situ", do not require to retrieve destructively samples and are nowadays an indispensable tool in cultural heritage for the characterization of materials, their degradation and weathering degree but also to assess the effectiveness of conservation interventions and to evaluate the compatibility of materials [5]. "In situ" NDT techniques on conservation studies are used to

i. estimate physical and chemical properties of building materials and their conservation state,

ii. characterize the "environmental" conditions of buildings, including water content (spatial distribution, salt content and so on) and stress field.

iii. monitor the evolution of the building, materials properties, crack opening and strain.

NDT techniques can be associated to destructive tests providing information that NDT cannot provide [6]. Literature concerning non-destructive techniques applied to built heritage conservation is very large. The goal of this chapter is to present, in a synthetic way, the different available techniques and show some examples of their applications on built cultural heritage. Many different techniques are presented from very simple ones using cheap technologies to more complex ones needing expensive high-technology equipment. Some NDT techniques deal with the superficial part of stones or other building materials, and others explore the inner parts of the building. The examination depth can go from millimetric to metric scale. We present studies realised on masonry, mainly stone, and also on bricks, as well as some studies concerning plaster renders, frescoes, statues and so on. We focus on methods to evaluate physical properties of materials, but chemical or mineralogical analytical methods are not discussed. This chapter is not a technical description about the different techniques but rather an introduction to the different possibilities of application. For each technique, a brief description of the physical basis is done, and the emphasis is put on the information they can provide and their interests in cultural heritage. Examples of applications are shown.

The first and maybe the most important NDT always applied is "expert" visual observation [7]. This observation allows to planning the study campaign and the choice of the suitable techniques to be used. Visual inspection is important but it is not enough for a deep characterization of the materials and present pathologies. One of the most important aspects conditioning the success of any NDT study is the observer's experience. Under good conditions, we can consider that non-invasive techniques are the most appropriate tool for the evaluation of the internal structure and materials quality of cultural heritage [8]. It is important to keep in mind that every studied case should be considered as unique. No strict general rules can be applied in cultural heritage studies.

2. Geometrical information and mapping

2.1. Observation and mapping

As mentioned in the introduction, the first and oldest non-contact technique applied to cultural heritage is observation. This observation can be done with naked eye or by means of optical devices and contribute to generate maps in which all relevant information are synthesised. Traditionally, naked eye observations are complemented by magnifying glass observation but without the possibility of taking pictures. Photography is the traditional way of reporting observations on site. Fibre optics microscope (FOM) can be used "in situ" to acquire magnified images of surfaces. High resolution (up to 600×) and contrasted images can be obtained and stored without any surface preparation. This kind of images can be employed in many different ways, including the identification of textures and composition of surfaces, the study of decay phenomena, the investigation of surface morphology and the evaluation of cleaning and consolidation inventions. Several examples of images taken by FOM on surfaces of monuments are shown in the work of Moropoulou et al. [5].

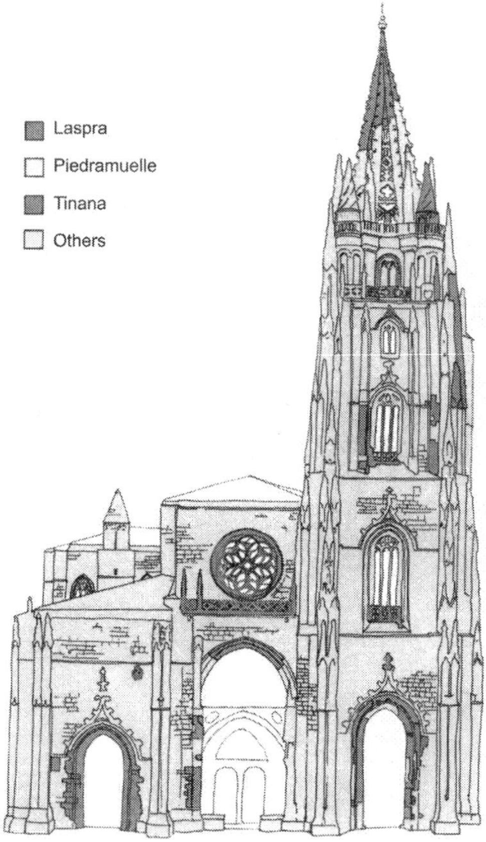

Figure 1. Lithological cartography of the main façade of the Cathedral of Oviedo (Spain) [9].

Esbert and Marcos [9] have mapped the different stones employed in the construction of the Cathedral of Oviedo (Spain). The cartography of one façade is shown in **Figure 1**. An example of weathering cartography of different parameters (weathering forms, rate and damage categories) of the Tomb no 778 of Petra site (Jordan) is shown in the work of Fitzner [10]. The classification of weathering forms used by these authors consists of four levels: Level I with four groups of weathering forms (loss of stone material, discoloration/deposits, detachment and fissures /deformation); Level II, each group is subdivided into main weathering forms (25); Level III with 75 individual weathering forms and Level IV with the differentiation of individual weathering forms according to intensities. More details are provided in the study of Fitzner and Heinrichs [11]. In **Figure 2**, we observe a map of the degradation forms presented in one façade of the Saint Nicolas Church of Maisons Laffitte (France). The chosen nomenclature and colours are the ones proposed in the ICOMOS-ISCS [12]

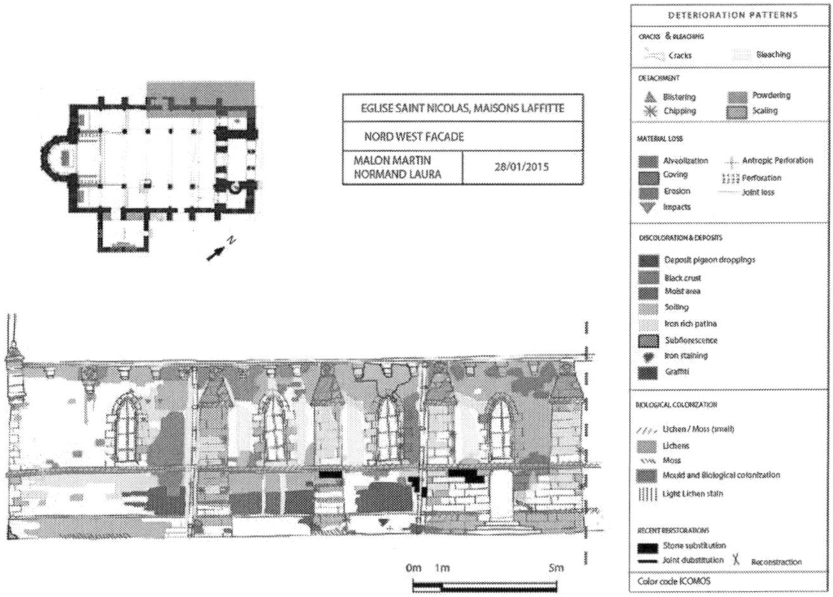

Figure 2. Mapping of weathering types of the Saint Nicolas church at Maisons-Laffitte (France).

2.2. Three-dimensional models

Photogrammetry is the science of making measurements from photographs. Usually, several pictures of the same object taken from different places are combined to obtain three-dimensional (3D) information of the object. Classically, we have the aerial photogrammetry, with pictures taken from an aircraft and usually pointed vertically towards the ground, and close-range photogrammetry, where the camera is close to the object. There is an established procedure about how to configure the position and orientation of the camera to obtain the most useful information. A video illustration of this technique can be seen in [13]. Digital cameras have facilitated the automatic treatment of such images, allowing also quantitative measurements [14]. For example, Solla et al. [15] applied photogrammetry to build the 3D structure of a bridge. To obtain a final 3D model with enough high resolution, they realised several small 3D models, each model was defined using three or four photographs, with consecutive models overlap of 40%. In the study of Arias et al. [16], photogrammetry is used for monitoring the evolution of a crack in the masonry of the Basilica da Ascensión in Galicia (Spain). A wall presenting a long vertical crack has been modelled by photogrammetry in 2001 and 2003, during this period an increase in crack size of 12% has been measured.

Terrestrial laser scanning allows a faster collection of the 3D coordinates of objects and can be applied at different scales for large surface on site of for smaller object with high resolution on site or in the laboratory. There exist different ways of processing the signal to obtain the 3D model; more detail about this technique can be found in the study of Yastikli [14]. Several authors have used this technique on cultural heritage with different objectives. In the work of Heinrichs and Azzam [17], high resolution 3D terrestrial laser scanning is used to study five rock-cut monuments in Petra. The obtained data were used to assess their original architecture, weathering damage, calculate the rock mass removed for monument creation, water run-off on the monument, etc. They installed an environmental survey network measuring temperature, humidity and electrical resistance. These measures have been done at surface and at different depths on the stone until 18 cm. Localisation of environmental data on a 3D model (orientation, slope and so on) allows a good interpretation of weathering processes.

2.3. Digital image analysis

Vazquez et al. [18] proposed to use digital image analysis of wall pictures to distinguish between unweathered limestone, weathered limestone and areas with more or less thick efflorescence. They applied this technique to the conservation study of the Chapel of Falla located in the Crypt of the Cadiz Cathedral (Spain). They distinguish six "classes" based on the grey level range, from 0 to 109: unweathered limestone; from 110 to 129: weathered limestone; from 130 to 159: efflorescence type 1; from 160 to 179: efflorescence type 2; from 180 to 190: efflorescence type 3 and from 191 to 255: efflorescence type 4. Efflorescence thickness increases from type 1 to 4. In this way, they can quantify the percentage of surface belonging to each type.

A similar study has been performed by Crespo et al. [19] in the ruins of Santo Domingo (Pontevedra, Spain) combining colour optical images with images obtained by a terrestrial laser scanning. They used not only the geometrical information given by laser scanning but also the intensity of the reflected signal. They combined the red channel signal of the colour image with backscattered laser intensity (sum, ratio) and they applied statistical clustering procedures to the pixel level of the resulting image. They obtained a good automatic classification of granite ashlar with remaining lime/mortar and granite ashlars affected by high moisture content. Meroño et al. [20] applied a combination of 3D model from a terrestrial laser scanner with multispectral images to the study of Santa Marina de Aguas church (Cordoba, Spain). They arrive to separate different areas with stone, plaster, wood and different kinds of degradations. As they use a 3D model, they can calculate not only 2D projections but also real surface distribution.

Cartography of lithology and pathologies needs as much as possible detailed support of the façade, wall or object to be studied. Even if traditional methods are still used as a first approach, more sophisticated and accurate results can be obtained by the use 3D acquisition techniques and by architectural lasergrammetry and photogrammetry associated with devoted software. There is also the possibility of using GIS software to integrate other information such as numerical data and detail pictures.

2.4. Measurements of stress and strain

Structural problems are very common on cultural heritage. Their origin can be very different and many times they are released or accelerated by external causes, natural as earthquake, dryness or flood or anthropic like works done in the building or in the surroundings. The first control devices were just glass or plaster plates placed on cracks to survey if the cracks continue to grow or not. Nowadays, there are many different ways to control and measure stress supported by building, the propagation and evolution of cracks or the deformation of the structure. A brief description of some of these methods will be presented here.

2.4.1. Stress measurement

The flat-jack test (**Figure 3**) is a slightly destructive test to measure the "in situ" stress conditions. Mortar joint is cut out to introduce a flat jack (deformable steel capsule made out of two cold-formed steel halves welded together where a fluid can be injected under pressure to open them). The flat jack is then inserted into the cut and the pressure is gradually increased until the original distance between both sides is reached. This pressure is equal to the original stress state on the masonry [21]. By using two parallel flat jacks, applying an increasing pressure and registering the induced deformation, stress–strain curves can be done and Young's modulus can be measured [22]. After data acquisition, the cuts are refilled with mortar. Several examples can be seen in the study of Simoes et al. [23]. Bartoli et al. [24] used single and double flat-jack tests in the characterization of the mechanical properties of the Torre Grosa in San Gimihna (Siena, Italy). The masonry of the tower was composed of an internal layer in bricks and mortar joints and an external side composed of calcareous ashlars and a filling between then with a heterogeneous material. They measured stress values and Young's modulus in both internal and external walls. These data were used in a 3D finite element model of the tower.

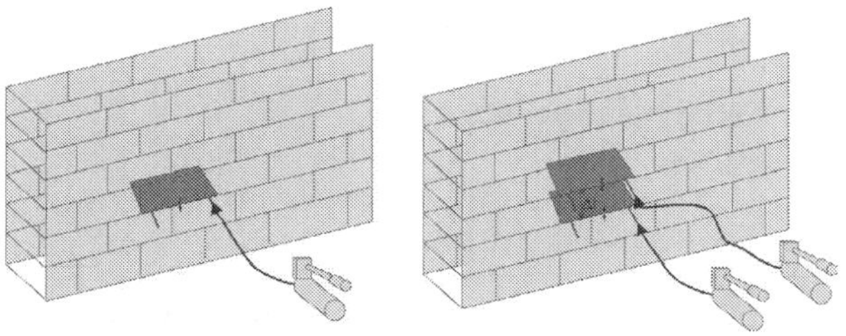

Figure 3. Simple and double flat jack techniques.

2.4.2. Strain measurement

Several systems can be used to measure strain, most of them are based on measuring distances and displacements: extensometers, inclinometers, callipers, linear variable differential transducers and so on. Monitoring can be done manually or automatically. Structural control can be static or dynamic [25]. Static monitoring consists in measuring displacement and inclination, without any external excitation source. The sources of deformation could be soil settlements, excessive load and temperature changes. Dynamic methods control the behaviour of the structure in response to an external vibration. These vibrations can be due to environmental vibrations, passive sources (traffic, wind and bell ring) or forced vibrations and active sources (hammering system or vibrodines). The response of the material is recorded by a network of accelerometers installed in chosen parts of the building. The vibration response is characteristic of the local and global behaviour of the building. Dynamic tests are used in the conception of the physical model of the structure.

Monitoring systems can be installed permanently or periodically and should be installed over a long period of time, more than 1.5 years to avoid seasonal variation errors. Sometimes instruments measuring deformation are installed in hardly accessible places, because of crack location and also to avoid vandalism. Nowadays, the information registered by strain gages or other instruments is tele-transmitted but, for example, in Castillo de San Marcos (St Agustine, Florida, USA) in 1987, the gage results were read using a tripod-mounted telescope [26]. Different movements can be monitored: relative movements of vertical structures, absolute horizontal movements, differential settlements and tilting.

For monitoring crack evolution, a detailed crack pattern survey should be carried out. The most frequently measured parameters to characterise crack relative movements are opening and sliding. In several Coptic monuments, more than 95 crack gauges and 10 temperature gauges have been installed in order to control the deformation behaviour of the monuments during restoration works [27].

In [28] a study is presented where the evolution of several cracks located in the Madara Horsemen, a rock bas-relief carved on the NW scap of the Madara Plateau (Bulgaria) included in the UNESCO World Heritage List have been followed during 10 years. Seven cracks cut the rock face that bears the bas-relief and three of them have been considered as dangerous. In 1990, a survey campaign started and three extensometers and five bench marks for calliper measure have been installed in different strategic locations. The obtained information allows one to model the dynamics of the rock blocks. For each extensometer, deformation has been measured in three directions: X, compression of crack; Y, moving of the rock slice/prims inside the massif and Z, uprising of the rock slice/prims. The authors concluded that movements along the rock face depend on seasonal and daily temperature fluctuations but movements in the periphery of the plateau, with higher velocities, are very sensitive to earthquakes.

Several applications of long-gauge sensors to monitor structural stability of different heritage building in Switzerland, Italy, Russia and Korea can be seen in the work of Glisic et al. [29]. They used SOFO ("Surveillance d'Ouvrages par Fibres Optiques" or structural monitoring by optical fibres) interferometric sensors. A description of the SOFO system can be found in the study of Inaudi et al. [30] and Inaudi [31].

3. Physical properties

We describe how to measure selected physical properties that are relevant for characterizing building materials and the different techniques that can be used.

3.1. Water absorption

Water is the most important weathering agent in buildings, and special attention needs to be paid to water transfer. Water contribute to stone degradation, among others, (i) by dissolving minerals, allowing the reaction of atmospheric pollutants with mineral grains and (ii) by the transport of soluble salt that will crystallize and become one of the most active ageing processes. The moisture distribution in buildings depends on the porosity of the materials, the pore size distribution and the environmental conditions. Two mechanisms are responsible for the introduction of salt in monuments, capillary rise of groundwater and infiltration of rainwater [32]. To know the kinetics of water absorption, several laboratory experiments are used: capillarity absorption, total immersion and so on. In monuments, different techniques have been employed to quantify the capacity of building materials to absorb water. Maybe the most common is the Karsten or RILEM tube. It consists of a cylindrical open reservoir in contact with the surface of the wall and connected to a graduated pipe. The opening of the cylinder can be vertical or horizontal to do measures in vertical or horizontal surfaces. The reservoir is filled through the pipe until a fixed level. Water starts to move to the porous material, and water height in the pipe is monitored as a function of time. Similarly to laboratory absorption tests, a coefficient is obtained as the ratio between the absorbed water in a time interval. The obtained results are very sensitive to the characteristics of the surface, roughness and weather conditions (temperature and relative humidity) [33]. This test is commonly used to measure the hydrophobic quality of restoration products, and it can be used also on joints to test the quality of the contact. A deep study about this test can be found in the study of Hendrickx [34] with a description of the experimental details, analytical models and examples of validation of these models. Mirowski pipe is similar to Karsten tube, but the pipe is closed and the reservoir is filled with a sponge [35]. The Italian pipette is another similar method. Common diameter of cylindrical reservoirs varies between 2 and 3 cm.

The contact sponge method [36] consists of a sponge enclosed in a contact plate. The sponge is higher than the vertical border of the plate. The sponge is filled with water, and the whole (sponge and plate) is weighed. The sponge is pressed manually against the stone surface until the vertical borders of the base touch the stone surface. After a selected time, the sponge is weighed again [35]. The amount of water added to the sponge, the applied pressure and the contact time depend on the material to be tested are fixed [3]. The permeability box consists of a rectangular box of 16 cm×34 cm filled with water and with an open surface in contact with the wall. The level of water in the box is maintained constant, and the amount of water that has penetrated into the wall is recorded. Another simple method, the water drop test, consists in placing a drop of water (or a non-polar liquid) on the surface of the material and observing the evolution of the water with time [37]. Depending on the behaviour of the water drop, we can determine whether the surface is water repellent or not, if the surface is not water repellent but underneath there is an impermeable layer, etc. This method cannot be applied on vertical surface. Schematic drawings of these different setups are shown in **Figure 4**.

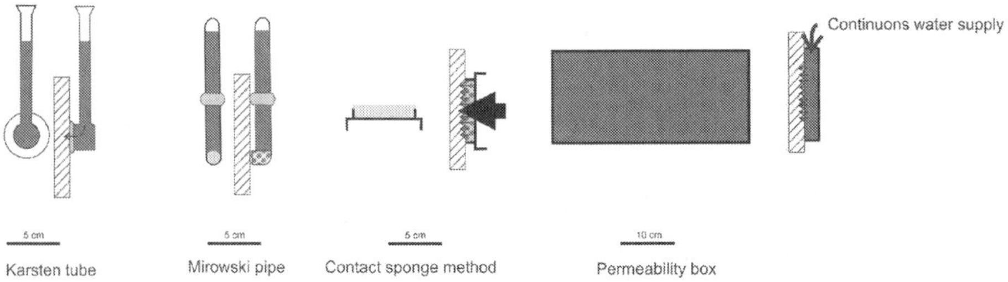

Figure 4. Different systems to measure water absorption "in situ".

In the work of Vandevoorde et al. [36], we can see a comparative study of the contact sponge methods, Karsten tube and Mirowski pipe in seven different stones with porosities between 10 and 30%. They compare these "in situ" measurements to laboratory capillarity rise method results. They found that contact sponge is capable of measuring initial water uptake in low porosity stones with high precision but it is not convenient for long time measurements or in high porosity stones due to the small amount of water supply. Karsten pipe is more convenient for longer time measurements and in high porosity materials. The results obtained by the Mirowski pipe were not convincing.

Another method has been proposed in the study of Drdácký et al. [38]. This method has the possibility of continuous measurement of water intrusion into the surface, allowing long-term measurements and recording. According to the designer, this method reduces the number of operators, it is more precise, effective and faster [39]. This pistol-like device measures the time necessary for the absorption of predefined volume of water and it can be used on inclined surfaces.

3.2. Permeability

Permeability is defined as the capacity for fluids (gas or liquid) to flow through a porous material. The term permeability is commonly employed for single fluid transport produced by a difference in the hydraulic head between two points in a completely saturated medium. Water absorption takes place in a unsaturated medium, and the driving forces are capillarity and gravity. There are two ways of measuring permeability (i) steady-state method, where a flow is imposed to the porous medium under constant hydraulic gradient and flow is measured and (ii) transient or pulse method where hydraulic head is instantaneously raised (or lowered) at one end of the sample and variation of pressure with time is recorded. The steady-state method corresponds to the permeability box test. The most common method employed for "in situ" measures uses the second experimental conditions (pulse method).

Brown and Smith [40] do a review of the different gas permeameters used in laboratory and field applications. They have developed a portable syringe air permeameter to be used on rock outcrops and cores. This permeameter induces a vacuum on the surface of the stone by suddenly increasing the volume of a chamber in contact with the stone surface. Air flows from the stone to the chamber gradually increasing the pressure on the chamber. Air pressure is registered as a function of time and from this pressure variation air permeability is estimated. This device was originally thought to measure rock permeability inhomogeneity in outcrops but it is actually used for "in situ" permeability measurements on buildings. This device is actually commercialized as TinyPerm by New England Research, Inc. (**Figure 5**). "In situ" air permeability measures have been done with this permeameter in two buildings of Paris (France) restored with lime mortars in order to check the compatibility of repair mortars with the original stone, the Euville limestone [41]. Similar measurements have been performed on laboratory samples of the same materials. It was found that the stone permeability is higher than the permeability of mortars by a factor between 1.5 and 15, despite the high scatter in the data. This method is a very convenient way of having 'in situ" estimation of materials permeability and its inhomogeneity.

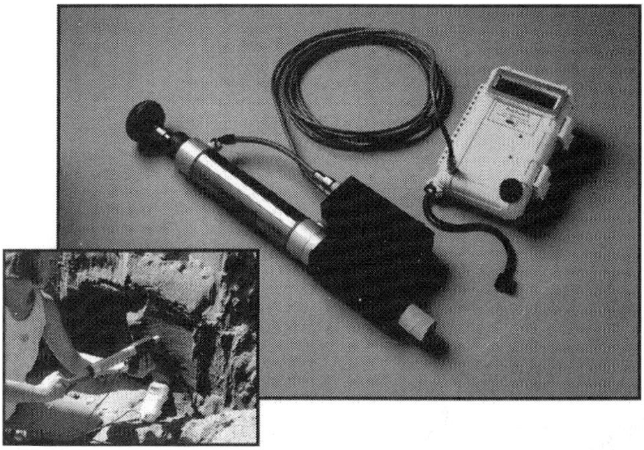

Figure 5. Tiny Perm (New England Research Inc.) taken with the permission from Brown and Smith [40].

Filomena et al. [42] compared air permeability measurements on 51 cylindrical and bedding parallel sandstone cores (∅ 2.54 cm, L 5cm) obtained with three different laboratory permeameters and the Tiny perm. They found that the permeability derived from Tiny perm measures in laboratory is about 37% lower than the ones obtained in confined samples. They concluded that probe permeameters have the advantage of providing closely spaced, non-destructive permeability data, which are mostly suitable to get 3D permeability value, to estimate the anisotropy effects and heterogeneity. Another system based on the same principal is the "Torrent permeability tester method" (Permea-TORR), but in this case vacuum is produced by a pump. Sena da Fonseca et al. [43] measured the permeability in 15 limestones and marbles from Portugal with a Permea-TORR and found very good relationships with the open porosity and the water absorption.

3.3. Humidity/water content

As "in situ" amount of water contained in a building material cannot be measured directly without sampling, different properties, which values change as a function of water content, can be used to estimate it. In order to validate the procedures, values of the measured properties in materials with well-known water content are used as standard for comparison. There are several kinds of moisture meters originally designed to estimate the humidity of wood that are now used on other built cultural heritage materials. Usually, they are not calibrated for stone but they can be used to estimate humidity variability in a site. The obtained results are affected by several factors such as the weather (especially humidity), the degradation of the surface, the presence of biological colonisation and so on [44]. There are two basic families of devices, one based on the measurement of the electrical resistivity of the materials, called "pin meters"; and a second group based on the measurement of the impedance, also called "pin-less meters" or electromagnetic wave meters (**Figure 6**). In pin meter, a small electrical current is passed between two pin tips, therefore, the measurement provides information only on a very thin line between both pins. Pinless moisture meters use a larger sensor pad that emits electromagnetic signals; therefore, the measures correspond to a larger area of the surface than in pin meters. The pin moisture meters need less access area that the pin-less one. In the same study, on Paris buildings previously described, we found better correlation between the nature of the studied material (mortar or stone) and water content with a pinless meter than with a pinless meter.

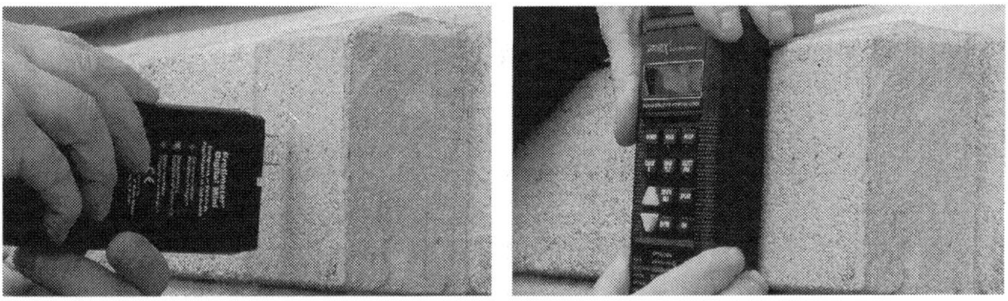

Figure 6. Left: Pin moisture meter (Protimeter), right: pinless moisture meter (Tramex).

3.4. Surface cohesion

Another very important property when quantifying the superficial weathering degree of buildings material is the cohesion of their surface. Several very simple slightly destructive tests are traditionally used.One consists in selecting a part of the surface on which a squared frame with an open surface of 10×10 cm is placed, then brushing this surface 10 times in the vertical and 10 times in the horizontal direction, and collecting the detached material. The collected material is dried and weighted. The results are expressed in grams per square centimetre [37]. Dividing the result of the test by the density of the obtained material, we can estimate the thickness of the removed layer [33]. The size of the collected grains can also be measured in the laboratory.

Another way to quantify the cohesion of surfaces is the peel-off resistance measured by the Power Strip test [45]. A Power Strip tape (20 mm x 50 mm) is applied and evenly pressed to the stone surface. A spring balance is attached to one end of the tape with a clamp. Afterwards, the strip is peeled off and the employed force recorded by the balance. The material stuck on the tape can be observed and weighted at the laboratory. If necessary the test can be repeated several times. The peel-off resistance is calculated from the force and the surface of the strip. The weather conditions have a great incidence on the obtained results [37]. Drdacky et al. [46] proposed a standardization of this method, also called Scotch Tape test, for assessing the cohesion and consolidation characteristics of historic stone surfaces.

Some authors used the roughness as a measurement of the surface quality. There are several ways of measuring it but most of them are not relevant for "in situ" measurements because the height range of the technique is too small to be applied on granular materials. In the study of Alvarez de Buergo et al. [47], a portable optical surface roughness meter has been used to select the best cleaning condition of the stone masonry at several areas of the Cathedral of Segovia (Spain). Hand profilometer can measure micro-relief amplitude from 0.5 to 6 mm [48]. Laser 3D scanners have also been used to characterise the 3D profile of stone walls and art object surfaces. Gomez-Heras et al. [49] used 3D laser scanner for monitoring the evolution of sample surface during salt crystallization test. This kind of technology can also be applied for "in situ" observations.

3.5. Surface hardness

Hardness of the surface is another property usually considered to determine the conservation degree of building materials. It is measured mainly by the Schmidt Hammer or sclerometer. The Schmidt Hammer was originally devised for carrying out "in situ", non-destructive tests on concrete [50]. It gives an immediate indication of the compressive strength of the material using the linear calibration curve supplied with each instrument [51]. It measures the height of rebound of a plunger impacting the surface at a defined pressure. There are different kinds of Schmidt hammer. The most employed in rock outcrops and building stones is the 'N' type with an impact energy of 2.207 Nm. Another type used in weak stones is the 'L' type with an impact energy of 0.735 Nm. More details about the use of Schmidt hammer and its correlation to uniaxial compressive strength can be found in the study of Aydin and Basu [52]. This test is very common on built cultural heritage studies. The Duroscope has a similar mechanism of operation but with a smaller and pointed plunger and a smaller spring-loaded mass. It can be used for surfaces with low strength. With Duroscope, the precision of the measure depends on the smoothness of the surface [53].

The Equotip is another device to test the mechanical properties of stone surfaces. In this case, a ball hits the surface with a fixed energy, and the velocity of the ball as it rebounds is recorded. There are different kinds of Equotip devices depending on the nature of the ball (tungsten carbide, ceramic and polished diamond), its size (2.8, 3 and 5 mm) and the impact energy (3, 11 and 90 Nm). The Equotip type D has been used to measure surface hardness changes of common Portland limestone, Cornish granite and a standard marine concrete exposed to coastal environment in the work of Coombes et al. [54]. They found that the durometer can detect changes on the surface hardness after short exposition period (months). This information indicates that it should also be useful in studying weathering in buildings environments. **Figure 7** shows pictures of a Schmidt Hammer, Duroscope and Equotip.

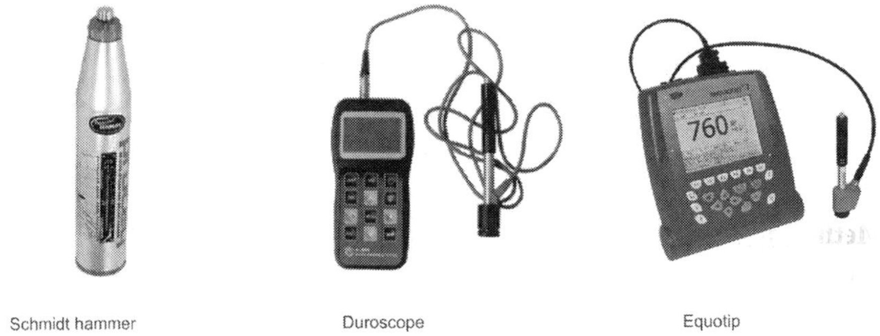

Figure 7. Different systems to measure surface hardness. Schmidt hammer (Proceq), Duroscope (Hitec Enterprises Inc) and Equotip (Proceq).

3.6. Color

Color is one of the most common properties measured in monumental stone, mainly when some preventive or restoration treatment will be applied or when stone ashlars should be replaced, as they can change the aesthetics of the building. Colour changes can be due to darkening by air pollution [55], to fire damage of stones [56],and so on. There are different systems to quantitatively express the colour. Traditionally, the Munsell color system has been used. The first edition of the Munsell Book Color has been edited in 1929. The Munsell system has three dimensions: HUE, which is the common name of the colour (red, green and yellow); VALUE, which represents the light strength and goes from 10 (white) to 0 (black) and CHROMA, which measures the degree of colour strength (intensity) and goes from neutral (grey) to maximum (there is not arbitrary end of the scale). So, colour is expressed as HUE/VALUE/CHROMA. Munsel colour chart served to determine the colour value of material by visual comparison between the objet and different sheets with colour samples. Nowadays, colorimeters and spectrocolorimeters measure the colour and can register the light spectrum in an automatic way. There are others systems to express colour as the CIELAB, values L*a*b*, where L* is lightness and goes from 0 (black) to 100 (white), a* axis is the red/green colour, with green at negative a* values and red at positive a* values; b* axis is the yellow/blue colour, with blue at negative b* values and yellow at positive b* values. All these systems are represented on "color solids". For example, **Figure 8** shows the Munsel and CIEL color solids and a spectrocolorimeter.

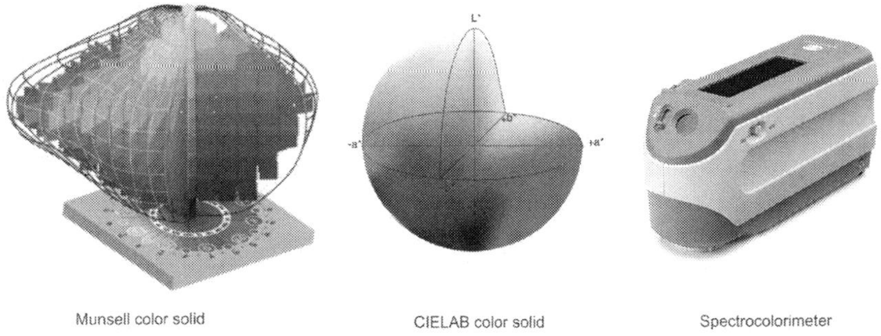

Figure 8. Color measurement systems and spectrocolorimeter (Konica Minolta).

4. Methods adapted from field geophysics

In this group, we use a classification depending on the employed techniques rather than on the measured property as before. We do a brief description of the physical principle of each technique, the different operating modes and we show some applications from bibliography.

4.1. Infrared thermography

Infrared thermography (IRT) is one of the non-destructive techniques, which is most frequently employed in civil and architectural inspections, especially on cultural heritage. It is used in the different phases of the studies: diagnosis, preventive maintenance and to verify the result of restorations. In cultural heritage, it has been applied for the study of very different aspects: detection of holes on masonry, plaster detachment from a wall [57], moisture location, material weathering, delamination between corroding reinforcement bars and surrounding materials [58], presence of cracks, biological colonization [33], evaluation of restoration materials [5] and the use of different kinds of mortars [59]. It is a non-contact method, which can be applied to wide surfaces of walls. Therefore, it can be used in hardly accessible spots or on irregular surface and/or with non-planar shapes [7]. Some IRT cameras have high spatial and thermal resolution. This technique presents a particular interest in the study of frescoes where it is not possible to use contact techniques [25]. One limitation is the limited penetration depth (some centimetres).

The camera receives infrared radiation emitted and reflected by the surface. The electromagnetic signal is transformed in an electrical signal and finally in a false colour image. Infrared waves have wavelengths ranging from 0.78 to 1000 μm but those between 7 and 14 μm are typically captured by thermographic camera sensors [60]. Thermal resolution can go from 0.08 to 2°C depending on the camera. The wavelength shifts with the temperature of the surface, decreasing as temperature increases. The signal arriving to the camera depends on the emissivity of the material (capacity of the material to emit energy), the surface colour, reflections on metal or glazed surfaces, meteorological conditions and the distance between the camera and the surface. Apparent different temperatures may be caused by variation in thermal diffusivity of materials that depends on thermal conductivity, specific heat and density of materials [15]. Technical aspects are also important as the characteristics of the camera and the calibration process needs to avoid biased measurements [57].

In cultural heritage, two kinds of thermography methods are employed: passive thermography and active thermography. Both need a temperature change during the test. Passive thermography used natural heating or cooling, the energy source is the sun. Thermography registers differences in temperature due to differences in heat transfer. Active thermography uses an external known source of heating or cooling to register the heat flux. In passive thermography, a single snapshot is used, whereas in active thermography a series of images (or a movie) is required. In "pulse thermography," the surface is heated briefly and then the temperature is recorded during cooling. In "lock-in thermal," a sine-modulated lamp heating is used, and the time dependence between reference and temperature signals is recorded [5]. Passive thermography is used for large areas and active thermography for the study of small areas of special interest. Generally, passive thermography is qualitative, whereas in active thermography some parameters as thermal conductivity can be calculated. APT methodology consisting of a combination of instrumentation and work flow providing fast, high resolution, low cost approach that automatically acquires, corrects and processes geo-located, temporally anchored, building scale thermal surveys has been used in the study of Hess et al. [60]. This method based on image mosaic permits the survey of whole buildings with high resolution.

The equipment consists of an infrared thermocamera with a specific wavelength range, different kinds of lenses and data treatment facilities. Active thermography also needs a heating source.

4.1.1. Application to detachments detection

Detachments create an air layer between the stone/bricks and the plaster. This air layer represents a thermal resistance to heat flow, preventing the heat flowing from the wall to the plaster during the cooling periods or from the plaster to the wall during heating. When the surface is in equilibrium with the environment (no heating or cooling), detachments cannot be observed. In passive thermography, during the day, detached areas are hotter than the areas without detachment; during the evening (cooling), they are cooler than the rest.

Thermography can detect not only plaster or other render detachments but also external bricks or stone detachment. de Freitas et al. [57] compared numerical simulation, physical model and "in situ" measurements of render detachments. They created a laboratory wall model with air bubbles between a concrete masonry and a polymeric render and heated the surface using an infrared heat source simulating solar radiation. They did also some measures on residential buildings façades in Porto (Portugal) on several areas where detachments were suspected and selected by the percussion method (which will be explained later). The results obtained by numerical simulations agree with the experimental data.

Jo and Lee [61] used active thermography for blistering detection on five-story pagoda located in the Magoksa Temple, Gongju (Korea). They tested different heating systems, and they concluded that lamps are unsuitable for large stone monuments and decided to use an infrared heater with a halogen lamp. **Figure 9** shows the obtained result on the four faces of the pagoda.

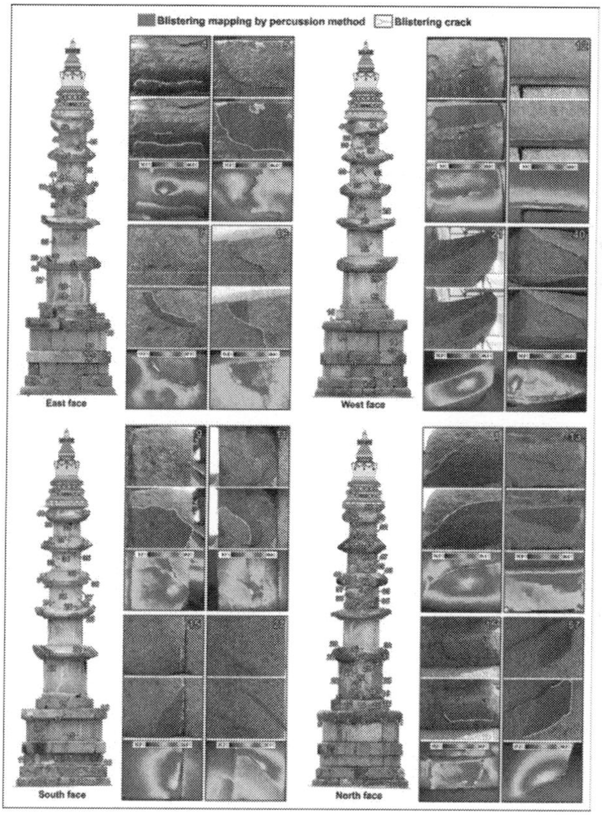

Figure 9. Active thermography applied to a pagoda from the Magoksa Temple (Korea) [61].

4.1.2. *Application to the detection of voids or cracks*

As for detachments, air in voids, pores or cracks produce lacks on thermal conductivity and are observed as detachments by thermal anomalies. Two thermographic campaigns on the Santa Maria ad Cryptas (Fossa, Italy) have been performed [59], one in 2007 and a second after the 6 April 2009 earthquake. During the first campaign, they found an incipient formation of cracks not visible with naked eye. After the earthquake, the analysis showed that some damage induced by the earthquake corresponds to thermal anomalies previously detected (**Figure 10**).

Figure 10. Passive thermography applied to the detection of cracks. (a) Lateral wall of the Church of Santa Maria ad Cryptas (Italy) after the 2009 earthquake; (b) aspect in 2007; (c) thermography image taken in 2007 showing traces of incipient splitting of the masonry. [59].

4.1.3. Application to the detection of materials contrast

Different types of stones, bricks and mortars have different thermal properties that result in different temperature response during cooling or heating phases.

Three different kinds of thermal tests have been applied in the Monastery of the Assumption of the Virgin (Ioannina, Greece) [62]: (i) "graduated heating thermography," where the increase of the surface temperature was registered during heating; (ii) lock-in thermography, with a sine-modulated lamps heating and (iii) the pulse phase infrared thermography, where the surface is heated briefly and the temperature decrease is recorded. The three techniques were applied to the donor's inscription and different characteristics have been evidenced: detection of delamination and detached areas, the presence of different depth and material type, different substances used in the paint mural and pre-existing murals.

4.2. Acoustic, sonic and ultrasonic methods

This group of techniques is largely the most employed in the study of cultural heritage. It goes from very simple and inexpensive ones to more sophisticated with a large number of sensors and advanced computation for data analysis.

4.2.1. Simple acoustic tests

A very old and common technique used to detect detachments on render, known as acoustic tracing, is to knock the surface with a finger phalange to check if it "sounds empty." In such a way, it is possible to locate the areas with likely rendering defects. The sound can be analyzed directly by the operator listening or it can be recorded and analyzed with electronic devices. Another technique consists in dragging a chain along the surface and listening to the emitted sound, or in the "natural percussion" method to analyze the sound generated by the rebound of a sphere dragged along the surface [63]. A semiautomatic variation is called AAT [64]. The material is excited by a hard rubber sphere hitting the surface, the location and the sound are recorded by a video camera. A special software allows the production of maps of detached areas. All these methods could substitute infrared thermography on polish surface. In order to do automatic discrimination of sounds, some acoustic analyses on the audible spectrum have been presented in the work of Sklodowski et al. [63]. They tested different tapping instruments, such as finger, rubber sphere a wire in a hard plastic coating. This last one has been considered as providing the best results. Bläuer et al. [37] used a 50 cm rod with a steel ball of 2 cm in diameter at one end but they also proposed to use a metal wire or a small Allen key.

Another technique consists in using a loudspeaker placed at the end of a duct in contact with frescoes surface at the other end [65]. Loudspeaker excites plane waves normal to the duct axis and the response of the portion of fresco in contact with the duct is measured with microphones. Collini and Garziera [66] use a contactless method based on a loudspeaker and a microphone but in this case they used acoustic absorbency and not frequency response. The loudspeaker and the microphone are placed, respectively, 1 m and 1 cm away from the fresco. They find a good correlation between the measured absorbency and the detachment thickness on artificial reproduction of frescoes with defects. Using also loudspeakers as acoustic excitation, the surface vibration can be measured by a scanning laser Doppler vibrometer [67].

4.2.2. Wave velocities measurement

It is the most simple acoustic measurement that can be used in cultural heritage material characterization. In consists in generating an elastic wave at a source point and registering its arrival at another point: if we know the distance between the two points, measuring the travel time of the wave which propagates from the source point to the receiver allows one to calculate the wave velocity (**Figure 11**). The wave can be generated by a percussion (sonic, frequency around 5 kHz) or by a piezoelectric transmitter (ultrasonic, frequency higher than 20 kHz). The receiver is usually a piezoelectric accelerometer converting vibration into electric signal. The simplest experimental procedure consists in measuring just the arrival time of the wave but the entire signal can be recorded and processed to calculate attenuation, energy or other wave parameters that can provide information about the elastic and mechanical properties of material along the wave path. There are three experimental setups to measure wave velocity "in situ" depending on the accessibility: (i) the direct method, where the wave source and the receiver are on opposite faces of the object (e.g., in columns), (ii) the semi-direct method, where emitter and receiver are located at a given angle (e.g., in building corners) and (iii) the indirect method, where the source and the receiver are placed on the same surface.

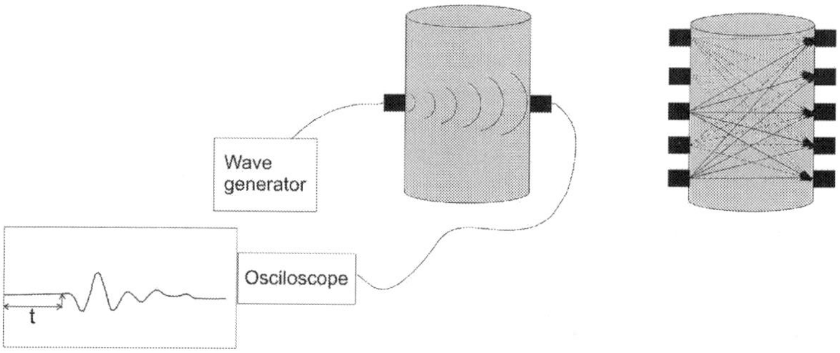

Figure 11. System for measuring wave velocity (left), to do acoustic tomography (right).

Two kinds of body waves are used on acoustic studies, primary (P) or compression waves and secondary (S) or shear waves. P waves can move through solid rock and fluids and the vibration direction is parallel to the direction of wave propagation. S waves are slower than P waves and can only propagate through solid rock, not through any fluid medium. S waves move particles perpendicular to the direction of wave propagation. In a homogeneous isotropic solid, the elastic properties of the material can be estimated from P and S wave velocities and density. Higher frequencies provide higher resolution but attenuation increases with frequencies. It is important to find the appropriate frequency as a function of the material and the wave travel distance. As a general rule, ultrasonic technique is suitable for small objects [68] and sonic for buildings.

These methods are used to estimate the density and elastic properties of materials (e.g. Young's modulus and Poisson ratio), it can indicate the presence of voids and cracks and effectiveness of repair by injection. In the case of inhomogeneous materials, the interpretation of the results is not simple [25]. Fort et al [69] in situ P wave velocities and Schmidt hammer surface strength and they found a direct relationship between them and an inverse correlation between P wave velocity and the decay of several granites of the Guadarrama mountains in Spain. In the study of Bromblet et al. [70], the P wave velocities of 62 marble columns of the cloister of the church Saint Trophime of Arles (France) have been compared with the values taken 16 years before (1993–2009) to estimate the evolution of their degradation over this time. They found a significant variation, sometimes the P wave velocity decreases but sometimes it increases. This increase could be due to the reduction of water content within the masonry because of the 2003/2004 restoration campaign. In the study of Pamplona and Simon [71], a similar work has been done on six marble sculptures but including acoustic tomography for some of them. They could study the weathering evolution and the effect of applied consolidation treatments.

Sklodowski [72] presented a method based on the propagation of superficial Rayleigh waves using edge probes rather than volume waves like with the previous techniques. Piezoelectric transducers are glued to the surface of steel edges providing only linear contact with the surface. Measurements are made using one transmitting and one or two receiving transducers at a distance of several centimetres without any contact interphase. The method provides wave velocities in a very superficial part of materials (few mm). In the work of Sklodowski [72] two studies in bricks and marble are presented, and differences in velocities between degraded and not degraded materials are observed.

4.2.3. Impact-echo technique

This method is based on impact-generated sound waves (about 2–50 kHz) that propagate through masonry and are reflected by internal discontinuities (cracks, detachments, reinforced structures and so on) [73]. It has been developed for the investigation of concrete structures, and it can determine the size and depth of these discontinuities. Waves are recorded, and the frequency spectrum is calculated. For each geometrical discontinuity, different waveforms and spectra are generated and dominant patterns (number and distribution of peaks in the spectra) can be recognized. Its main advance is that it does not need to couple the transducer with the surface but, on the other side, it is tedious in the sense that it requires many measurements without the possibility to determine the dimensions of defects filled with water. A robot to automatise the measurements is in development [74]. **Figure 12** shows the principle and the equipment required for the impact-echo technique.

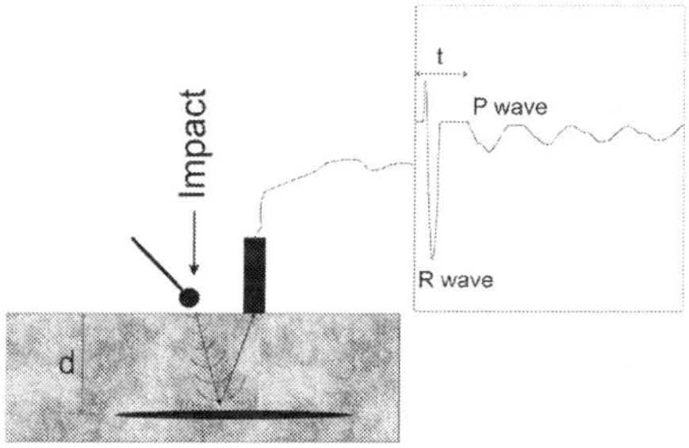

Figure 12. Impact-echo system.

Even if the impact-echo technique was originally designed for concrete, it has also been applied for the study of stone masonry [75] with good results. The required equipment includes a hammer or steel sphere, a wave receiver, amplifier and signal analysis facilities. If several measurements are done with increasing distances between impact and receiver [76], a mini-seismic method is developed, which can be interpreted as a classical seismic test.

4.2.4. Sonic or ultrasonic tomography

It consists in obtaining a 3D distribution of wave velocities in the interior of the studied object. In this case, a large number of measurements needs to be done with many different wave paths between emitter and receiver covering as much as possible the volume to be investigated

(Figure 11). Several transducers can be used but the number of points can also be increased by repeating the measurements at different locations. Both sonic and ultrasonic waves can be used. Signals are recorded, and time flight and amplitude are generally used in calculations. The tomographic technique can be used in 2D or 3D mode. To obtain good results, a precise knowledge of the geometrical and structural model is necessary. In a single block object, only the geometrical model is needed, but in multiple element objects, the inner structure and composition should be known. We can assimilate this tomography to the one applied in seismology because it uses the same inversion methods. The equipment consists of survey tools to precisely determine the position of the measure points, sonic or ultrasonic equipment and tomography software for inversion calculations.

One example of acoustic tomography in the sculpture of Leonora d'Aragone (Francesco Laurana) is presented in the study of Capizzi et al. [68]. This sculpture (22 cm x 40 cm x 43 cm) is carved from a unique block of microcrystalline marble and presents possible crack problems. High-resolution ultrasonic tomography was the most appropriate technique to monitor the structural continuity of a possible fracture and to investigate the internal marble conditions. A 3D ultrasonic tomography was obtained from 157 measurement points spaced from 2 to 5 cm (1832 raypaths). Extensive signal processing has been done before inversion. **Figure 13** shows some of the results obtained.

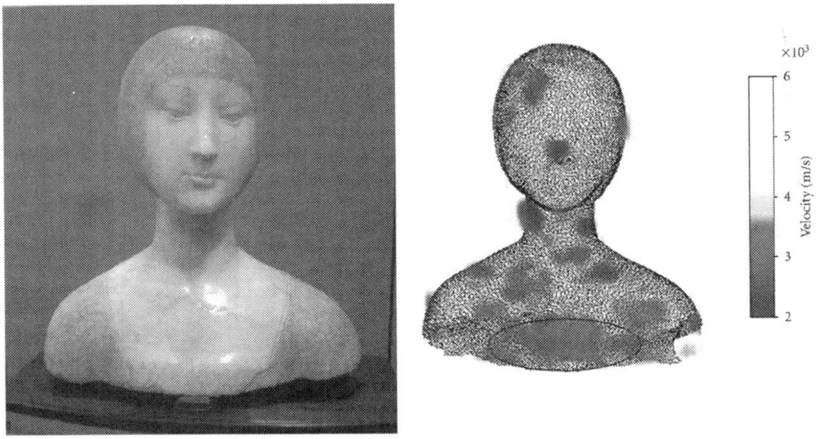

Figure 13. 3D ultrasonic travel time tomography on the Eleonora d'Aragona statue. Adapted from Cosentino et al. [4].

This technique gave also good results in a granitic megalithic of Axeitos (La Coruna, Spain) [77] where the heterogeneous structure in terms of conservation could be determined (**Figure 14**). Another example of a map of ultrasonic velocities as indicator of stone degradation level is found in the study of Fitzner [10] where the results for the Lion horoscope of the Nemrut Dag in Turkey are shown. Velocities go from 300 to 4500 m/s clearly showing the most degraded areas.

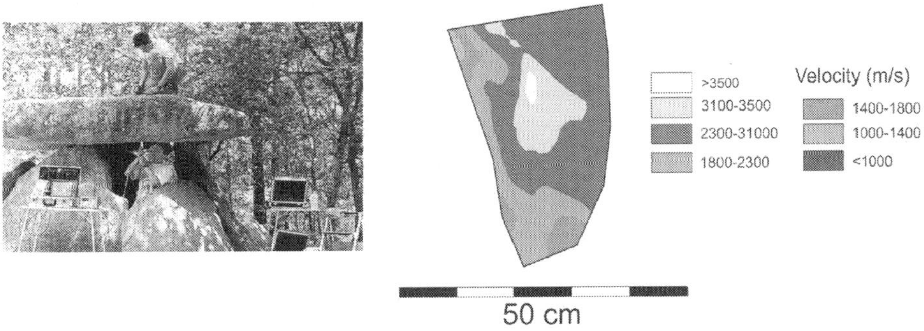

Figure 14. 3D ultrasonic tomography of the megalith d'Axeitos (Spain). Reproduced with the permission from Ebert et al. [77].

4.2.5. Monitoring using acoustic emission techniques

Acoustic emission (AE) is not an inspection technique but a monitoring one. When a crack is created or propagates, elastic energy is released through elastic wave propagation (similar to earthquakes). Recording signals generated by these waves is an indicator that cracking processes are active. The acoustic emission technique records continuously the elastic energy released during micro-crack generation, crack propagation or crystallization events. This elastic energy propagates in the sample as elastic waves, which can be detected and converted into electrical signal by piezoelectric acoustic emissions transducers. Several parameters are of interest like the number of recorded events (number of wave trends), and for each event the number of counts (number of times amplitude crosses a predefined threshold), the duration of each event, etc. In a continuous homogenous material, the recording of acoustic emissions by several transducers located at different positions can lead to the localization of the source of the acoustic emission, similarly to the localization of earthquakes. In buildings, it is quite difficult to localize the source of the events. High-frequency events propagate in masonry with greater attenuation. Based on experimental results, some authors [78] found that for a distance of 10 m in a tower of complex masonry structure, acoustic emission only with frequency components lower than 100 kHz are detectable.

This technique is currently used in rock mechanics but only few examples can be found of its application to cultural heritage. Two examples in the study of salt crystallization in the laboratory are presented in the works of Grossi et al. [79] and Menéndez and David [80] Grossi et al. [79] registered acoustic emissions during classical crystallization tests with sodium sulphate; and Menéndez and David [80] registered AE during non-classical crystallization tests with gypsum. An example of the use of acoustic emission techniques to monitor the crack growth in three medieval towers in Alba (Italy) can be found in the study of Anzani et al. [81]. They used the number of counts as an indicator of the energy released. They monitored two towers during around 1500 hours and another during 3500 hours. During this time, several seismic events took place and they were reflected in a peak of acoustic emission signal. A continuous background of acoustic emissions has also been registered indicating a permanent crack growth. A masonry building (Casa Capello, Rivoli, Italy) has been monitored to evaluate the status of cracks that spread out after the collapse of a breast wall [82]. An 800-hour campaign showed that acoustic emission activity corresponds to highest velocity of crack advancement. The crack growth and the associated acoustic emission are shown in **Figure 15**, based on the study of Carpinteri and Lacidogna [82].

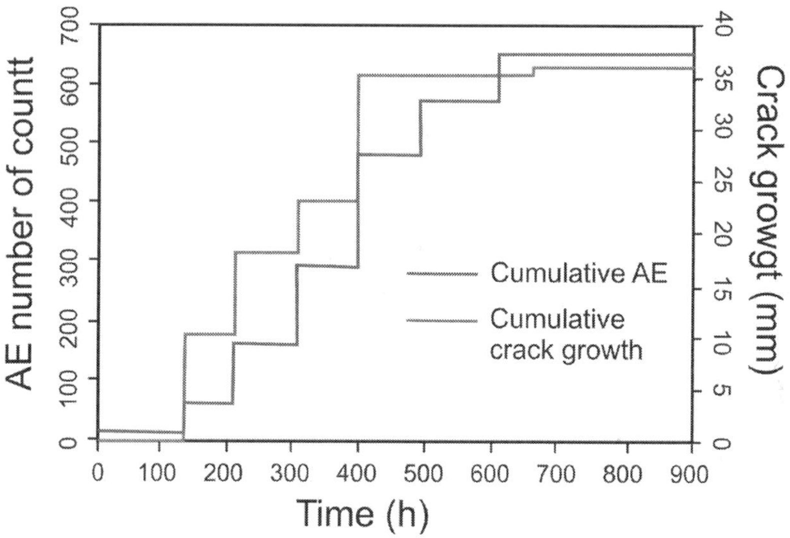

Figure 15. Cumulative crack growth and associated acoustic emission activity in Casa Capello (Rivoli, Italy). Adapted from Carpinteri and Lacidogna [82].

Another example can be found in the study of Suarez del Rio et al [83], where the acoustic emission close to several cracks of the Cathedral of Palma de Mallorca (Spain) has been monitored. They also measured displacements in those cracks. They conclude that the origin of cracks in this building is mainly due to thermal expansion more than mechanical problems because AE activity was much higher during day time than at night, was more important in the external part than in the internal one of the cathedral and also because displacements measured in the South façade were more important than in the Nord one.

4.3. Electromagnetic techniques

4.3.1. Electrical methods

This method consists in injecting an electrical current between two electrodes on the material and measuring the resulting voltage difference on another couple of electrodes (**Figure 16**). By changing the distance between electrodes, electric profiles are obtained. The electrical current can be direct or alternating. The goal is to obtain the resistivity distribution of the interior of the material. The resistivity of materials depends on the mineralogy, fluid content, porosity and water saturation degree [8].

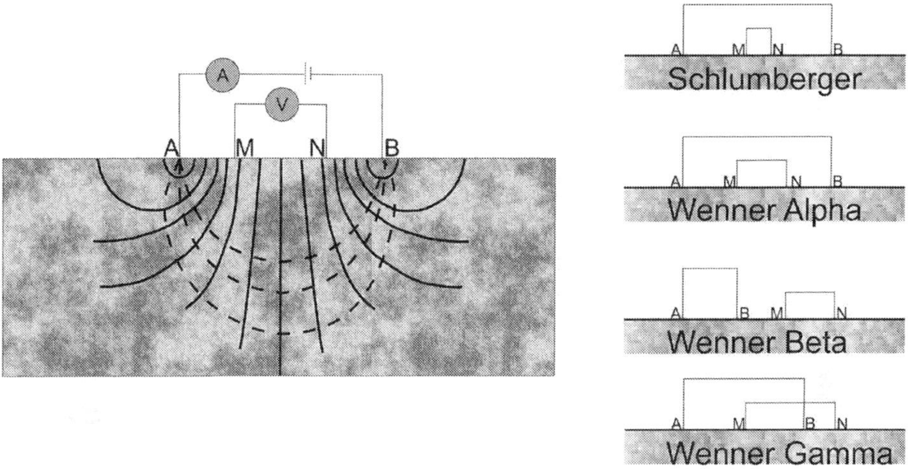

Figure 16. Principle of the electrical method (left) and different electrode arrays (right).

4.3.1.1. Methods using injected current

The electrodes can be spatially distributed in different ways. There are different types of spatial electrodes array: if all the electrodes are aligned, several arrays are possible such as Schlumberger, Wenner (alpha, beta or gamma), Lee and dipole; if they are not aligned, square or rectangular arrays can be used [84]. A 2D and 3D resistivity distribution can be obtained by changing the distance between electrodes. Two-dimensional parallel profiles in one or two direction can allow obtaining 3D representation of the material [8]. In the electrical resistivity tomography, the potential field distribution within an irregular shaped object is difficult to model [68].

Numerous 2D and 3D electrical tomographic surveys carried out on walls, columns and floors using direct electric current are shown in the study of Cosentino et al. [85]. The data were interpreted by back-projection inversion software or by 2D and 3D complete inversion procedure. They used a 256-channel instrument but there are instruments with larger number of channels. A problem was the choice of the electrodes; small electrodes with similar contact resistance are required. They used disposable electrocardiogram electrodes with external adhesive strip. An example of their results is shown in **Figure 17** for an ancient wall (~1.5 m thick) covered by a mosaic in the Fountain Room of Zisa Palace in Palermo. This mosaic presented some moisture problems. The 3D inversion in a 2 m×3 m area permitted to locate a water leakage area at 70–80 cm inside the wall, which was related to an unknown pipe collecting water from the roof.

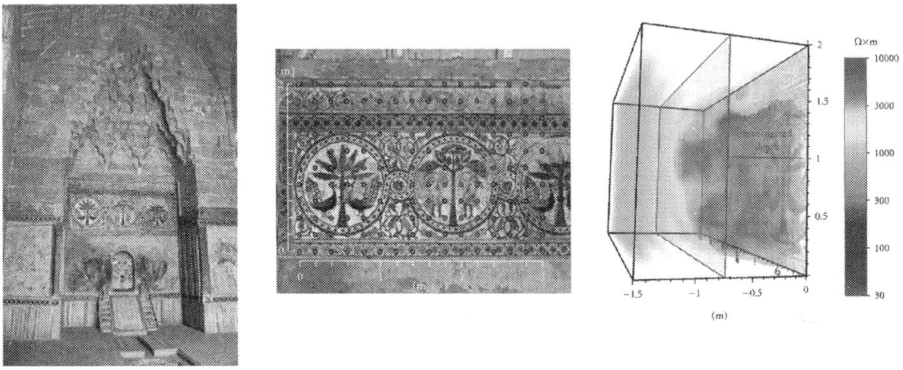

Figure 17. Frontal wall of the Fountain Room of the Arabian Zisa Palace in Palermo (left); position of the 176 potential electrodes (centre) and 3D inversion model of the acquired data (right). Adapted from Cosentino et al. [4].

Martinho et al. [86] combined resistivity tomography data with soluble salt analysis in three stone bas-reliefs located in the ground floor of the cloister of the Santa Cruz Monastery in Coimbra (Portugal). The results indicate that lower resistivity values are located in the areas of lower salt concentration. Moisture and salt distribution pointed to more than one source of moisture. The lower resistivity values correspond to severe decay areas. Electric resistivity tomography made possible to estimate the thickness of the panels.

Alternating current has been used in the study of Biernat et al. [87] to identify damp distribution in a wall. They used several current and voltage electrodes and independent power sources to find conductivity distribution by inversion of the data (electric impedance tomography). They concluded that this method can be of great importance in controlling the quality of damp proofing solutions.

4.3.1.2. Self-potential method

In this method, no current is injected, only the natural potential difference is measured. Differences in electrical potential can be due to electrofiltration, thermoelectrical, electrochemical or mineralization potentials [8]. Self-potential signals in cultural heritage material can be related to redox reactions and interstitial humidity within stones and mortars [88]. This method can be used, mainly, for determination of moisture content. Self-potential tomography is useful for practical application on flat surfaces [88]. Martinho et al. [89] combined this technique with classical studies (macroscopic stone description, qualitative visual assessment of stone deterioration, quantification of salts, etc), seismic refraction method and infrared thermography on a Gothic tomb located in the Igreja da Graca church (Santarem, Portugal). They found a direct linear relationship between self-potential and temperature.

http://dx.doi.org/10.5772/62408

4.3.2. Ground penetrating radar

This technique is similar to the acoustic technique but with another kind of waves. This technique is also an adaptation of a geophysical on site technique. In this case, electromagnetic pulses (frequency 500–2500 MHz) are injected into the material by an antenna (**Figure 18**). The most common experimental setup is the reflection mode with the antenna moving in one direction on the surface and the receiving antenna moving at the same time as the transmitter. Both antennas are usually placed on the same device. In transmission mode, the receiver is placed on an opposite surface of the objet. This method is less common and generally is only used to measure the signal velocity for calibration of the test.

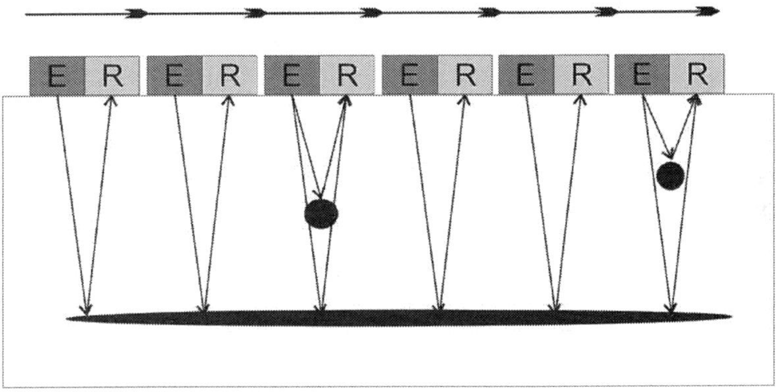

Figure 18. Principle of ground penetrating radar (GPR).

The propagation of the radar signal depends on the conditions and dielectric properties of the materials. For example, for geologic profiling at depths between 0 and 30 m, the antenna can have a frequency of about 100 MHz; in archaeology, with depths between 0 and 10 m, antennas of frequency 200–300 MHz can be used and for shallower objects to be detected, frequencies of 1000–2500 MHz can be employed. Penetration depth decreases with the frequency of the signal but the spatial resolution of the method increases with frequency.

The measured parameters are the signal velocity and the attenuation. The signal is reflected by an interface separating two media with different dielectric constants and the travel time corresponds to a distance twice the depth of the interface. As in acoustic methods, the choice of the antenna depends on the material and on the size and location of the defects.

This technique gives good results in detecting and locating fractures and discontinuities into the materials but does not allow estimating the mechanical parameters. Furthermore, it is difficult to apply in small objects or irregular sculpted surfaces [68]. In **Figure 19**, we can observe the reconstruction of the layers along the cross section of Monte di Pieta (Naples, Italy) façade [90].

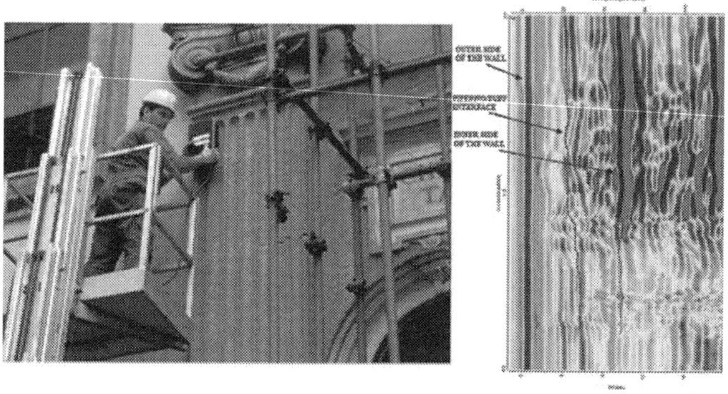

Figure 19. GPR applied to Monte de Pieta (Naples, Italy): pilasters and radar section with the construction layers[90].

Ground penetrating radar (GPR) has been used for a structural evaluation of several ancient stone masonry arch bridges in the NW of Spain still in use [91]. They realized two parallel profiles through the bridges at 1 m distance between them. The final goal was to elaborate finite difference time domain models of the bridges. They could detect ancient restorations, reconstruction of arches or all along the pathway; differences in building materials in the same stonework; presence of possible hidden arches or different historical shape of the structure; identification of reinforcement solid piers; thickness of the stone ring and nature of foundations.

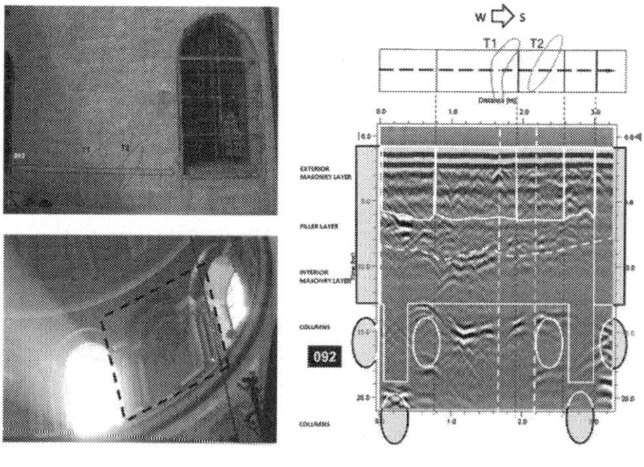

Figure 20. Upper left: External view of the southwest masonry of the Dome of the Catholic, Church of the Holy Sepulchre: position of the GPR scan and cracks T1 and T2. Lower left: Interior view of the back side of the examined area. Right: GPR scan with the interpretation of the interior structure [5].

In **Figure 20**, we can observe a typical result obtained by processing data for a GPR survey [5]. The profile corresponds to the exterior masonry of the Dome of the Catholic part in the Church of the Holy Sepulchre. Two cracks were observed in the exterior and in the GPR data and we can observe how they penetrate the complete thickness of the ashlars. The position of the filler layer between the external and internal ashlars can be determined.

4.3.2.1. Application: GPR 3D reconstruction

It consists in obtaining a 3D volume of the radar data by acquiring a dense subset of parallel 2D radar profiles [92]. The position of traces and distance between them should be accurately obtained in order to produce good 3D reconstructions. The shorter the distance between traces the more precise the 3D reconstruction. The volume rendering can be done by special softwares. Orlando and Slob [93] applied this technique to monitor cracks in a building with structural problems probably induced by movements of terrain. They used a GPR with 2 GHz bipolar antennas, with dipoles placed in a rectangular arrangement and doing four profiles at the same time: between parallel x-direction antennas, between parallel y-direction antennas, perpendicular with x direct source and y receiver and with y antenna as direct and x as receiver. Several 10 m profiles were done with spacing 0.1 m, four times a year to follow the crack evolution on a floor. They did not find any significant difference between the different periods of observation and conclude that this method is partially suitable for direct crack detection.

4.3.2.2. Application: Radar tomography

As in acoustic tomography, GPR tomography is generally used to map the interior of objects that can be accessed from at least two sides [92]. Tomography uses the direct transmission method. Transmitter and receiver antennas are separated and located successively in various positions to entirely cover the area under investigation. For every measurement, the velocity is calculated, and an inversion method applied to reconstruct the internal structure. An application of this technique to the study of piers in the cathedral of Noto (Portugal) can be seen in the study of Binda et al. [94]. Antennas were placed at different heights of the pier, one every 10 cm (46 acquisitions, 4.6 m) at each height a single signal was recorded, holding the transmitter and moving the receiver in all the positions at the opposite sides of the pier. Examples of application of radar tomography to study of internal structure of lapideous balcony corbels and the juxtaposition of patinas on urban environments are shown in the work of Cosentino et al. [85].

5. Combination of several NDT

A unique NDT technique cannot give all the necessary information to understand the conservation problem of the building. Information is needed about the nature of the materials, their properties, the 3D structure of the building and the evolution of all these characteristics with time. A combination of several techniques is always employed, sometimes only "simple" techniques are enough to cover the goals of the study but sometimes these simple techniques should be combined with one or more complex techniques. For example, before doing acoustic tomography, GPR will be very helpful to have a first approach of the internal structure and allows the construction of the geometrical model to carry out the numerical inversion. Acquisition methods should be referred to the same geometric system to correlate the information provided by the different methods [15].

Many examples can be found in the literature showing the interest of combining information coming from different non-destructive techniques. Solla et al. [15] used photogrammetry, thermography and GPR to study the Lubian bridge (Galicia Spain). This bridge has a very complicated history from the XV century with different reconstructions, rebuilt and restoration campaigns using different materials: original masonry (granite ashlar), granite masonry with clay mortars, cement, pavement of granite flagstones and cement. Photogrammetry allows one to do a detailed 3D reconstruction of the bridge; thermographic contributes to the identification of different materials and GPR is used to establish the internal structure of the bridge with the different areas of original and restoration/reconstruction parts.

A combination of GPR and acoustic techniques has been used to study the structure and the conservation degree of a column proceeding from the Hospital of Saint Pau I la Santa Creu in Barcelona (Spain) [95]. The column has been moved to the laboratory for investigation. They performed series of GPR vertical and horizontal profiles with spacing 5 and 3 cm, respectively, vertical profiles height was 110 cm. Acoustic tomography has been performed using 13 sensors placed around the column and with a hammer as vibration source. After these non-destructive tests, mechanical compression test has been done and the interior of the column could be observed after breakage. GPR vertical and radial profiles give different and complementary information about the internal array of bricks, presence or not of reinforcement elements but differentiating between voids or cracks and changes in material is difficult just with GPR data. Seismic data inform about the presence of damaged areas. In this example, NDT techniques show the presence of a metallic pipe in the interior of the column, the pipe was corroded in some areas, which could be related to the most damaged zones.

Kilic [58] applied a combination of GPR and thermal survey to the study of Urla primary school (Turkey), an Ottoman structure of XVI century. The results show how an integrated approach using a combination of NDT methods can detect defects affecting the structural condition, both visible and hidden, of historical buildings. This is especially interesting in cases of lack of written documentation on the original construction (position of pipes, internal structure and so on).

In the Non-Destructive Testing in Civil Engineering (NDTCE'09) conference, a very large set of different examples of application of NDT and minor destructive techniques in different sites in Italy has been presented by Binda and Saisi [7]. These examples concern the survey of structural damage associated with cracks, the detection of multiple leaf walls or detachments, presence of voids or inclusions, determination of the masonry "quality" and moisture detection. Each example presented has different problems and the employed techniques vary to solve them. In each example, a combination of two or more techniques is employed: cartography of materials and cracks, georadar, sonic or ultrasonic tests, thermography, single or double flat jacks and power drilling method (partially destructive). As a general conclusion, the authors stated that the use of NDT in the diagnostic of historic buildings is not an easy task, the problem to be solved and the appropriate techniques should be well known.

Bergamo et al. [96] presents an application of several NDT and minor destructive techniques to the study of arch bridged in use in Italy. They concluded that georadar is useful to investigate this type of structures; vibrational analysis, flat-jack and penetration tests are useful to investigate the causes of damage and to calibrate the finite element models of the bridges. They found that thermography is the most reliable technique to detect moisture, discontinuities and the presence of different building materials.

The choice of the techniques will depend on the historical interest of the object, the actual danger and the financial possibilities. A table with the different NDT techniques used in cultural heritage is presented in the study of McCann and Forde [97], with for each technique the measured parameter(s), the advantage and disadvantage and the relative cost. This table can be a useful tool for a first approach to select the methods that can be applied for a particular building.

In the study of Meneely et al. [98], a toolkit to monitor historical buildings in order to know their evolution with time and with the change of environmental condition (climatic change and atmospheric pollution) is presented. This kit is based on an initial 3D laser approach to spatially map data collected from other techniques. They proposed to use/measure (i) high quality digital photography (ii) colorimeter, (c) permeability, (iv) ground penetrating radar, (v) thermography, (vi) X-ray fluorescence to analyse the surface chemistry and others. All this information can be stored in a Geographical Information System to obtain spatial distribution of different parameters and the relationships between them. The frequency of the survey depends not only on financial and logistical constraints but also on the rate of change of the site.

6. Conclusions

Many different "in situ" non-destructive techniques are available for the study of built cultural heritage. Some of them are very old techniques of almost a century ago, whereas others are based on the most recent technology and are still in development. As an example of a technique under development, we can cite a system using microwaves to measure water and salt contents on frescoes and mural paintings based on the relationships between these parameters and the dielectric properties of the materials [99].

NDT techniques are of especial interest to construct analytical or numerical models of buildings. Carino [73] constructed a finite element model of the main façade of the Mote di Pieta de Naples (Italy) that could simulate the construction and its historical phases in order to justify the existing major damages and to estimate the current state of stress. In the construction of the model, they used data from 3D laser scanning, GPR and flat-jack techniques.

As a general conclusion, we can say that every study on built cultural heritage used a combination of NDT. The number and the complexity of this technique depend on the interest of the building, its conservation degree, the goal of the study and the available budget. Different NDT techniques are complementary, and the combination of data provided by each technique is crucial to understand the past behaviour or the building and to predict the future one as a function of new conditions affecting the building (mechanical, climatic and so on). Most of the time, NDT are the only way to have "access" to the inner part of the building. Great progress has been done in this field in the last decades, and the research is still very active.

Acknowledgements

I like to thank Christian David for his thorough reading and clever suggestions about the organization and the writing of this chapter.

Author details

Beatriz Menéndez

Address all correspondence to: beatriz.menendez@u-cergy.fr

Geosciences and Environment Cergy (GEC), University of Cergy-Pontoise, Cergy-Pontoise, France

References

[1] Fitzner B. Damage diagnosis on stone monuments – in situ investigation and laboratory studies. In: Proceeding of International Symposium of the Conservation of the Bangudae Petroglyph, Ulsan City/Korea, 15 July 2002. Ulsan City Korea: Stone Conservation Laboratory, Seoul National University, 29–71.

[2] Fitzner B. Documentation and evaluation of stone damage on monuments. In: Proceedings of the 10th International Congress on Deterioration and Conservation of Stone, 27 June–2 July 2004. Stockholm, Sweden, 667–690.

[3] Svahn H. Non-Destructive Field Tests in Stone Conservation: Final Report for the Research and Development Project: Literature Study. 2006. Riksantikvarieämbetet. Stockholm, 63 p.

[4] Cosentino OL, Capizzi P, Martorana R, Messina P, Schiavone S. From geophysics to microgeophysics for engineering and cultural heritage. International Journal of Geophysics. 2011; 2011: 8, ID: 428412. DOI: 10.1155/2011/428412

[5] Moropoulou A, Labropoulos KC, Delegou ET, Karoglou M, Bakolas A. Non-destructive techniques as a tool for the protection of built cultural heritage. Construction and Building Materials. 2013; 48: 1222-1239. DOI: 10.1016/j.conbuildmat.2013.03.044

[6] Bosiljkov V, Uranjek M, Zarnic R, Bokan-Dosihkov V. An integrated diagnostic approach for the assessment of historic masonry structures. Journal of Cultural Heritage. 2010; 11: 239-240. DOI: 10.1016/j.culher.2009.11.007

[7] Binda L, Saisi A. Application of NDT to the diagnosis of historic structures. In: Proceedings of the Non-Destructive Testing in Civil Engineering NDTCE'09, Non-Destructive Testing in civil Engineering Conference, June 30–July 3 2009. Nantes, 27 p.

[8] Martinho E, Dionisio A. Main geophysical techniques used for non-destructive evaluation in cultural built heritage: a review. Journal of Geophysics and Engineering. 2014; 11: 1-15. DOI: 10.1088/1742-2132/11/5/053001

[9] Esbert RM, Marcos RM. Las piedras de la Catedral de Oviedo y su deterioración. Colegio Oficial de Aparejadores y Arquitectos Técnicos del Principado de Asturias, 1983. Oviedo, 147 p.

[10] Fitzner B. Diagnosis of weathering damage on stone monuments. Macla.2014; 18: 21:28. 18: 21-28.

[11] Fitzner B, Heinrichs K. Damage diagnosis on stone monuments – weathering forms, damage categories and damage index. In: Prikryl R, Viles HA Ed., Abstracts of the International Conference "Stone Weathering and Atmospheric Pollution Network (SWAPNET)", 7–11 May 2001. Prachov Rocks (Czech Republic).

[12] ICOMOS-ISCS. Vergès-Belmin Ed. Illustrated glossary on stone deterioration patterns. 2008. ISBN: 978-2-918086-00-0. Champigny sur Marne, France

[13] Photogrammetry. http://culturalheritageimaging.org/Technologies/Photogrammetry/

[14] Yastikli N. Documentation of cultural heritage using digital photogrammetry and laser scanning. Journal of Cultural Heritage. 2007; 8: 423-427. DOI: 10.1016/j.culher. 2007.06.003

[15] Solla M, Asorey-Cacheda R, Nuñez-Nieto X, Conde-Carnero B. Evaluation of historical bridges through recreation of GPR models with the FDTD algorithm. NDT&E international. 2016; 77: 19-27. DOI: 10.1016/j.ndteint.2015.09.003

[16] Arias P, Armesto J, Ordonez C, Lorenzo H, Caparrini H. Damage quantification and monitoring in masonry monuments through digital photogrammetry. Key Engineering Materials. 2007; 347: 291-296. DOI: 10.4028/www.scientific.net/KEM.347.291

[17] Heinrichs K, Azzam R. Investigation of salt weathering on stone monuments – the 'petraSalt' research project. Proceedings of the 19th Conference on Engineering Geology and of the Forum for young Engineering Geologists, 13–15 March 2013, Munich: Kurosch Thuro.

[18] Vazquez MA, Galan E, Guerrero MA, Ortiz P. Digital image processing of weathered Stone caused by efflorescences: a tool for mapping and evaluation of stone decay. Construction and Building Materials. 2011; 25: 1603-1611. DOI: 10.1016/j.conbuildmat.2010.10.003

[19] Crespo C, Aemesto J, Gonzalez-Aguilera D, Arias P. Damage detection on historical buildings using unsupervised classification techniques. International Archives of Photogrammetry, Remote Sensing and Spatial Information Sciences. 2010; 38(5): 184-188.

[20] Meroño JE, Perea AJ, Aguilera MJ, Laguna AM. Recognition of materials and damage on historical buildings using digital image classification. South African Journal of Science. 2015; 111(1/2): 68-76. DOI: 10.17159/sajs.2015/20140001

[21] RILEM. Recommendation MDT. D. 4: In-situ stress tests based on the flat jack. Materials and Structures. 2004; 37: 491-496. DOI: 10.1617/14119

[22] RILEM. Test method recommendations of RILEM TC 177-MDT 'Masonry durability and on-site testing' - D.5: In-situ stress - strain behaviour tests based on the flat jack. Materials and Structures. 2004; 37: 497-501. DOI: 10.1617/14120

[23] Simoes A, Gago A, Lopes M, Bento R. Characterization of old masonry wall: flat-jack method. Proceedings of the 15th World Conferènce on Earthquake Engineering, 24–28 September, 2012. Lisbon, Portugal, 10 p.

[24] Bartoli G, Betti M, Giordano S. In situ and dynamic investigations of the "Torre Grossa" masonry tower. Engineering Structures 2013; 52: 718-733. DOI: 10.1016/j.engstruct.2013.01.030

[25] Binda L, Saisi A, Tiraboschi C. Investigation procedures for the diagnostic of historic masonries. Construction and Building Materials. 2000; 14: 199-233. DOI: 10.1016/S0950-0618(00)00018-0

[26] Fenn DB, Rutenbeck T, Knab LI, Clifton JR, Nistal-Moret BBG. Structural monitoring of Castillo de San Marcos national monument. Program report of Old Cultures in New Worlds, 8th ICOMOS General Assembly and International Symposium, 1987. Washington, 400-407.

[27] Rossi PP, Rossi C. Surveillance and monitoring of ancient structures: recent developments. In: Structural Analysis of Historical Constructions II. Possibilities of Numerical and Experimental Techniques - Proceedings of the IVth International Seminar on Structural Analysis of Historical Constructions, 10–13 November 2004. Padova, Italy: Taylor & Francis, Technology & Engineering, 1466.

[28] Dovrev N, Avramova-Tacheva E, Kostack B. Monitoring of the cracks affecting the "Madara Horseman" rock bas-relief, North-East Bulgaria. In: St Ivan Rilski, Kostov RI, Gaydarska B, Gurova M Eds., Geoarcheology and Archeomineralogy. Proceedings of the International Conference, 29–30 October 2008. Sofia: Publishing House, 385-390.

[29] Glisic B, Posenato D, Casanova N, Inaudi D, Figini A. Monitoring of heritage structures and historical monuments using long-gage fiber optic interferometric sensor – an overview. In: Baidar Bakht and Aftab Mufi Ed. Proceeding of the 3rd International Conference on Structural Health Monitoring of Intelligent Infrastructure, 1–16 November 2007. Vancouver.

[30] Inaudi D, Casanova N, Glisic B. Long-term deformation monitoring of historical constructions with fiber optic sensors. In: Lorenço PB, Roca P, Eds., Historical Constructions 2001. Guimaraes.

[31] Inaudi D. SOFO sensors for static and dynamic measurements. Proceedings of the 1st FIG International Symposium of Engineering Survey for Constructions and Structural Engineering. 28 June–1 July 2004, Nottingham, United Kingdom. 10 p.

[32] Charola AE. Salts in the deterioration of porous materials: an overview. Journal of the American Institute for Conservation. 2000; 39: 327-343. DOI: 10.1179/019713600806113176

[33] Menezes A, Gloria Gomes M, Flores-Colen I. In-situ assessment of physical performance and degradation analysis of rendering walls. Construction and Building Materials. 2015; 75: 283-292. DOI:10.1016/j.conbuildmat.2014.11.039

[34] Hendrickx R. Using the Karsten tube to estimate water transport parameters of porous buildings materials. Materials and Structures. 2013; 46: 1309-1320. DOI: 10.1617/s11527-012-9975-2

[35] Vandevoorde D, Cnudde V, Dewanckele J, Boone M. Comparison of non-destructive techniques for analysis of water absorbing behavior of stone. In: Proceeding of the 12th International Congress on the Deterioration and Conservation of Stone; 22–26 October 2012; New York. 10 p.

[36] Vandevoorde D, Cnudde V, Dewanchele J, Brabant L, de Bouw M, Meynen V, Verhaeven E. Validation of in situ applicable measuring techniques for analysis of the water absorption by stone. Procedia Chemistry. 2013; 8: 317-327. DOI: 10.1016/j.proche.2013.03.039

[37] Bläuer C, Frazen C, Vergès-Belmin V. Simple tests in stone conservation. In: 12th International Congress on the Deterioration and Conservation of Stone, 22–26 October 2012. 8 p.

[38] Drdácký M, Černý M, Slížková Z, Zíma P. Microtube device for innovative water uptake measurements. In: Krüger M, Ed., Proceedings of the European Workshop on Cultural Heritage Preservation, Berlin 26–28 September 2011. Fraunhofer IRB Verlag: Stuttgart, 126-130. ISBN: 978-3-8167-8560-6

[39] Drdácký M, Hasníková H, Lesák J, Zíma P. Innovated water uptake measurements on historic stone surfaces. In: Proceeding of the 12th International Congress on the Deterioration and Conservation of Stone, 22–26 October 2012. New York, 10 p.

[40] Brown S, Smith L. A transient-flow syringe air permeameter. Geophysics. 2013; 78 (5): D307-D313. DOI: 10.1190/geo2012-0534.1

[41] Menéndez B, Lopez-Arce P, Mertz JD, Tagnit-Hamou M, Aggoun S, Kaci A, Guiavarch M, Cousture A. Stone-mortar interaction of similar weathered stone repair mortars used in historic building. Accepted In: Proceedings 13th International Congress on the Deterioration and Conservation of Stone, 6–10 September 2016. Glasgow.

[42] Filomena CM, Hornung J, Stollhofen H. Assessing accuracy of gas-driven permeability measurements: a comparative study of diverse Hassler-cell and probe permeameter devices. Solid Earth. 2014; 5 (1): 1–11. DOI: 10.5194/se-5-1-2014

[43] Sena da Fonseca B, Castela AS, Duarte RG, Neves R, Montemor MF. Non-destructive and on site method to assess the air-permeability in dimension stones and its relationship with other transport-related properties. Materials and Structures. 2015; 48: 3795-3809. DOI: 10.1617/s11527-014-0440-2

[44] Duarte R, Flores-Colen I, de Brito J. In situ testing techniques for evaluation of water penetration in rendered facades - the portable moisture meter and Karsten tube. In: Proceedings of the XII International Conference on Durability of Building Material and Components, 12–15 April 2011. Porto, 8 p.

[45] Meinhardt J, Snethlage R, Auras M. Natural stone monitoring – Investigation methods for a reliable evaluation of the effectiveness of conservation measures. In: 12th International Congress on the Deterioration and Conservation of Stone, 22–26 October 2012. 14 p.

[46] Drdacky L, Lesak J, Rescic S, Slızkova Z, Tiano P, Valach J. Standardization of peeling tests for assessing the cohesion and consolidation characteristics of historic stone surfaces. Materials and Structures. 2012; 45:505–520. DOI: 10.1617/s11527-011-9778-x

[47] Alvarez de Buergo M, Vazquez-Calvo C, Fort R. The measurement of surface roughness to determine the suitability of different methods for stone cleaning. Geophysical Research Abstracts Vol. 13, EGU2011-6443 EGU General Assembly, 2011.

[48] Dabski M. Rock surface micro-roughness, Schmidt hammer rebound and weathering rind thickness within LIA Skálafellsjökull foreland, SE Iceland. Polish Polar Research. 2014; 35(1): 99–114, 2014 DOI: 10.2478/popore−2014−0008

[49] Gomez-Heras M, Smith BJ, Viles HA, Meneely J, McCabe S. HD Laser scanning for the evaluation of salt decay laboratory simulations of building limestone. In: Proceedings of the Salt Weathering on Buildings and Stone Sculptures Conference, 22–24 October 2008. Copenhagen, 149-158.

[50] Viles J, Goudi A, Grab S, Lalley J. The use of the Schmidt Hammer and Equotip for rock hardness assessment in geomorphology and heritage science: a comparative analysis. Earth Surface Processes and Landforms. 2011; 36: 320-333. DOI: 10.1002/esp.2040

[51] Christaras B. Non destructive methods for investigation of some mechanical properties of natural stones in the protection of monuments. Bulletin of the International Association of Engineering Geology. 1996; 54: 59-63.

[52] Aydin A, Basu A. The Schmidt hammer in rock material characterization. Engineering Geology. 2005; 81: 1–14. DOI: 10.1016/j.enggeo.2005.06.006

[53] Torok A. Surface strength and mineralogy of weathering rusts on limestone buildings in Budapest. Building and Environment. 2003; 38: 1185-1192. DOI: 10.1016/S03360-1323(03)00072-6

[54] Coombes MA, Feal-Perez A, Naylor LA, Wilhem K. A non-destructive tool for detecting changes in the hardness of engineering materials: application of the Equotip durometer in the coastal zone. Engineering Geology. 2013; 167: 14-19. DOI: 10.16/j.enggeo.2013.10.003

[55] Grossi C, Esbert RM, Diaz-Pache F, Alonso FJ. Soiling of building stone in urban environment. Building and Environment. 2003; 38: 147-159.

[56] Torok A. In situ methods of testing stone monuments and the application of nondestructive physical properties testing in masonry diagnosis. In Materials, Technologies and Practice in Historic Heritage Structures, 2010. Netherlands: Springer. DOI: 10.1007/978-90-481-2684-2_10

[57] de Freitas S, de Freitas VP, Barreira E. Detection of façade plater detachments using infrared thermography – A nondestructive technique. Construction and Building Materials. 2014; 70: 80-87. DOI:10.1016/j.conbuildmat.2014.07.094

[58] Kilic G. Using advanced NDT for historic buildings: Towards an integrated multidisciplinary health assessment strategy. Journal of Cultural Heritage. 2015; 16: 526-535. DOI: 10.1016/j.culher.2014.09.010

[59] Paoletti D, Ambrosini D, Sfarra S, Bisegna F. Preventive thermographic diagnosis of buildings for consolidation. Journal of Cultural Heritage. 2013; 14: 116-121. DOI: 10.1016/j.culher.2012.05.005

[60] Hess M, Vanoni D, Petrovic V, Kuester F. High-resolution thermal images methodology for non-destructive evaluation of historic structures. Infrared Physics & Technology. 2015; 73: 219-225. DOI: 10.1016/j.infrared.2015.09.020

[61] Jo YH, Lee CH. Quantitative modeling and mapping of blistering zone of the Magoksa Temple stone pagoda (13th century, Republic of Korea) by graduated heating thermography. Infrared Physics & Technology. 2014; 65: 43-50. DOI: 10.1016/j.infrared.2014.02.011

[62] Kordatos EZ, Exarchos DA, Stavrakos C, Moropoulou A, Matikas TE. Infrared thermographic inspection of murals and characterization of degradation in historic monuments. Construction and Building Materials. 2013; 48: 1261-1254. DOI: 10.1016/j.conbuildmat.2012.06.062

[63] Sklodowski R, Drdacky M, Sklodowski M. Identifying subsurface detachment defects by acoustic tracing. NDT&E International. 2013; 56: 56-64. DOI: 10.1016/j.ndteint.2013.02.002

[64] Drdacky M, Lesac J. Non-invasive survey of detachment of historic rendering. In: FortR, Alvarez de BuergoM, Gomez-HerasM, Vazquez-CalvoC, Eds. Heritage, Weathering and Conservation, 2006. London: Taylor & Francis Group, 591-597.

[65] Del Vescovo D, Fregolent A. Assessment of fresco detachments through a non-invasive acoustic method. Journal of Sound and Vibration. 2005; 284: 1015-1031.

[66] Collini L, Garziera R. A contact-less diagnosis system for frescoes. Part 2: acoustic excitation-acoustic response. NDT&E International. 2013; 56: 76-81. DOI: 10.1016/j.ndteint.2013.02.006

[67] Castellini P, Esposito E, Legoux V, Paone N, Stefanaggi M, Tomasini EP. On field validation of non-invasive laser scanning vibrometer measurement of damaged frescoes: experiments on large walls artificially aged. Journal of Cultural Heritage. 2000; 1 (Supplement 1): S349-S356. DOI: 10.1016/S1296-2074(00)00145-X

[68] Capizzi P, Cosentino PL, Schiavone S. Some tests of 3D ultrasonic travel time tomography on the Eleonora d'Aragona statue (F. Laurana, 1468). Journal of Applied Geophysics. 2013; 91: 14-20. DOI: 10.1016/j.jappgeo.2013.01.012

[69] Fort R, Alvarezde Buergo M, Perez-Monserrat EM. Non-destructive testing for the assessment of granite decay heritage structures compared to quarry stone. International Journal of Rock Mechanics & Mining Sciences. 2013; 61: 296–305. DOI: 10.1016/j.ijrmms.2012.12.048

[70] Bromblet P, Vergès-Belmin V, Simon S. Ultrasonic velocity measurements for the long-term monitoring of the degradation of marble columns in the cloister of the church Sait-Trophine in Arles (France). In: 12th International Congress on the Deterioration and Conservation of Stone, 22–26 October 2012, 11 p.

[71] Pamplona M, Simon S. Long-term condition survey by ultrasonic velocity testing of outdoor marble sculptures. In: 12th International Congress on the Deterioration and Conservation of Stone; 22–26 October 2012, 11 p.

[72] Sklodowski M. Application of ultrasonic edge probes to on-site testing of mechanical properties of historical materials. In Tiano P, Pardini C, Eds., Proceedings of the International Workshop – SMW08 In situ Monitoring of Monumental Surfaces, 27–29 October 2008. Florence, Italy: Edifir, 177-186. ISBN: 978-88-7970-390-1

[73] Carino NJ. The impact-echo method; an overview. In Chang PC, Ed. Proceedings of the 2001 Structures Congress & Exposition, Washington, D.C., 21–23 May 2001, 2001. Reston: American Society of Civil Engineers, 18 p.

[74] Hola J, Schabowicz K. State-of-the-art non-destructive methods for diagnosis testing of building structures – anticipated development trends. Archives of Civil and Mechanical Engineering. 2010; 10(3): 5-18. DOI: 10.1016/S1644-9665(12)60133-2

[75] Sadri A. Application of impact-echo technique in diagnoses and repair of stone masonry structures. NDT&E International. 2003; 36: 195–202. DOI: 10.1016/S0963-8695(02)00064-6

[76] Miranda LF, Rio J; Miranda Guedes J, Costa A. Sonic impact method – a new technique for characterization of stone masonry walls. Construction and Building Materials. 2012; 36: 27–35. DOI: 10.1016/j.conbuildmat.2012.04.018

[77] Ebert RM, Ordaz J, Alonso FJ, Montoto M. Manual de diagnostico y tratamiento de materiales pétreos y cerámicos, 1997. Col legi d'aparelladors i Arquitectes Tècnics de Barcelona: Barcelona.

[78] Carpinteri A, Lacidogna G. Damage evaluation of three masonry towers by acoustic emission. Engineering Structures. 2007; 29: 1569-1579. DOI: 10.1016/j.engstruc.2006.08.008

[79] Grossi CM, Esbert RM, Suarez del Rio LM, Montoto M, Larenzi-Tabasso M. Acoustic emission monitoring to study sodium sulphate crystallization in monumental porous carbonate stones. Studies in Conservation. 1997; 42: 115-125.

[80] Menéndez B, David C. The influence of environmental conditions on weathering of porous rocks by gypsum; a non-destructive study using acoustic emissions. Environmental Earth Sciences. 2013; 68: 1691-1706. DOI: 10.1007/s12665-012-1861-2

[81] Anzani A, Binda L, Carpinteri A, Invernizzi S, Lacidogna G. A Multilevel approach for the damage assessment of historic masonry tower. Journal of Cultural Heritage. 2010; 11: 459-470.

[82] Carpinteri A, Lacidogna G. Monitoring a masonry building of the 18th century by acoustic emission technique. WIT Transactions on the Built Environment. 2001; 55: 327–337. DOI: 10.2495/STR010321

[83] Suarez del Rio LM, Gomez Ruiz de Argandona V, Calleja L, Rodriguez Rey A, Grossi-Sampedro CM, Montoto M. Acoustic emission monitoring of the Cathedral of Palma de Mallorca (Spain). In Materials, Technologies and Practice in Historic Heritage Structures, 2010 Netherlands: Springer, 351-365. DOI: 10.1007/978-90-481-2684-2_18

[84] Chapellier D. Prospect ion électrique de surface. Geophysics on line course 2002. Université de Lausanne, Institut Français du Pétrole.

[85] Cosentino PL, Capizzi P, Fiandaca G, Martorana R, Messina P. Advances in Microgeophysics for engineering and cultural heritage. Journal of Earth Science. 2009; 20(3): 626-630. DOI 10.1007/s12583-009-0052-x

[86] Martinho E, Alegria F, Dionisio A, Grangeia C, Almeida F. 3D-resistivity imaging and distribution of water soluble salts in Portuguese Renaissance Stone bas-reliefs. Engineering Geology. 2012; 141-142: 33-44. DOI: 10.1016/j.enggeo.2012.04.010

[87] Biernat K, Idziaszek-Gonzalez A, Nita K, Sikora J, Wojtowicz S. Nondesructive impedance method of brickwork damp identification. Proceedings of the 42 International Conference and NDT Exhibition, NDE for Safety/DEFEKTOSKOPIE 2012, October 30–November 1 2012. Sec u Chrudimi (Czech Republic), 7-12.

[88] Cammarano F, Di Fiore B, Mauriellon P, Patella D. Examples of application of electrical tomographies and radar profiling to cultural heritage. Annali di Geofisica. 2000; 43(2): 309-324. DOI: 10.4401/ag-3631

[89] Martinho E, Dionisio A, Almeida F, Mendes M, Grangeia C. Integrated geophysical approach for Stone decay diagnosis in cultural heritage. Construction and Building Materials. 2014; 52: 345-352. DOI: 10.1016/j.conbuildmat.2013.11.047

[90] Lignola GP, Manfredi G. A combination of NDT methods for the restoration of monumental façades: The case study of Monte di Pieta (Naples, Italy). Journal of Cultural Heritage. 2010; 11: 360-364. DOI: 10.1016/j.culher.2009.11.010

[91] Solla M, Lorenzo H, Rial FI, Novo A. Ground-penetrating radar for the structural evaluation of masonry bridges: Results and interpretation tools. Construction and Building Materials. 2012; 29: 458-465. DOI: 10.1016/j.conbuildmat.2011.10.001

[92] Topczewski L, Fernandes FM, Cruz PJS, Lourenço PB. Practical implications of GRP investigation using 3D data reconstruction and transmission tomography. Journal of Building Appraisal. 2007; 3: 59-76. DOI: 10.1057/palgrave.jba.2950060

[93] Orlando L, Slob E. Using multicomponent GPR to monitor cracks in historical buildings. Journal of Applied Geophysics. 2009; 67:327-334.

[94] Binda L, Saisia A, Tiraboschia C, Vallea S, Collab C, Fordec M. Application of sonic and radar tests on the piers and walls of the Cathedral of Noto. Construction and Building Materials. 2003; 17: 613–627. DOI: 10.1016/S0950-0618(03)00056-4

[95] Santos-Assunçao S, Perez-Garcia V, Caselles O, Clapes J. Assessment of complex structures with GPR compared to other non-destructive testing studies. Remote Sensing. 2014; 6: 8220-8237. DOI: 10.3390/rs6098220

[96] Bergamo O, Campione G, Donadello S, Russo G. In-situ testing procedure as an integral part of failure analysis of historical masonry arch bridges. Engineering Failure Analysis. 2015; 57: 31–55. DOI: 10.1016/j.engfailanal.2015.07.019

[97] McCann DM, Forde MC. Review of NDT methods in the assessment of concrete and masonry structures. NDT&E International. 2001; 34: 71-84. DOI: 10.1016/S0963-8695(00)00032-3

[98] Meneely JD, Smith BJ, Curran J, Ruffell A. Developing a 'non-destructive scientific toolkit' to monitor monuments and sites. Proceedings of the ICOMOS Scientific

Symposium Changing World, Changing Views of Heritage, 7 October 2009. Valletta, Malta. 9 p.

[99] Olmi R, Bini M, Ignest A, Priori S, Riminesi C, Felici A. Diagnoses and monitoring of frescoes using evanescent-field dielectrometry. Measurement Science and Technology. 2006; 17: 2281-2288

Chapter 9

Microwave Non-Destructive Testing of Non-Dispersive and Dispersive Media Using High-Resolution Methods

Cédric Le Bastard, Khaled Chahine, Yide Wang,
Vincent Baltazart, Nicolas Pinel,
Christophe Bourlier and Xavier Derobert

Additional information is available at the end of the chapter

http://dx.doi.org/10.5772/62410

Abstract

This chapter discusses the principle and application of two model-based algorithms for processing non-dispersive and dispersive ground penetrating radar (GPR) data over layered medium under monostatic antenna configuration. Both algorithms have been selected for their super-time resolution capability and reduced computational burden; they allow GPR to measure a layer thickness smaller than the fraction of the dominant wavelength. For non-dispersive data, the ESPRIT algorithm is generalized to handle different kinds of data models encountered in experiments and in the literature. For dispersive data, the proposed adaptation of the MPM algorithm allows recovering the full-time resolution and jointly estimating the time delays and quality factors of a layered medium with reduced bias. Both processing techniques are applied to probe-layered roadways for NDT&E purposes.

Keywords: ground penetrating radar, high-resolution signal processing methods, non-dispersive media, dispersive media, civil engineering

1. Introduction and context

Ground penetrating radar (GPR) is a non-destructive testing tool widely used in the fields of defense, agriculture, and mainly in geological applications and civil engineering [1–8]. It allows measuring different surveyed media parameters or detecting and localizing buried objects [9–15 19]. In this chapter, we focus on horizontally stratified media such as roadways, walls, and soils. The inner structure of the media can be deduced from GPR profiles by means of echo detection and amplitude estimation. Some processing approaches to obtain such geometrical information can use the full-waveform inversion [6, 7] or migration processing [8]. In this chapter, we focus on estimation-based methods of low computational burden, which allow estimating only one, two, or three parameters.

Echo detection provides time-delay estimation (TDE) associated with each layer [15]. TDE is usually performed using conventional FFT-based methods (inverse FFT or cross-correlation methods), for which the time resolution depends on system GPR frequency bandwidth. To afford enhanced time resolution, e.g., for measuring thin thickness of layered medium, advanced model-based signal processing techniques are required.

This chapter shows that subspace-based methods are relevant for GPR applications for improving the performance of classical TDE techniques. These techniques use simplified assumptions, namely monostatic radar configuration at nadir probing layered non-dispersive media. Under these hypotheses, the data are conventionally modeled as the sum of shifted and attenuated copies of the Tx pulse. When dealing with experiments, the latter simplified data model is restricted to either lossless or weakly lossy materials, e.g., pavement, and to small depth-layered structures. Subspace-based techniques then provide unbiased TDE and super-time resolution capability as opposed to the conventional FFT-based techniques. Among the subspace-based techniques, the ESPRIT algorithm has been selected, thanks to its accuracy and its reduced computational burden.

For dispersive media, the above simplified data model does not hold because the wavelets undergo time spread and distortion with depth. Conventional subspace techniques then provide biased TDE and highly degraded time resolution. A few specific advanced processing techniques have been then developed and tested to extend the application of subspace techniques to dispersive media. Given a particular dispersion–absorption model, namely the Jonscher's parameterization of the medium permittivity, the MPM algorithm has been adapted to dispersive media to estimate unbiased TDE and recover the super-time resolution performance.

Both subspace methods are based on the frequency domain data model. In this case, each echo is mathematically represented by a complex pole from which the time delay can be readily estimated. For non-dispersive media, the pole is purely imaginary, i.e., $e^{-2j\pi f T_k}$; for dispersive media, the complex pole includes both real and imaginary parts of which the ratio depends on the Q factor, which characterizes the dispersion of the medium.

Therefore, this chapter develops two model-based high-resolution methods to estimate the different parameters of both non-dispersive and dispersive media, such as time delay, thickness, permittivity, and quality factor Q. For non-dispersive media, the ESPRIT algorithm is extended to the signal model taking into account the radar pulse and any noise characteristics. Furthermore, six special cases of the ESPRIT method are introduced and discussed. The influence of interface roughness and medium heterogeneity inside the medium is also discussed. For dispersive media, the matrix pencil method is extended to two variants of signal model: depth invariant (DI; Q does not vary from one layer to another) and depth variant (DV; Q varies from one layer to another). The performance of the proposed methods in terms of time resolution, accuracy, and root mean square error is tested with simulated data. The obtained results show that the two proposed methods succeed in improving the time resolution in the case of non-dispersive data and in jointly estimating the time delays and quality factors in the case of dispersive data.

2. Non-dispersive media

2.1. Data model

In this section, we present the forward model used to process radar data with high-resolution methods. First, numerous physical considerations are discussed. Then, the forward model is described. The high-resolution methods are applied on a non-dispersive medium like the roadway [15]. The medium considered here is assumed to be a horizontally stratified medium with K layers, as shown in **Figure 1**.

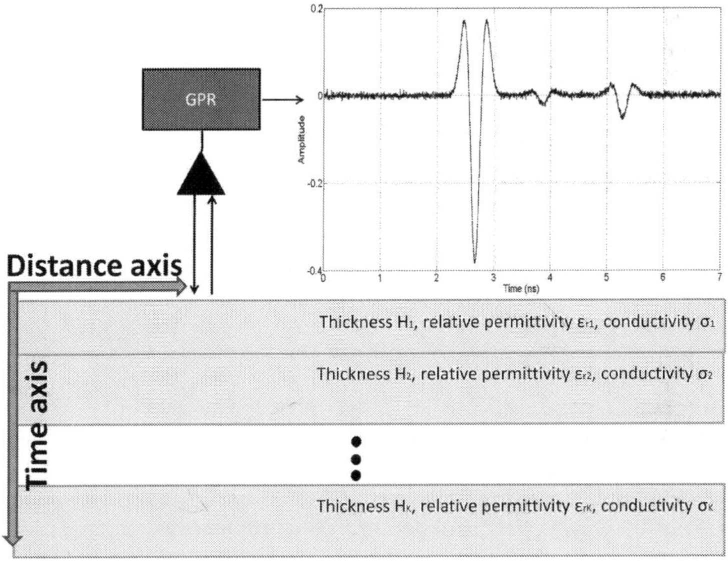

Figure 1. Horizontally stratified medium with K layers probed by Ground Penetrating Radar and the associated radar profile (A-scan) with three primary echoes.

A monostatic radar antenna is positioned in the far-field zone of the probed medium (antenna beyond the Fraunhoffer limit) at nadir. A vertical radar profile (function of time axis, called A-scan) is then obtained, as shown in **Figure 1**. When a sequence of A scans is acquired along the scanning direction by collecting radar data at different locations, the radargram image (which is also called B-scan) is obtained, as shown in **Figure 2**.

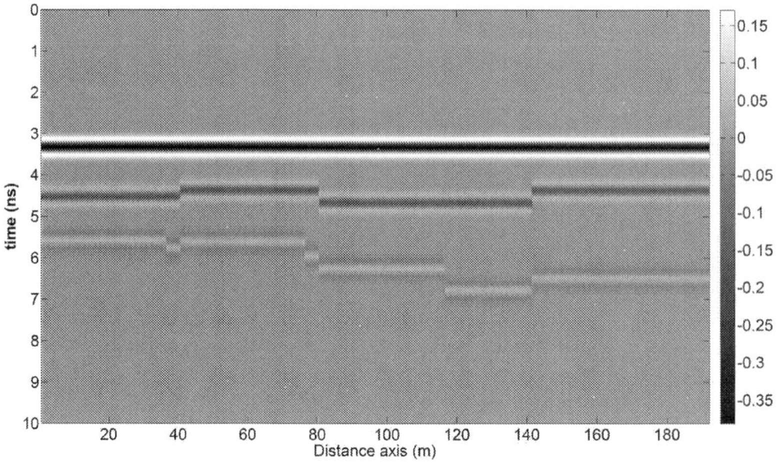

Figure 2. Example of B-scan of GPR.

Each layer is composed of aggregates, which are embedded within the hosting material. The size of the aggregates (1–2 cm in size) influences the scattering of waves depending on the dominant radar wavelength. When the aggregates are smaller than the radar wavelength, each layer can be considered as a homogeneous medium with smooth interfaces, which is characterized by its permittivity, conductivity, and thickness. However, when the aggregates are bigger or of the same order of size as the radar wavelength, each layer must then be considered as a heterogeneous medium comprising random rough interfaces. In such case, each layer is characterized not only by its effective permittivity, conductivity, and thickness but also by the two random rough interfaces delimiting each layer. These two rough interfaces are defined by their height probability density function and their surface height autocorrelation function. Note that, following Koudogbo et al. [17] and Pinel et al. [18], at nadir and within studied GPR band, surface scattering dominates in comparison with volume scattering. This is an important simplification that allows us to avoid modeling each aggregate individually, which would involve volume scattering. Instead, we can model these aggregates by considering an effective permittivity. Besides, as thoroughly analyzed by Pinel et al. [18], the roughness of the interfaces plays a non-negligible role in the backscattering process only for large bands, or in other words, for high frequencies: typically, from about 3 GHz. To probe the roadway, the pulse GPRs are more often used, and the typical frequency is between 0.1 and 3 GHz. Nowadays, some GPRs of "stepped frequency radar" kind are also employed. They work at higher frequencies, but the main drawback is the measurement time, which remains important.

Then, as a first approximation, the roadway structure is considered as a stratified medium consisting of K homogeneous layers with small dielectric contrast, for which the interface roughness can be neglected at GPR frequencies (1–2 GHz). The aggregates, which are included within the material, are supposed to be small enough compared to the dominant GPR wavelength. For example, the GPR pulse at 1.5 GHz has a dominant wavelength of 20 cm in

http://dx.doi.org/10.5772/62410

air and about 9 cm within the pavement material, whereas the maximum aggregate size is about 2 cm. Thus, each layer can therefore be considered as a homogeneous medium, which is characterized by an effective dielectric constant and thickness.

According to Adous [20], the dielectric constant of pavement layer is generally between 4 and 8. The dispersivity of the medium is usually neglected because the conductivity is low (between 10^{-3} and 10^{-2} S/m [21]). Roadway thicknesses typically range from 1 to 40 cm. For the pavements, the thickness varies from 1 to 10 cm. For ultrathin asphalt surfacing (UTAS) pavement [8], the thickness decreases to 1–3 cm. In the latter case, the backscattered echoes from adjacent interfaces overlap with each other, and it becomes very difficult even impossible to correctly estimate the thickness by conventional FFT-based methods. We also assume that the GPR measurements are performed in far field (i.e., the antennas are located beyond the Fraunhoffer limit), so that the received signal model simplifies to a one-dimensional model [13], and an analytical expression of the signal can be deduced from the solutions of Maxwell's equations. In standard radar configuration, i.e., monostatic radar in far field at nadir, the received signal is composed of d backscattered echoes, which are reflected from the medium interfaces.

Providing the latter assumptions on the medium and the antenna configuration, the echoes are simply the time-shifted and attenuated copies of the transmitted pulse $e(t)$. The echo amplitudes c_k are given by the Fresnel coefficients; they are reinforced by a large dielectric contrast between the layers. There are two kinds of echoes: the so-called "primary" echoes and the "multiple" echoes. The first ones (primary echoes) encounter only one reflection from the interface, whereas the multiple echoes result from several reflections. The multiple echoes show negligible amplitude if the layered medium shows small dielectric contrasts, as previously assumed. Thus, the received signal $s(t)$ is the sum of d echoes $e(t-T_k)$, whose amplitudes are proportional to the Fresnel coefficients of each medium interface. The received signal, also called signal model, can then be written as follows:

$$s(t) = \sum_{k=1}^{d} c_k e(t - T_k) + b(t) \tag{1}$$

where $e(t)$ is the transmitted signal; d is the number of backscattered echoes, which are mostly composed of K primary echoes; c_k is the amplitude of the kth echo; T_k is the time delay of the kth echo; and $b(t)$ is an additive noise, which is assumed to represent the measurement uncertainties.

The dispersivity of the medium is supposed negligible. In frequency domain, the echo amplitudes c_k are therefore constant over the frequency bandwidth, so the received signal is a linear combination of complex exponentials that are modulated by the radar pulse $e(f)$ plus an additive noise. The received signal in frequency domain can then be written as follows:

$$\tilde{s}(f) = \tilde{e}(f) \sum_{k=1}^{d} c_k e^{-2j\pi f T_k} + \tilde{b}(f) \tag{2}$$

where the $\dot{}$ symbol represents the corresponding Fourier transform of the time function (Equation 1), and f stands for the frequency. Suppose that N frequency samples are chosen within the frequency bandwidth B of the GPR. Thus, the received signal can be written in matrix form as follows:

$$\tilde{\mathbf{s}} = \mathbf{\Omega A c} + \tilde{\mathbf{b}} \tag{3}$$

where:

- $\tilde{\mathbf{s}} = [\tilde{s}(f_1) \quad .... \quad \tilde{s}(f_N)]^T$ is the $(N \times 1)$ data vector, which may represent either the Fourier transform of the pulse GPR signal or the measurements from a stepped frequency radar. The superscript ^{T} indicates the transpose operator;
- $\mathbf{\Omega} = \mathrm{diag}(\tilde{e}(f_1) \quad ... \quad \tilde{e}(f_N))$ is the $(N \times N)$ diagonal matrix whose diagonal elements are the amplitudes of the Fourier transform of the radar pulse. The radar pulse is supposed known; it can be measured by the signal backscattered from a metallic plane;
- $\mathbf{A} = [\mathbf{a}(t_1) \quad ... \quad \mathbf{a}(t_d)]^T$ is the $(N \times d)$ mode matrix;
- $\mathbf{a}(t_k) = [e^{-2j\pi f_1 T_k} \quad ... \quad e^{-2j\pi f_N T_k}]^T$ is the $(N \times 1)$ mode vector or steering vector;
- $\mathbf{c} = [c_1 \quad \dots \quad c_d]^T$ is the $(d \times 1)$ vector of echoes amplitudes;
- $\tilde{\mathbf{b}} = [\tilde{b}(f_1) \quad ... \quad \tilde{b}(f_N)]^T$ is the $(N \times 1)$ zero mean noise vector;
- $f_n = f_1 + (n-1)\Delta f$, where f_1 is the beginning of the band, and Δf is the frequency difference between two adjacent samples.

The covariance matrix is defined as $\mathbf{\Gamma} = \langle \tilde{\mathbf{s}}\tilde{\mathbf{s}}^H \rangle$, where the $\langle \cdot \rangle$ operator denotes the ensemble average, and the superscript H indicates the complex conjugate transpose. In practice, the covariance matrix $\mathbf{\Gamma}$ is estimated from M independent snapshots of the data vector $\tilde{\mathbf{s}}$ as follows:

$$\hat{\mathbf{\Gamma}} = \frac{1}{M} \sum_{m=1}^{M} \tilde{\mathbf{s}}_m \tilde{\mathbf{s}}_m^H \tag{4}$$

According to signal model (3) and assuming the noise to be independent of the echoes, the covariance matrix $\mathbf{\Gamma}$ of $\tilde{\mathbf{s}}$ can be written as follows:

$$\mathbf{\Gamma} = \mathbf{\Gamma}_{sig} + \sigma^2 \Sigma_0 = \mathbf{\Omega A \Gamma}_c \mathbf{A}^H \mathbf{\Omega}^H + \sigma^2 \Sigma_0 \tag{5}$$

where $\mathbf{\Gamma}_{sig}$ and $\mathbf{\Sigma}_0$ are the covariance matrices of the signal and noise, respectively. The matrix $\mathbf{\Gamma}_c$ is the $(d \times d)$ dimensional covariance matrix of the vector of echo amplitudes. The noise covariance has been normalized such that $tr(\mathbf{\Sigma}_0) = N$[22] and *tr(.)* is the trace operators.

2.2. High-resolution method ESPRIT applied to TDE

When enhanced time resolution is required for separating overlapping echoes (e.g., probing thin layered media), conventional FFT-based methods are not efficient. Advanced signal processing methods, such as subspace-based methods, have emerged in the 1980s to elegantly solve this problem. These methods are called subspace-based methods because they exploit the properties of the eigenstructure of the covariance matrix $\mathbf{\Gamma}$ [22]. Among the existing methods, ESPRIT has been selected because it affords direct parameter estimation with a lower computational load.

The ESPRIT algorithm was firstly proposed for array signal processing by Roy et al. [23]. Several extensions of the ESPRIT algorithm have been proposed for TDE in the literature. N-ESPRIT by Saarnisaari [24] can handle the colored noise but assumes an ideal radar pulse, whereas P-ESPRIT by Swindlehurst [25] takes into account the radar pulse but supposes a temporally white Gaussian noise. Compared to the latter extensions, the M-ESPRIT algorithm, which is presented in this section, is matched to the general data model in Equation (3) by taking both the radar pulse shape and any noise into account [16].

The principle of the ESPRIT method is to exploit some properties of rotation invariance [23]. The first step of the algorithm consists in dividing the data vector $\tilde{\mathbf{s}}$ into two overlapping subvectors. Each subvector has $N-1$ elements and overlaps with each other by $N-2$ elements. Thus, the $((N-1)\times d)$ dimensional mode matrices of each sub-band, $\mathbf{A}_1$ and $\mathbf{A}_2$, are related to each other by the $(d\times d)$ diagonal matrix $\mathbf{\Psi}$, whose elements depend on the time-delay parameter to be estimated as follows:

$$\mathbf{A_2} = \mathbf{A_1}\Psi \tag{6}$$

such that:

$$\mathbf{A} = \begin{pmatrix} \mathbf{A}_1 \\ - \end{pmatrix} = \begin{pmatrix} - \\ \mathbf{A}_2 \end{pmatrix} \tag{7}$$

and

$$\Psi = \mathrm{diag}(e^{-2j\pi\Delta f T_1} \;\ldots\; e^{-2j\pi\Delta f T_d}) \tag{8}$$

The matrix $\mathbf{\Psi}$ defined in Equation (8) cannot be estimated from data. On the basis of the generalized singular value decomposition (GSVD) of the covariance matrix $\mathbf{\Gamma}$, the diagonal elements of $\mathbf{\Psi}$ can be retrieved from the similar matrix $\mathbf{\Phi}$ defined thereafter in Equation (17c). From the GSVD of the covariance matrix $\mathbf{\Gamma}$, we have the following equation:

$$\mathbf{\Gamma}_{sig}\mathbf{V}_{sig} = \mathbf{\Sigma}_0\mathbf{V}_{sig}\Upsilon \tag{9}$$

with

$$\Upsilon = diag(\lambda_1 - \sigma^2, \quad \ldots \quad , \lambda_d - \sigma^2) \tag{10}$$

where the generalized eigenvectors $\mathbf{v}_k$ associated with the signal subspace are arranged in the matrix $\mathbf{V}_{sig}$ as columns; λ_i is the ith generalized eigenvalue of covariance matrix $\mathbf{\Gamma}_{sig}$; and σ^2 is the smallest generalized eigenvalue of the noise subspace. ESPRIT exploits the linear relation in Equation (6) within the GSVD of $\mathbf{\Gamma}$. Thus, the GSVD of Equation (9) can be written in each data sub-band as follows:

$$\mathbf{\Gamma}_{sig,1}\mathbf{V}_{sig} = \Sigma_{0,1}\mathbf{V}sig\Upsilon \tag{11}$$

$$\mathbf{\Gamma}_{sig,2}\mathbf{V}_{sig} = \Sigma_{0,2}\mathbf{V}sig\Upsilon \tag{12}$$

where $\mathbf{\Gamma}_{sig,1}$ and $\mathbf{\Gamma}_{sig,2}$ are the two $(N-1\times N)$ dimensional matrices defined as follows:

$$\mathbf{\Gamma}_{sig,1} = \mathbf{\Omega}_1\mathbf{A}_1\mathbf{\Gamma}_s\mathbf{A}^H\mathbf{\Omega}^H \tag{13}$$

$$\mathbf{\Gamma}_{sig,2} = \mathbf{\Omega}_2\mathbf{A}_2\mathbf{\Gamma}_s\mathbf{A}^H\mathbf{\Omega}^H \tag{14}$$

where $\mathbf{\Omega}_1$=diag($\breve{e}(f_1) \quad \ldots \quad \breve{e}(f_{N-1})$) and $\mathbf{\Omega}_2$=diag($\breve{e}(f_2) \quad \ldots \quad \breve{e}(f_N)$) are the two $((N-1)\times(N-1))$ diagonal matrices. $\mathbf{\Sigma}_{0,1}$ and $\mathbf{\Sigma}_{0,2}$ are the two overlapping band matrices of size $((N\times 1)\times N)$, and defined as the $N-1$ upper lines and the $N-1$ lower lines of the noise covariance matrix as follows:

$$\mathbf{\Sigma}_0 = \begin{pmatrix} \mathbf{\Sigma}_{0,1} \\ - \end{pmatrix} = \begin{pmatrix} - \\ \mathbf{\Sigma}_{0,2} \end{pmatrix} \tag{15}$$

Equations (6), (12), and (14) are used to obtain the following expression:

$$\mathbf{\Sigma}_{0,2}\mathbf{V}_{sig} = \mathbf{\Omega}_2\mathbf{A}_1\mathbf{\Psi}\mathbf{T} \tag{16}$$

where $\mathbf{T}=\boldsymbol{\Gamma}_c\mathbf{A}^H\boldsymbol{\Omega}^H\mathbf{V}_{sig}\boldsymbol{\Upsilon}^{-1}$ a $(K\times K)$ invertible matrix. Then, Equations. (11), (13), and (16) are used, and after some mathematical manipulations, we obtain the following equation:

$$\boldsymbol{\Sigma}_{0,2}\mathbf{V}_{sig}=\boldsymbol{\Theta}\boldsymbol{\Sigma}_{0,1}\mathbf{V}_{sig}\boldsymbol{\Phi} \tag{17a}$$

with

$$\boldsymbol{\Theta}=\boldsymbol{\Omega}_2\boldsymbol{\Omega}_1^{-1} \tag{17b}$$

$$\boldsymbol{\Phi}=\mathbf{T}^{-1}\boldsymbol{\Psi}\mathbf{T} \tag{17c}$$

Equation (17c) means that the matrices $\boldsymbol{\Phi}$ and $\boldsymbol{\Psi}$ are similar, and thus they have the same eigenvalues, which are expressed as $e^{-2j\pi\Delta f T_k}$. As opposed to $\boldsymbol{\Psi}$, $\boldsymbol{\Phi}$ can be retrieved from data. Thus, the d time delays of the echoes can be retrieved from the argument of the d eigenvalues of the matrix$\boldsymbol{\Phi}$ In practice, the covariance matrix is estimated, and only the matrices $\hat{\boldsymbol{\Sigma}}_{0,1}$, $\hat{\boldsymbol{\Sigma}}_{0,2}$, and $\hat{\mathbf{V}}_{sig}$ are estimated. The matrices $\boldsymbol{\Omega}_1$ and $\boldsymbol{\Omega}_2$ are obtained by the measurements of the radar pulse over a metallic plane. From Equation (17a), there exists two ways to calculate the matrix $\boldsymbol{\Phi}$. The first one uses the least square (LS) solution as follows:

$$\hat{\boldsymbol{\Phi}}=\left[\left(\hat{\boldsymbol{\Sigma}}_{0,1}\hat{\mathbf{V}}_{sig}\right)^H\hat{\boldsymbol{\Sigma}}_{0,1}\hat{\mathbf{V}}_{sig}\right]^{-1}\left(\hat{\boldsymbol{\Sigma}}_{0,1}\hat{\mathbf{V}}_{sig}\right)^H\boldsymbol{\Theta}^{-1}\hat{\boldsymbol{\Sigma}}_{0,2}\hat{\mathbf{V}}_{sig} \tag{18}$$

The second way is the total least square (TLS) solution, which is known to improve the robustness of the solution especially at low signal-to-noise ratio (SNR) [26]. The TLS solution is computationally more demanding and requires performing the GSVD of the couple of matrices $[\hat{\boldsymbol{\Sigma}}_{0,2}\hat{\mathbf{V}}_{sig},\ \boldsymbol{\Theta}\hat{\boldsymbol{\Sigma}}_{0,1}\hat{\mathbf{V}}_{sig}]$.

2.3. M-ESPRIT applied to specific data models

M-ESPRIT is a general algorithm, which can handle different data configurations encountered in the literature and in GPR experiments. This section reviews the simplifications in Equation (17a), which corresponds to the six specific data models in **Table 1**. The assumptions on the data model imply the modifications of both the noise metric and the signal subspace. The corresponding equations to be solved for matrix $\boldsymbol{\Phi}$ are also derived in **Table 1**, column 6. The LS and the TLS solutions for matrix $\boldsymbol{\Phi}$ can then be readily established.

Case	Assumptions	Data model ($\tilde{\mathbf{s}}$) and Covariance matrix ($\mathbf{\Gamma}$)	Signal subspace	Noise metric	Equation (17c) to solve for K
1 Conventional ESPRIT	Dirac pulse and white noise	$\tilde{\mathbf{s}} = \mathbf{Ac} + \tilde{\mathbf{b}}$ $\mathbf{\Gamma} = \mathbf{A\Gamma}_c\mathbf{A}^H + \sigma^2\mathbf{I}$	$\mathbf{A}$	$\mathbf{I}$	$\mathbf{V}_{sig,2} = \mathbf{V}_{sig,1}\mathbf{\Phi}$ $\mathbf{V}_{sig} = \begin{pmatrix} \mathbf{V}_{sig,1} \\ - \end{pmatrix} = \begin{pmatrix} - \\ \mathbf{V}_{sig,2} \end{pmatrix}$
2 M-ESPRIT	Radar pulse and any noise	$\tilde{\mathbf{s}} = \mathbf{\Omega}\,\mathbf{Ac} + \tilde{\mathbf{b}}$ $\mathbf{\Gamma} = \mathbf{\Omega}\mathbf{A}\mathbf{\Gamma}_c\mathbf{A}^H\mathbf{\Omega}^H + \sigma^2\mathbf{\Sigma}_0$	$\mathbf{\Omega}\mathbf{A}$	$\mathbf{\Sigma}_0$	$\mathbf{\Sigma}_{0,2}\mathbf{V}_{sig} = \mathbf{\Theta}\mathbf{\Sigma}_{0,1}\mathbf{V}_{sig}\mathbf{\Phi}$ $\mathbf{\Sigma}_0 = \begin{pmatrix} \mathbf{\Sigma}_{0,1} \\ - \end{pmatrix} = \begin{pmatrix} - \\ \mathbf{\Sigma}_{0,2} \end{pmatrix}$ $\mathbf{\Theta} = \mathrm{diag}\left(\frac{\grave{e}(f_2)}{\grave{e}(f_1)} \cdots \frac{\grave{e}(f_N)}{\grave{e}(f_{N-1})} \right)$
3 P-ESPRIT	Radar pulse and white noise	$\tilde{\mathbf{s}} = \mathbf{\Omega}\,\mathbf{Ac} + \tilde{\mathbf{b}}$ $\mathbf{\Gamma} = \mathbf{\Omega}\mathbf{A}\mathbf{\Gamma}_c\mathbf{A}^H\mathbf{\Omega}^H + \sigma^2\mathbf{I}$	$\mathbf{\Omega}\mathbf{A}$	$\mathbf{I}$	$\mathbf{V}_{sig,2} = \mathbf{\Theta}\mathbf{V}_{sig,1}\mathbf{\Phi}$
4 N-ESPRIT	Dirac pulse and any noise	$\tilde{\mathbf{s}} = \mathbf{\Omega}\,\mathbf{Ac} + \tilde{\mathbf{b}}$ $\mathbf{\Gamma} = \mathbf{A\Gamma}_c\mathbf{A}^H + \sigma^2\mathbf{\Sigma}_0$	$\mathbf{A}$	$\mathbf{\Sigma}_0$	$\mathbf{\Sigma}_{0,2}\mathbf{V}_{sig} = \mathbf{\Sigma}_{0,1}\mathbf{V}_{sig}\mathbf{\Phi}$
5 Variant of case 4	Whitened data and any noise	$\tilde{\mathbf{s}} = \mathbf{Ac} + \mathbf{\Omega}^{-1}\tilde{\mathbf{b}}$ $\mathbf{\Gamma} = \mathbf{A\Gamma}_c\mathbf{A}^H + \sigma^2\mathbf{\Sigma}^w$ With $\mathbf{\Sigma}^w = \mathbf{\Omega}^{-1}\mathbf{\Sigma}_0\mathbf{\Omega}^{-H}$	$\mathbf{A}$	$\mathbf{\Sigma}^w$	$\mathbf{\Sigma}^w_{0,2}\mathbf{V}_{sig} = \mathbf{\Sigma}^w_{0,1}\mathbf{V}_{sig}\mathbf{\Phi}$ $\mathbf{\Sigma}^w = \begin{pmatrix} \mathbf{\Sigma}^w_{0,1} \\ - \end{pmatrix} = \begin{pmatrix} - \\ \mathbf{\Sigma}^w_{0,2} \end{pmatrix}$
6 Variant of case 5 to deal with specific pre-processing	In case of sub-band averaging techniques (correlated echoes)	$\tilde{\mathbf{s}}^w_m = \mathbf{A}_1\mathbf{\Psi}^{m-1}\mathbf{c} + \mathbf{\Omega}^{-1}_m\tilde{\mathbf{b}}_m$ $\mathbf{\Gamma}_\mathrm{M} = \mathbf{A}_1\mathbf{\Gamma}^M_c\,\mathbf{A}_1{}^H + \sigma^2\mathbf{\Sigma}^M$	$\mathbf{A}_1$	$\mathbf{\Sigma}^M$	$\mathbf{\Sigma}^M_{0,2}\mathbf{V}_{sig} = \mathbf{\Sigma}^M_{0,1}\mathbf{V}_{sig}\mathbf{\Phi}$ $\mathbf{\Sigma}^M = \begin{pmatrix} \mathbf{\Sigma}^M_{0,1} \\ - \end{pmatrix} = \begin{pmatrix} - \\ \mathbf{\Sigma}^M_{0,2} \end{pmatrix}$

Table 1. Expressions of 6 data models, the corresponding covariance matrices and signal subspaces, and the associated solutions to Equation (17a).

Case 1 refers to the conventional ESPRIT algorithm [23] with the ideal data model. In comparison with the data model in Equation (3), both the radar pulse matrix $\mathbf{\Omega}$ and the covariance matrix of noise $\mathbf{\Sigma}_0$ reduce to the $(N \times N)$ identity matrix. In this case, $\mathbf{V}_{sig,1}$ and $\mathbf{V}_{sig,2}$ are the two overlapping sub-band matrices of size $((N \times 1) \times N)$, and they are defined as the $N-1$ upper lines and the $N-1$ lower lines of the matrix $\mathbf{V}_{sig}$. Case 2 in **Table 1** describes the general data model, which is supported by M-ESPRIT presented in Section 2.2.

In comparison, Cases 3 and 4 describe intermediate solutions for data modeling. For Case 3, the radar pulse is taken into account in the data model by matrix $\mathbf{\Omega}$. The data model in Case 4 is composed of time-shifted Dirac pulses with any noise ($\mathbf{\Sigma}_0$).

The two last data models, Cases 5 and 6, have a practical use when further preprocessing is performed. The whitening of the data by the radar pulse in Case 5 makes it possible to recover the situation of ideal Dirac pulses like for conventional ESPRIT, but in this case, the noise is modified. In this case, the Case 4 can be used.

In **Table 1**, the sixth data model provides a way to deal with the situation of totally correlated backscattered echoes. When the echoes are totally correlated, the full rank property of the covariance matrix of the sources $\mathbf{\Gamma}_c$ does not hold anymore, and the ESPRIT algorithm fails to work. This situation is likely to be encountered within a multipath environment because the source vector c in Equation (3) represents the different paths from the same original source. To resolve this problem, the covariance matrix may be estimated by using an averaging preprocessing technique [27]. This technique first whitens the data by the radar pulse and then organizes the frequency bandwidth (N samples) into M sub-bands of L data each, overlapping with each another by $L - 1$ samples. Consequently, the covariance matrix to be processed by the subspace algorithm is reduced to the size $L \times L$. L will be referred to below as the effective frequency bandwidth; obviously, it must remain larger than the number of echoes d. In **Table 1**, matrix $\mathbf{A}_1$ is defined as the L first lines of the mode matrix in Equation (3). The matrices $\mathbf{\Gamma}_c^M$ and $\mathbf{\Sigma}^{\mathbf{M}}$ are the covariance matrices of sources and noise, respectively, after using a preprocessing method.

Furthermore, to decrease the computational burden of the ESPRIT method, linear versions have also been introduced by Zha [28] and Le Bastard et al. [29]. These methods use linear operations on the data and do not require the costly eigendecomposition of the covariance matrix. Further details are available in Refs. [28–31].

2.4. Simulations, experiments, and discussion

2.4.1. Simulation

This section illustrates with simulated data the performance of the M-ESPRIT method, in comparison with N-ESPRIT, P-ESPRIT, and the conventional ESPRIT. The performance of the algorithms has been evaluated with respect to both the time resolution capability and the relative root mean square error (RRMSE) on the estimated thickness. RRMSE on the estimated thickness is defined as follows:

$$RRMSE(\%) = 100 \times \frac{\sqrt{\frac{1}{I}\sum_{i=1}^{I}(\hat{H}_i - H)^2}}{H} \tag{19}$$

where H is the true value of a parameter, and $\hat{H}_i$ is the corresponding estimated value in the ith simulation. I is the number of Monte Carlo trials. The simulations have been carried out for the thickness measurement of a two-layered stratified medium. The time resolution describes the processing capability to separate two echoes within a limited band. It is thereafter defined as the $B\Delta\tau$ product, where B is the frequency bandwidth of the radar device, and $\Delta\tau$ is the smallest time shift between two echoes that the processing technique is able to distinguish. In this section, different data sets have been generated by varying the thickness from 1 to 41.2 mm, such that the corresponding $B\Delta\tau$ product spans the interval [0.027; 1], as shown in **Figure 4**.

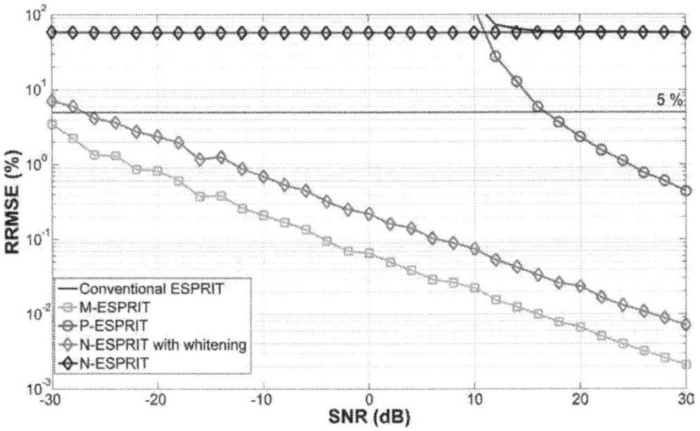

Figure 3. RRMSE variation of the estimated thickness vs. SNR after 100 Monte Carlo simulations with two uncorrelated echoes; $B\Delta\tau = 0.5$.

The performance is established by a Monte Carlo process, which consists of $I = 100$ independent runs of algorithms. For each run, the covariance matrix was estimated from 100 independent snapshots. To analyze the influence of the noise, the simulations were carried out with a noise whose covariance matrix is non-diagonal. The radar pulse $e(t)$ was estimated in practice by the signal backscattered from a metallic plane. For the simulations, it is modeled as a Ricker pulse defined as the second derivative of a Gaussian pulse [31]. The parameters of the pulse have been chosen according to the characteristics of conventional GPR: central frequency of 1.5 GHz with frequency bandwidth of 2 GHz. The data set is composed of $N = 41$ equi-spaced frequency samples within a 2 GHz bandwidth. According to Adous [20], the dielectric constants of the first and second layers were set to 4 and 7, respectively, and the conductivity of the two layers was neglected. The SNR was fixed with respect to the second echo and defined as the ratio between the power of the second echo and the noise variance.

Figure 3 shows the RRMSE variations of the estimated thickness versus SNR in the context of overlapping uncorrelated echoes. It can be seen that the RRMSE continuously decreases with increasing SNR for all algorithms (except for N-ESPRIT). M-ESPRIT allows obtaining the best performance for all SNRs. In contrast, the algorithms N-ESPRIT and conventional ESPRIT provide a large and constant RRMSE meaning unrealistic biased TDE; both algorithms are not adapted to the application because they do not take into account the radar pulse or/and the noise metric. In this simulation, P-ESPRIT performs better than the two latter algorithms but is expected to reach a constant RRMSE at higher SNR, meaning an asymptotically biased TDE.

To further analyze the performance, N-ESPRIT was used after data whitening by the radar pulse, referred to in **Figures 3** and **4** as N-ESPRIT with whitening. This modification of the original algorithm proposed by Saarnisaari [24] strongly improves the performance, which comes closer to that of M-ESPRIT. The latter result means that it is more important to take into account the radar pulse in the mode vector than in the noise metric.

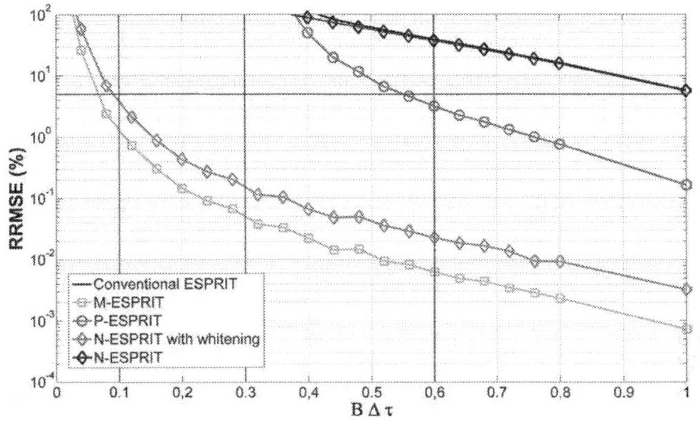

Figure 4. RRMSE variation of the estimated thickness vs. BΔτ product after 100 Monte Carlo simulations with two uncorrelated echoes; SNR=15dB

Figure 4 shows the RRMSE variations of the estimated thickness versus $B\Delta\tau$ product at medium SNR for overlapping uncorrelated echoes. It can be seen that the RRMSE increases continuously with decreasing $B\Delta\tau$ product. Like in **Figure 3**, M-ESPRIT shows the best performance followed by N-ESPRIT with whitening. In contrast, the other algorithms (conventional ESPRIT, N-ESPRIT, and P-ESPRIT) show poorer performance.

2.4.2. Experiments

The performance of the algorithms was also tested in laboratory on smooth materials. A wideband exponential tapered slot antenna (ETSA) [32] is combined with a broadband vector network analyzer (VNA) to provide the experimental ultra-wide band (UWB) stepped-frequency radar (SFR).

2.4.2.1. Experiment 1

First, the algorithms presented in the previous sections were tested on a virtual two-layered medium filled in with air. This virtual medium was created with the backscattered data recorded by the SFR from the metallic plates at two different heights. The data from the virtual two-layered medium is then considered as the sum of the two backscattered data as follows: $\tilde{s}(f)=\tilde{s}_{hi}(f)+\tilde{s}_{hj}(f)$, where $\tilde{s}_{hi}(f)$ and $\tilde{s}_{hj}(f)$ are the backscattered data from the metallic plate at the heights $h_i$and h_j, respectively, as shown in **Figure 5**.

First, the null correlation case is considered to provide the best performance of the algorithms. The full correlation case will be presented in the following subsection. Thus, the null correlation matrix $\boldsymbol{\Gamma}^{null}$ was formed as the sum of the covariance matrices of each individual echo as follows: $\boldsymbol{\Gamma}_{null} = \langle \tilde{\mathbf{s}}_{hi}\tilde{\mathbf{s}}_{hi}{}^{H}\rangle + \langle \tilde{\mathbf{s}}_{hj}\tilde{\mathbf{s}}_{hj}{}^{H}\rangle$. The following four heights are considered to form three two-layered virtual media: h_1, $h_2 = h_1 + 2.25$ cm, $h_3 = h_1 + 4.5$ cm, $h_4 = h_1 + 6.75$ cm. They span three $B\Delta\tau$ products: $B\Delta\tau = 0.3$, $B\Delta\tau = 0.6$, and $B\Delta\tau = 0.9$, respectively. The frequency band (2–4) GHz is spanned with $N = 201$ samples. The measurements are carried out at high SNR (about 40 dB). The transient signal in the antennas is filtered with a time window, as shown in **Figure 5** (grey curves).

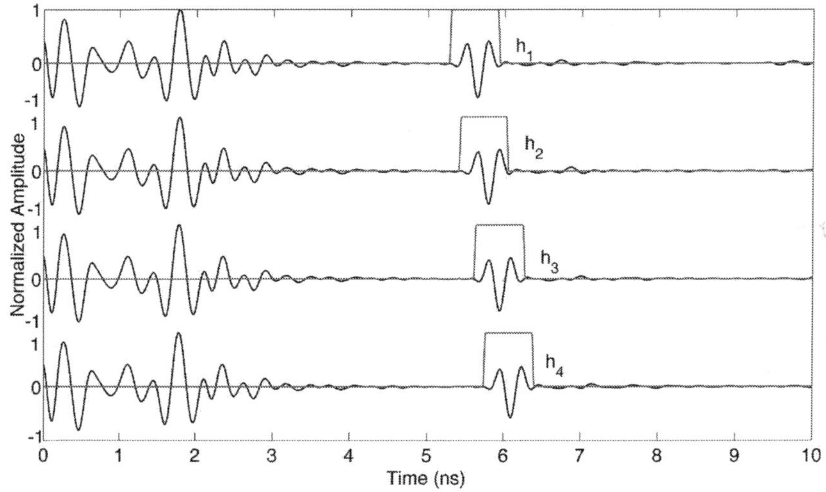

Figure 5. Backscattered raw data from the metallic plate at different heights, h_1, h_2, h_3, and h_4.

	$B\Delta\tau = 0.3$	$B\Delta\tau = 0.6$	$B\Delta\tau = 0.9$
ESPRIT	76.5	16.41	2.46
M-ESPRIT	1.92	0.34	0.04
N-ESPRIT with whitening	3.24	0.11	0.06

Table 2. Relative bias (in %) on the thickness for 3 $B\Delta\tau$ products with conventional ESPRIT, M-ESPRIT, and N-ESPRIT with whitening.

Table 2 shows that the relative bias decreases with increasing $B\Delta\tau$ product. The conventional ESPRIT algorithm provides the largest error on the thickness estimation, followed by N-ESPRIT and M-ESPRIT. At high SNR, M-ESPRIT performs better than N-ESPRIT at small $B\Delta\tau$ product; both algorithms give similar results for $B\Delta\tau$ beyond 0.6.

http://dx.doi.org/10.5772/62410

2.4.2.2. Experiment 2

Second, SFR comprising two ETSA antennas (whose transmitter and receiver are very close to each other, about 20 cm) is tested on a smooth PVC slab. The PVC slab has a dielectric constant of ε_r = 2.97+0.015j and a thickness of 4 cm. The height of the antenna is about 70 cm (far-field condition). The frequency band is (1-2.6) GHz with N = 81 samples. Thus, $B\Delta\tau$ product is 0.728. In this case, the echoes are totally correlated, thus the Case 6 of the **Table 1** is applied. We used the MSSP preprocessing [8, 20] with a sub-band size of about 90% of the whole frequency bandwidth (L=73). The bias on the estimated thickness is 3.11%, which highlights the good performance of the M-ESPRIT algorithm.

3. Dispersive media

3.1. Constant-Q data model for dispersive media

In Ref. [33], Bano modeled the frequency dependence of permittivity for radar wave propagation in constant-Q media using Kjartansson's formulation for seismic wave propagation in viscoelastic media [34]. This formulation is based on a complex power law of frequency for the effective dielectric permittivity and results in a frequency-independent quality factor Q of the form: $\frac{1}{Q} = \frac{\varepsilon''_e}{\varepsilon'_e} = \tan\left[\frac{\pi}{2}(1-n)\right]$, where ε'_e and ε''_e are the real and imaginary parts of the effective dielectric permittivity, respectively, and the index n ($0<n<1$) is related to Q as follows: $n = \frac{2}{\pi}\arctan(Q)$. The case $n=1$ corresponds to lossless propagation ($Q=\infty$, i.e., no attenuation) presented in Section 2.

A backscattered complex signal, represented as a linear combination of d echoes, results from probing a horizontally stratified medium, under normal incidence of plane waves. An echo originates from the interface between two horizontally superposed layers. Each layer is assumed to be homogeneous and having a thickness e, a constant quality factor Q (or, equivalently, a dispersion index n), and a dielectric constant ε^0. The stratified structure is said to be DI (Depth Invariant) if Q does not vary from one layer to another; otherwise, it is called a DV (Depth-variant) structure [35]. Considering Kjartansson's absorption–dispersion model and substituting for each layer of a DV structure the expression of the corresponding wavenumber k in the equation of a plane wave exp{-2jke} the following backscattered signal is obtained:

$$s(f) = \sum_{m=1}^{d} c_m \prod_{i=1}^{m} \exp\left\{-2\pi f_r \tau_l \left(j + \tan\left[\frac{\pi}{4}(1-n_l)\right]\right)\left(\frac{f}{f_r}\right)^{(1+n_l)/2}\right\} + b(f) \quad (20)$$

where $c_m = a_m \exp\{j\theta_m\}$ is the frequency-independent complex amplitude of the mth primary echo, $\tau_l = 2e_l / V_{rl}$ is the traveling time of the wave in the lth layer, V_{rl} is the reference phase velocity at an arbitrary reference angular frequency, and $b(f)$ is a complex white Gaussian noise with zero mean and variance $2\sigma^2$. After sampling, the frequency variable f is replaced by $f_k = k f_s$, where f_s is the sampling period or frequency shift. The backscattered signal becomes:

$$s(k) = \sum_{m=1}^{d} c_m \Pi_{l=1}^{m} z_l^{k^{(1+n_l)/2}} + b(k) \quad k = 1,2,...,N \tag{21}$$

where the pole of the lth layer is given by the following equation:

$$z_l = \exp\left\{-2\pi f_r \tau_l \left(j + \tan\left[\frac{\pi}{4}(1-n_l)\right]\right)\left(\frac{f_s}{f_r}\right)^{(1+n_l)/2}\right\}. \tag{22}$$

The product in the signal model represents the cumulative effect of the $(m-1)$ traversed layers of the stratified medium on the mth echo.

The assumption of a DI structure allows rewriting Equation (21) in the following compact form:

$$s(k) = \sum_{m=1}^{d} c_m \breve{z}_m^{k^{(1+n)/2}} + b(k) \tag{23}$$

where the pole of the mth primary echo is:

$$\breve{z}_m = \prod_{l=1}^{m} z_l = \exp\left\{-2\pi f_r T_m^{eq}\left(j + \tan\left[\frac{\pi}{4}(1-n)\right]\right)\left(\frac{f_s}{f_r}\right)^{(1+n)/2}\right\} \tag{24}$$

with

$$T_m^{eq} = \sum_{l=1}^{m} \tau_l \tag{25}$$

is the equivalent time delay of the mth echo. Note that, compared with Equation (21), the subscript of n disappears in Equation (23). From Equations (21) and (23), it can be seen that the DI case is a special case of the more general DV one. Under matrix form, the signal model of both structures is expressed by the following equation:

$$\mathbf{s} = \mathbf{A}\mathbf{c} + \mathbf{b} \tag{26}$$

with the following notational definitions:

$$\begin{aligned}
\mathbf{s} &= \left[s(1)\ s(2)\ \ldots\ s(N)\right]^T : N \times 1 \text{ data vector} \\
\mathbf{A} &= \left[\mathbf{a}_1\ \mathbf{a}_2\ \ldots\ \mathbf{a}_d\right] : N \times d \text{ matrix of mode vectors} \\
\mathbf{a}_m &= \left[\prod_{l=1}^{m} z_l^{1^{(1+n_l)/2}}\ \prod_{l=1}^{m} z_l^{2^{(1+n_l)/2}}\ \cdots\ \prod_{l=1}^{m} z_l^{N^{(1+n_l)/2}}\right]^T : N \times 1 \text{ DV mode vector} \\
\mathbf{a}_m &= \left[\breve{z}_m^{1^{(1+n)/2}}\ \breve{z}_m^{2^{(1+n)/2}}\ \ldots\ \breve{z}_m^{N^{(1+n)/2}}\right]^T : N \times 1 \text{ DI mode vector} \\
\mathbf{c} &= \left[c_1\ c_2\ \ldots c_d\right]^T : d \times 1 \text{ vector of complex amplitudes} \\
\mathbf{b} &= \left[b(1)\ b(2)\ \ldots\ b(N)\right]^T : N \times 1 \text{ noise vector.}
\end{aligned}$$

Note that because of the DV structure, the product of the poles cannot be put in the form of $\breve{z}_m$ as shown in Equation (24) and, consequently, mode vectors of different echoes have different forms.

3.2. Nonlinearity compensation through predistortion linearization

Using the first and second extension theorems in [36, Section 4] yields, the following geometric progression of data samples for a single echo is obtained [36]:

$$\left\{\tilde{s}(k)\right\}_{k=1}^{N} = \left\{c\breve{z}, c\breve{z}^{2(1+n)/2},\ c\breve{z}^{2\times 2^{(1+n)/2}-1},\ \ldots,\ c\breve{z}^{(N-1)\times 2^{(1+n)/2}-(N-2)}\right\} \tag{27}$$

Evidently, any Hankel data matrix constructed from the above data sequence admits Vandermonde decomposition. Moreover, we note that the geometric progression arises from probing the medium with non-uniform frequencies given by the following equation:

$$v(k) = \left[(k-1)2^{(1+n)/2} - (k-2)\right]^{2/(1+n)} f_s,\ k = 1, 2,\ \ldots,\ \mathrm{N}. \tag{28}$$

Hence, $v(k)$ can be considered as the transfer function or the nonlinearity of a predistorter. For a non-dispersive medium ($n=1$), we point out that $v(k)$ becomes equal to kf_s. Probing a dispersive medium with v in lieu of f results in a linear frequency dependency of the phase.

3.3. Depth-invariant method

In this section, we develop interpolation-based MPM for the problem of parameter estimation of a DI structure characterized by the data model in Equation (23). For the DI structure, the parameter estimation problem can now be stated as follows: *Given the data sequence* $\{s(k)\}_{k=1}^{N}$ *estimate the time delays* $\{T_m^{eq}\}_{m=1}^{d}$ *and the index n (or, equivalently, the quality factor Q) of the stratified medium.*

The required data sequence is obtained by spline interpolation. This is achieved by first determining the spline interpolant from the uniform data sequence provided by the measurement device and then evaluating the spline interpolant at non-uniform frequencies $\{v(k)\}_{k=1}^{N}$ obtained for a given value of n. Assuming a priori knowledge of n and neglecting the interpolation errors, the discrete backscattered signal becomes as follows:

$$\tilde{s}(k) = \sum_{m=1}^{d} c_m \tilde{z}_m^{(k+1)2^{(1+n)/2}-(k-2)} + \tilde{b}(k). \tag{29}$$

Then $\tilde{\mathbf{s}} = \tilde{\mathbf{A}}\mathbf{c} + \mathbf{b}$ with the following definitions:

$$\begin{aligned}
\tilde{\mathbf{s}} &= \left[\tilde{s}(1) \quad \tilde{s}(2) \quad \ldots \quad \tilde{s}(N)\right]^T \\
\tilde{\mathbf{A}} &= \left[\tilde{\mathbf{a}}_1 \quad \tilde{\mathbf{a}}_2 \quad \ldots \quad \tilde{\mathbf{a}}_d\right] \\
\tilde{\mathbf{a}}_m &= \left[\tilde{z}_m \quad \tilde{z}_m^{2^{(1+n)/2}} \quad \tilde{z}_m^{2\times 2^{(1+n)/2}-1} \quad \ldots \quad \tilde{z}_m^{(N-1)\times 2^{(1+n)/2}-(N-2)}\right]^T \\
\tilde{\mathbf{b}} &= \left[\tilde{b}(1) \quad \tilde{b}(2) \quad \ldots \quad \tilde{b}(N)\right]^T.
\end{aligned}$$

It is readily verified that the transformed mode vector $\tilde{\mathbf{a}}_m$ admits a Vandermonde structure (its samples follow a geometric progression with $\tilde{z}_m^{2^{(1+n)/2}-1}$ as the common ratio). As a consequence, the matrices $\tilde{\mathbf{S}}_1$ and $\tilde{\mathbf{S}}_2$, obtained from the Hankel matrix of $\tilde{\mathbf{s}}$ by deleting its last and first columns, respectively, admit the following decomposition:

$$\tilde{\mathbf{S}}_1 = \tilde{\mathbf{Z}}_1\mathbf{C}\tilde{\mathbf{Z}}_2 \quad \tilde{\mathbf{S}}_2 = \tilde{\mathbf{Z}}_1\mathbf{C}\tilde{\mathbf{Z}}_0\tilde{\mathbf{Z}}_2 \tag{30}$$

where

$$\tilde{\mathbf{Z}}_1 = \begin{bmatrix} \breve{z}_1 & \breve{z}_2 & \cdots & \breve{z}_d \\ \breve{z}_1^{2^{(1+n)/2}} & \breve{z}_2^{2^{(1+n)/2}} & \cdots & \breve{z}_d^{2^{(1+n)/2}} \\ \vdots & \vdots & \ddots & \vdots \\ \breve{z}_1^{(N-L-1)\times 2^{(1+n)/2}-(N-L-2)} & \breve{z}_2^{(N-L-1)\times 2^{(1+n)/2}-(N-L-2)} & \cdots & \breve{z}_d^{(N-L-1)\times 2^{(1+n)/2}-(N-L-2)} \end{bmatrix}$$

$$\tilde{\mathbf{Z}}_2 = \begin{bmatrix} 1 & \breve{z}_1^{2^{(1+n)/2}-1} & \cdots & \breve{z}_1^{\left(2^{(1+n)/2}-1\right)(L-1)} \\ 1 & \breve{z}_2^{2^{(1+n)/2}-1} & \cdots & \breve{z}_2^{\left(2^{(1+n)/2}-1\right)(L-1)} \\ \vdots & \vdots & \ddots & \vdots \\ 1 & \breve{z}_d^{2^{(1+n)/2}-1} & \cdots & \breve{z}_d^{\left(2^{(1+n)/2}-1\right)(L-1)} \end{bmatrix}$$

$$\tilde{\mathbf{Z}}_0 = \mathrm{diag}\left\{\breve{z}_1^{2^{(1+n)/2}-1}, \breve{z}_2^{2^{(1+n)/2}-1}, \ldots, \breve{z}_d^{2^{(1+n)/2}-1}\right\}$$

$$\mathbf{C} = \mathrm{diag}\{c_1, c_2, \ldots, c_d\}.$$

As in conventional MPM, each value of $\lambda = \breve{z}_m^{2^{(1+n)/2}-1}$ is a rank reducing number of the pencil $\tilde{\mathbf{S}}_2 - \lambda\tilde{\mathbf{S}}_1$. The estimates of the index n and the time delays $\{T_m^{eq}\}_{m=1}^d$ are then:

$$n = 1 - \frac{4}{\pi}\arctan\left[\frac{\Re\left(\log \breve{z}_m^{2^{\frac{1+n}{2}}-1}\right)}{\Im\left(\log \breve{z}_m^{2^{\frac{1+n}{2}}-1}\right)}\right] \tag{31}$$

$$T_m^{eq} = -\left[\frac{\Im\left(\log \breve{z}_m^{2^{(1+n)/2}-1}\right)}{2\pi f_r\left(2^{(1+n)/2}-1\right)\left(\frac{f_s}{f_r}\right)^{(1+n)/2}}\right] \tag{32}$$

These estimates allow determining the complex amplitudes using a LS fit whose solution is given by the following equation:

$$\mathbf{c} = \left(\tilde{\mathbf{A}}^H \tilde{\mathbf{A}}\right)^{-1} \tilde{\mathbf{A}}^H \tilde{\mathbf{s}}. \tag{33}$$

In practice, however, the value of n is unknown and needs to be estimated. To solve this problem, we propose a recursive approach summarized in the following steps [37]:

1. **Step 1.** Use the data sequence $\tilde{\mathbf{s}}$ provided by the spline interpolant at frequencies selected according to Equation (28) to construct a Hankel matrix $\tilde{\mathbf{S}}$. Start with $n=1$ as initial value, which corresponds to the initial uniform data sequence provided by real measurements.

2. **Step 2.** Run MPM and determine n and $\{T_m^{eq}\}_{m=1}^d$ by using Equations (31) and (32).

3. **Step 3.** Compare the estimated value of n with the value used in Step 1. If they are different, i.e., $\Delta n \neq 0$, substitute the new estimate of n in Equation (28) to obtain a new set of non-uniform frequencies at which the spline interpolant is to be evaluated. Repeat Steps 1–3 until convergence, i.e., $\Delta n = 0$. This termination condition might never be satisfied for noisy data, and so it will be replaced by a predefined practical threshold, e.g., 10^{-6}.

4. **Step 4.** Use the estimates of n and $\{T_m^{eq}\}_{m=1}^d$ obtained in the last iteration to estimate the complex amplitudes using Equation (33).

3.4. Depth-variant method

This section is concerned with the problem of parameter estimation of a DV structure characterized by the data model in Equation (21). The proposed algorithm can be seen as an extension or a generalization of the interpolation-based MPM described in the above section to a DV data model. It first unifies the form of the mode vectors through an equivalent-medium approximation. Indeed, the forward Q filtering effect experienced by each of the d echoes of a DV Q structure in a horizontally stratified earth can be approximated by the effect of a one-layered medium characterized by its equivalent parameters $(T_m^{eq},\ n_m^{eq})$ T_m^{eq} is given by Equation (25), and n_m^{eq} is given by the following equation [31]:

$$n_m^{eq} = 1 - \frac{4}{\pi}\arctan\frac{\sum_{l=1}^{m} \tau_l \tan\left[\frac{\pi}{4}\left(1 - n_l\right)\right]}{T_m^{eq}}. \tag{34}$$

The equivalent-medium approximation then allows writing:

http://dx.doi.org/10.5772/62410

$$\mathbf{A} \approx \left[\hat{\mathbf{a}}_1 \quad \hat{\mathbf{a}}_2 \quad \dots \quad \hat{\mathbf{a}}_\mathbf{d}\right] \tag{35}$$

where

$$\hat{\mathbf{a}}_\mathrm{m} = \left[\hat{\mathrm{z}}_\mathrm{m}^{1^{(1+n_m^{eq})/2}} \quad \hat{\mathrm{z}}_\mathrm{m}^{2^{(1+n_m^{eq})/2}} \quad \dots \quad \hat{\mathrm{z}}_\mathrm{m}^{N^{(1+n_m^{eq})/2}}\right]^\mathrm{T} \tag{36}$$

$$\hat{\mathrm{z}}_m = \exp\left\{-2\pi f_r T_m^{eq}\left(j + \tan\left[\frac{\pi}{4}\left(1 - n_m^{eq}\right)\right]\right)\left(\frac{f_s}{f_r}\right)^{(1+n_m^{eq})/2}\right\}. \tag{37}$$

The above approximation becomes exact for a DI structure, i.e., $\hat{\mathrm{z}}_\mathrm{m} = \breve{z}_\mathrm{m}$. Note also that upon setting $\{n_l = 1\}_{l=1}^{d}$, the above model reduces to the undamped exponential model. Second, it uses a projection scheme to strip the mode vectors alternately. The DV parameter estimation problem can be stated as follows: *Given the data sequence* $\{s(k)\}_{k=1}^{N}$ *estimate the time delays* $\{\tau_l\}_{l=1}^{d}$ *and the dispersion indices* $\{n_l\}_{l=1}^{d}$ *of the stratified medium.*

The proposed algorithm, called recursive and alternately projected MPM (RAP-MPM), is developed for the case of two echoes [38]. The extension for $d>2$ is straightforward; however, it requires more effort and notations and is not considered in this work. Suppose, for now, that n_1^{eq} and n_2^{eq} are known (n_1^{eq} is practically equal to n_1) and consider the application of the nonlinear transformation in Equation (28) to the uniform frequencies. Setting $n = n_1^{eq}$ in $\nu(k)$ and probing the medium with $\{\nu_k\}_1^N$ give rise to the following discrete backscattered signal:

$$\tilde{s}(k) = c_1 \hat{z}_1^{\left[\frac{\nu_k}{f_s}\right]^{(1+n_1^{eq})/2}} + c_2 \hat{z}_2^{\left[\frac{\nu_k}{f_s}\right]^{(1+n_2^{eq})/2}} + \tilde{b}(k) \tag{38}$$

Nevertheless, from a practical point of view, as it is impossible to probe the medium with non-uniform frequencies, the data sequence in Equation (38) is obtained, as for interpolation-based MPM, by cubic spline interpolation. Neglecting the interpolation errors, the data model in matrix form becomes:

$$\tilde{\mathbf{s}} = \tilde{\mathbf{A}}\mathbf{c} + \tilde{\mathbf{b}} \tag{39}$$

with the following notational definitions:

$$\tilde{\mathbf{s}} = \left[\tilde{s}(1) \;\; \tilde{s}(2) \;\; \dots \;\; \tilde{s}(N)\right]^T$$

$$\tilde{\mathbf{A}} = \left[\tilde{\mathbf{a}}_1 \;\; \tilde{\mathbf{a}}_2\right]$$

$$\tilde{\mathbf{a}}_1 = \left[\hat{z}_1 \;\; \hat{z}_1^{2\left(1+n_1^{eq}\right)/2} \;\; \dots \;\; \hat{z}_1^{(N-1)\times 2^{\left(1+n_1^{eq}\right)/2} - (N-2)} \right]^T$$

$$\tilde{\mathbf{a}}_2 = \left[\hat{z}_2 \;\; \hat{z}_2^{2\left(1+n_2^{eq}\right)/2} \;\; \dots \;\; \hat{z}_2^{\left[(N-1)\times 2^{\left(1+n_1^{eq}\right)/2} - (N-2)\right]^{\left(1+n_2^{eq}\right)/\left(1+n_1^{eq}\right)}} \right]^T$$

$$\tilde{\mathbf{b}} = \left[\tilde{b}(1) \; \tilde{b}(2) \;\; \dots \;\; \tilde{b}(N)\right]^T$$

It is readily verified from the equation of $\tilde{\mathbf{a}}_1$ that the transformed mode vector of the first echo admits a Vandermonde structure, and so its data matrix $\tilde{\mathbf{S}}_1$ becomes of rank one. Consequently, the matrices $\tilde{\mathbf{S}}_{11}$ and $\tilde{\mathbf{S}}_{12}$ admit the following decomposition:

$$\tilde{\mathbf{S}}_{11} = c_1 \tilde{\mathbf{z}}_{11} \tilde{\mathbf{z}}_{21} \;\; \tilde{\mathbf{S}}_{12} = c_1 \tilde{\mathbf{z}}_{11} \hat{\mathbf{z}}_1^{2^{\left(1+n_1^{eq}\right)/2} - 1} \tilde{\mathbf{z}}_{21} \tag{40}$$

where

$$\tilde{\mathbf{z}}_{11} = \left[\hat{\mathbf{z}}_1 \;\; \hat{\mathbf{z}}_1^{2^{\left(1+n_1^{eq}\right)/2}} \;\; \dots \;\; \hat{\mathbf{z}}_1^{(N-L-1)\times 2^{\left(1+n_1^{eq}\right)/2} - (N-L-2)} \right]^T$$

$$\tilde{\mathbf{z}}_{21} = \left[1 \;\; \hat{\mathbf{z}}_1^{2^{\left(1+n_1^{eq}\right)/2} - 1} \;\; \dots \;\; \hat{\mathbf{z}}_1^{\left(2^{\left(1+n_1^{eq}\right)/2} - 1\right)(L-1)} \right]$$

The corresponding matrix pencil can then be written as follows:

$$\tilde{\mathbf{S}}_{12} - \lambda \tilde{\mathbf{S}}_{11} = c_1 \tilde{\mathbf{z}}_{11} \left[\hat{\mathbf{z}}_1^{2^{\left(1+n_1^{eq}\right)/2} - 1} - \lambda \right] \tilde{\mathbf{z}}_{21} \tag{41}$$

As in conventional MPM, the value of $\lambda = \hat{\mathbf{z}}_1^{2^{\left(1+n_1^{eq}\right)/2} - 1}$ reduces the rank of the pencil in Equation (41) to zero and ought to be a GEV of the total matrix pair containing both echoes. However, due to the DV structure of the dispersive medium, the Vandermonde structure can be restored for one mode vector at a time. In such a case, estimating λ as the GEV of the total matrix pair is valid provided that the contribution of other echoes is filtered out after each restoration. Therefore, λ will be first estimated coarsely as a GEV of the total matrix pair. The estimation will be refined in later steps of the algorithm. The estimates of n_1^{eq} and T_1^{eq} can then be deduced from λ as follows:

$$n_1^{eq} = 1 - \frac{4}{\pi} \arctan \left[\frac{\mathrm{R}\left(\log \hat{z}_1^{2^{\left(1+n_1^{eq}\right)/2}-1}\right)}{\mathrm{I}\left(\log \hat{z}_1^{2^{\left(1+n_1^{eq}\right)/2}-1}\right)} \right] \tag{42}$$

$$T_1^{eq} = -\left[\frac{\mathrm{I}\left(\log \hat{z}_1^{2^{\left(1+n_1^{eq}\right)/2}-1}\right)}{2\pi f_r \left(2^{\left(1+n_1^{eq}\right)/2} - 1\right)\left(\frac{f_s}{f_r}\right)^{\left(1+n_1^{eq}\right)/2}} \right] \tag{43}$$

Note that we know a rough estimate of the pole, we can filter out the contribution of the associated mode vector to the Hankel data matrix $\tilde{\mathbf{S}} = \tilde{\mathbf{S}}_1 + \tilde{\mathbf{S}}_2$ through an orthogonal projection as in the work of Chen et al. [39]. However, contrary to Chen et al. [39] that used a QR decomposition of the known mode vector to determine the orthogonal subspace, we carry out a singular value decomposition of the Hankel data matrix constructed from $\tilde{\mathbf{a}}_1$ as $\tilde{\mathbf{S}}_1 = \mathbf{U}_1 \mathbf{\Sigma}_1 \mathbf{V}_1^H$. Using MATLAB notation, the orthogonal complement is then given by $\mathbf{U}_1^{\mathrm{orth}} = \mathbf{U}_1(:\ , \mathrm{rank}(\mathbf{\Sigma}_1) + 1 : \mathrm{end})$ The definition of $\mathbf{U}_1^{\mathrm{orth}}$ reflects the fact that the rank of $\tilde{\mathbf{S}}_1$ or, equivalently, $\mathbf{\Sigma}_1$ would be overestimated, unless $\tilde{\mathbf{a}}_1$ admits a Vandermonde structure guaranteed by setting $n = n_1^{eq}$ in Equation (28). More importantly, such a definition of $\mathbf{U}_1^{\mathrm{orth}}$ enables, once n_1^{eq} and T_1^{eq} are estimated, to filter out the contribution of $\tilde{\mathbf{a}}_1$ even if the rank is overestimated or, equivalently, $n \neq n_1^{eq}$. To summarize, the Vandermonde structure of the mode vector is essential only to parameter estimation as seen in Equations (42) and (43) but not to filtering as shown by the definition of the orthogonal complement. Left multiplying $\tilde{\mathbf{S}}$ by the conjugate transpose of $\mathbf{U}_1^{\mathrm{orth}}$ gives $\tilde{\mathbf{S}}^{\mathrm{pr}}$:

$$\tilde{\mathbf{S}}^{\mathrm{pr}} = \left(\mathbf{U}_1^{\mathrm{orth}}\right)^{\mathrm{H}} \tilde{\mathbf{S}} = \left(\mathbf{U}_1^{\mathrm{orth}}\right)^{\mathrm{H}} \left(\tilde{\mathbf{S}}_1 + \tilde{\mathbf{S}}_2\right) = \left(\mathbf{U}_1^{\mathrm{orth}}\right)^{\mathrm{H}} \tilde{\mathbf{S}}_2 \tag{44}$$

As pointed out in Ref. [40], the multiplication from the left destroys the shift-invariance property in the column space but maintains it in the row space. This means that $\tilde{\mathbf{S}}^{\mathrm{pr}}$ can still be used to estimate the parameters of the second echo. When the processed data sequence is noisy, the orthogonal projection is only an approximation due to the perturbation introduced by noise in the signal subspace, and thus only the principal left singular vectors are to be considered when determining the orthogonal complement. The number of these vectors must be chosen in a way that ensures the best orthogonal projection on the one hand and maximum filtering of the first mode vector on the other hand. Obviously, these requirements are contradictory; orthogonal projection improves with decreasing number of singular vectors because the largest singular values are less affected by noise, while the filtering effect improves with increasing number of vectors. The optimum number of principal singular vectors as to minimizing estimation errors is hence found heuristically to be two, i.e., $\mathbf{U}_1^{\mathrm{orth}} = \mathbf{U}_1(:\ , 2+1:\mathrm{end})$. Similar to Equation (38), setting $n = n_2^{eq}$ in Equation (28) restores the Vandermonde structure, only this time for the second mode vector. The estimates of n_2^{eq} and T_2^{eq} are then determined from the GEV of the matrix pair $\left[\tilde{\mathbf{S}}_2^{\mathrm{pr}}, \tilde{\mathbf{S}}_1^{\mathrm{pr}}\right]$ using Equations (42) and (43). The following steps of this alternately projected approach are to filter out the contribution of the second mode vector and re-estimate the parameters of the first and so on until practical convergence is achieved. The estimated values of $\{n_m^{eq}\}_{m=1}^{d}$ and $\{T_m\}_{m=1}^{d}$ then allow determining the complex amplitudes by means of a LS fit having the following solution:

$$\mathbf{c} = (\tilde{\mathbf{A}}^{\mathrm{H}} \tilde{\mathbf{A}})^{-1} \tilde{\mathbf{A}}^{\mathrm{H}} \tilde{\mathbf{s}}. \tag{45}$$

Note that the development of the algorithm has till now been based on the a priori knowledge of n_1^{eq} and n_2^{eq}. In practice, however, the values of the indices are unknown and need to be estimated. To solve this problem, we propose a recursive scheme of the previously described approach. The RAP-MPM (**R**ecursive and **A**lternately **P**rojected **M**atrix **P**encil **M**ethod) can now be summarized in the following steps for the case of $d = 2$ [38]:

Step 1. Form a Hankel data matrix $\tilde{\mathbf{S}}$ from a data sequence corresponding to frequencies given in Equation (28). If the model order $M < d$ (d is known), left multiply $\tilde{\mathbf{S}}$ by the conjugate transpose of $\mathbf{U}_{\mathrm{m}}^{\mathrm{orth}}$. Start with $M = d$ and $n = 1$ as initial values, which correspond to the exact uniform data sequence provided by real measurements.

Step 2. Run MPM and determine $\{n_m^{eq}\}_{m=1}^{d}$ and $\{T_m^{eq}\}_{m=1}^{d}$ by using Equations (42) and (43).

Step 3. Compare the estimated value of n_m^{eq} with the value used in step 1. If they are different, i.e., $\Delta n \neq 0$ substitute the new estimate of n_m^{eq} in Equation (28) to obtain a new set of non-uniform frequencies at which the spline interpolant is to be evaluated. Repeat steps 1, 2 and 3 until convergence, i.e., $\Delta n = 0$. This termination condition might never be satisfied for noisy data, and so it will be replaced by a predefined practical threshold, e.g., 10^{-6}

Step 4. Retain the estimates of n_m^{eq} and T_m^{eq} of the last iteration and use them each time $\mathbf{U}_{\mathrm{m}}^{orth}$ is to be determined from the mode vector $\tilde{\mathbf{a}}_m$ for a given value of n.

Step 5. Set the model order M to $d-1$ and repeat steps 1 to 4 until the relative change of the estimates between two consecutive alternated projections is smaller than a predefined value. Two to three projections were found to be sufficient.

Step 6. Use the estimates of $\{n_m^{eq}\}_{m=1}^{d}$ and $\{T_m\}_{m=1}^{d}$ to determine the complex amplitudes using Equation (45).

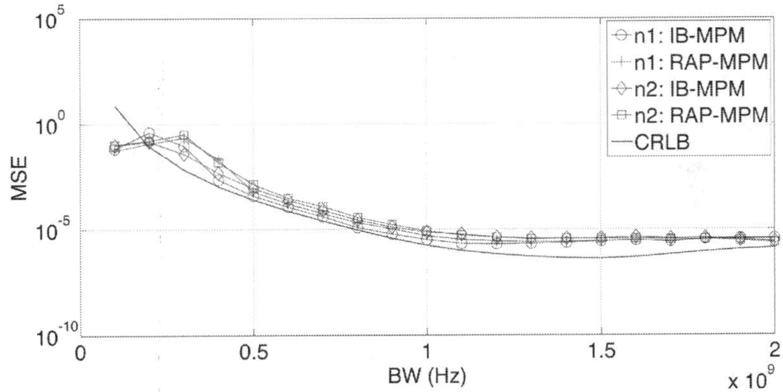

Figure 6. MSE of n_1 and n_2 versus bandwidth.

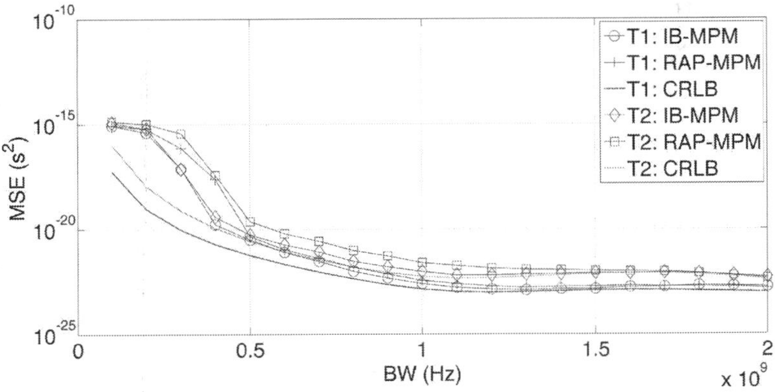

Figure 7. MSE of τ_1 and τ_2 versus bandwidth.

3.5. Performance evaluation

This section first examines the performance of both algorithms for two simulated backscattered echoes of a DI structure under scenarios of varying bandwidth, SNR, and $B\tau_2$. Then, it examines the performance of RAP-MPM in the case of a DV structure. We consider a reference frequency f_r=1.5 GHz and a frequency shift f_s=10 MHz. The frequency-independent complex amplitudes are $c_1=0.5\exp(j0.7)$ and $c_2=1-c_1^2$. Unless otherwise defined, SNR represents the average SNR over the data interval. The convergence threshold for both algorithms is fixed at 10^{-6}, and the number of alternate projections for RAP-MPM is set to three. For all simulations, the estimates of n and τ are computed for 500 independent trials. The mean squared errors (MSEs) are compared with the Cramer-Rao lower bounds (CRLBs).

3.5.1. *MSE versus bandwidth*

This simulation is concerned with the ability of the developed algorithms to handle wideband signals. The bandwidth is varied symmetrically about f_r. **Figures 6** and **7** show the MSEs of the parameters for two ($n_1=n_2=0.83$, τ_1 = 3.2 ns, τ_2 = 1.2 ns and SNR=30 dB) backscattered echoes along with the corresponding CRLBs as a function of bandwidth. It can be seen that the MSEs of both algorithms are comparable and mirror the CRLB as the bandwidth varies. This is in contrast with the behavior of conventional algorithms where their MSEs deviate from the CRLB beyond a certain bandwidth. Hence, the obtained results prove that the developed algorithms do not suffer from the bandwidth limitation of conventional estimation methods. A similar conclusion can be drawn for the DV structure ($n_1=0.87$, $n_2=0.83$, τ_1= 3.2 ns, τ_2 = 1.2 ns and SNR=30 dB), as **Figures 8** and **9** show that the estimates of RAP-MPM have MSEs that decrease as bandwidth increases.

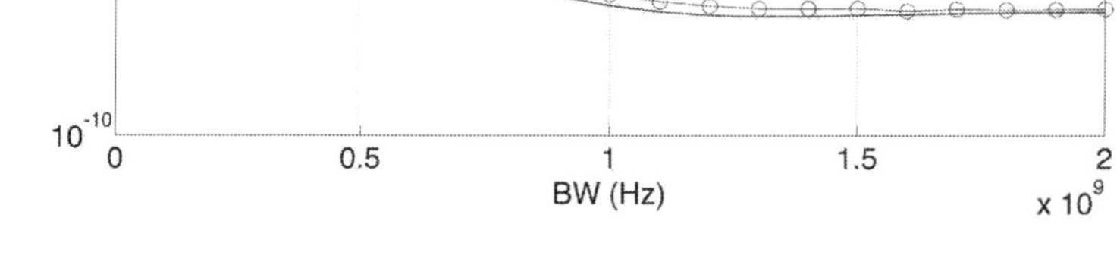

Figure 8. RAP-MPM MSEs of n_1 and n_2 versus bandwidth.

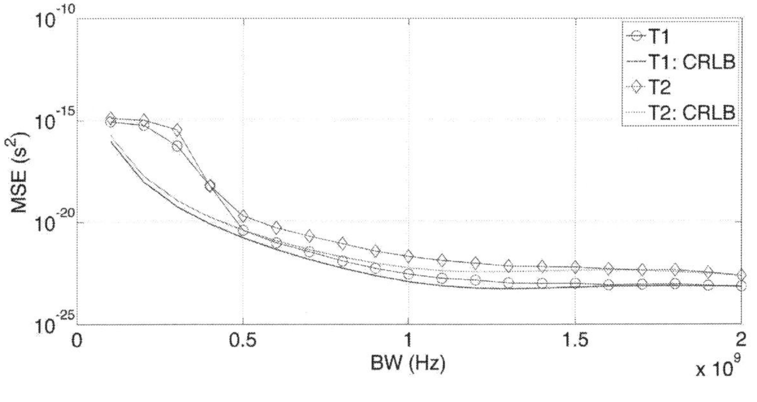

Figure 9. RAP-MPM MSEs of τ_1 and τ_2 versus bandwidth.

3.5.2. *MSE versus SNR*

This simulation investigates the robustness of the algorithms with respect to noise. **Figures 10** and **11** depict the MSEs of the parameters for two backscattered echoes ($n_1 = n_2 = 0.83$, τ_1 = 3.2 ns, τ_2 = 1.2 ns, and BW = 2 GHz) along with the corresponding CRLBs as a function of SNR. It can be seen that both algorithms are unbiased and robust to noise, as their MSEs approach the CRLBs even for low SNRs.

For the DV structure ($n_1 = 0.87$, $n_2 = 0.83$, $\tau_1 = 3.2$ ns, $\tau_2 = 1.2$ ns, and BW = 2 GHz), the MSEs of the indices and the time delays approach their CRLBs starting, respectively, from SNR = 5 and 10 dB as shown in **Figures 12** and **13**.

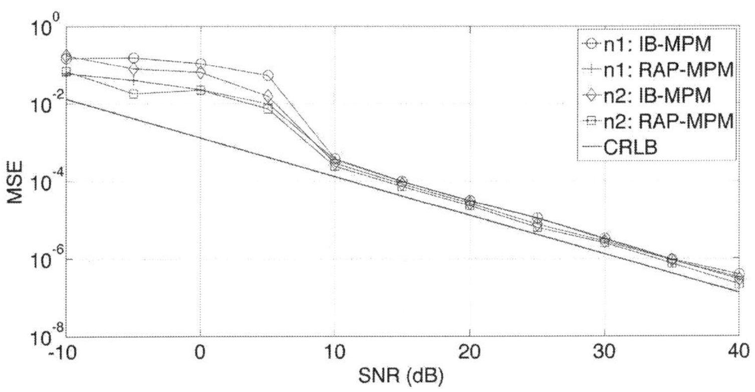

Figure 10. MSE of n_1 and n_2 versus SNR.

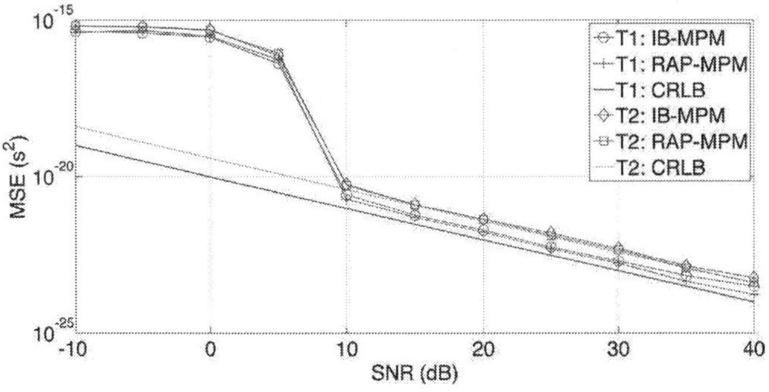

Figure 11. MSE of τ_1 and τ_2 versus SNR.

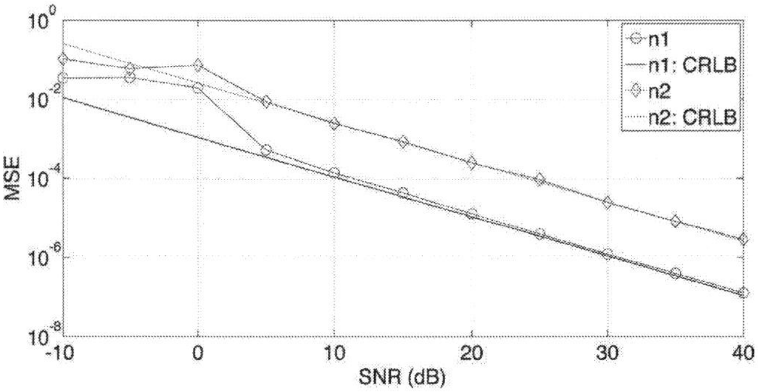

Figure 12. RAP-MPM MSEs of n_1 and n_2 versus SNR.

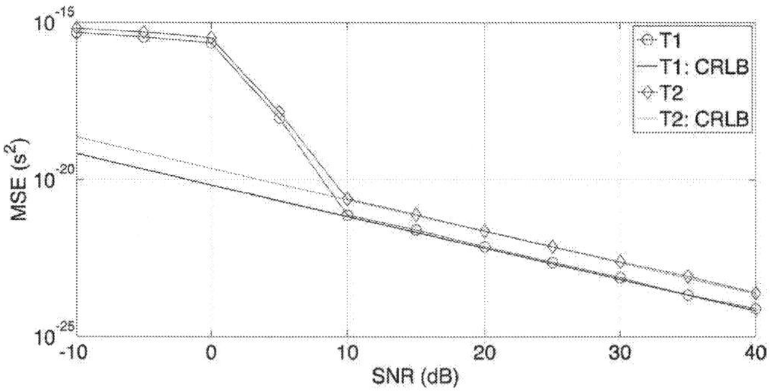

Figure 13. RAP-MPM MSEs of τ_1 and τ_2 versus SNR.

3.6. RRMSE versus $B\tau_2$

This simulation examines the resolution capability of the algorithms by varying τ_2 at BW = 2 GHz so that $0.2 \leq B\tau_2 \leq 2$. **Figure 14** shows the RRMSE of the estimates of τ_2 as a function of $B\tau_2$ along with the CRLB and a 5% error threshold for $n_1 = n_2 = 0.83$, τ_1 = 3.2 ns, and SNR = 30 dB. We note here that defining SNR with respect to both echoes leads to a drastic decrease in the SNR of the second echo as $B\tau_2$ increases. SNR is hence fixed with respect to the second echo. The obtained results illustrate that the algorithms afford less than 5% error for $B\tau_2 < 1$, which indicates their high-resolution capability.

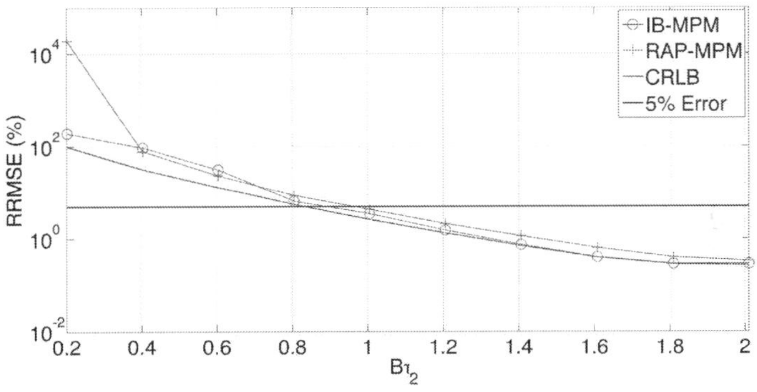

Figure 14. The RRMSE versus $B\tau_2$.

4. Conclusion

This chapter deals with model-based high-resolution signal processing algorithms applied to non-dispersive and dispersive GPR data. For non-dispersive data, a modified version of the conventional ESPRIT algorithm, called M-ESPRIT, is presented for estimating the time delays. M-ESPRIT allows taking the radar pulse and any noise signature into account and improves the time resolution of the GPR for thin pavement thickness estimation with conventional GPR of 2 GHz bandwidth. Contrary to the other versions of ESPRIT, M-ESPRIT enables processing raw data directly without any preprocessing. Simulation and experimental results showed that M-ESPRIT provides better performance. For dispersive data, a wideband extension of the matrix pencil method to the joint parameter estimation of dispersive media obeying the constant-Q model is presented. The first and second extension theorems were used to derive a data sequence that enables the singular extension of the Hankel data matrix and, as a consequence, restores its effective rank. The desired data sequence, however, was found to correspond to a set of non-uniform frequencies which are impossible to acquire from real measurements. To solve this problem, two MPM-based algorithms were then developed. IB-

MPM and RAP-MPM are based on interpolating the data sequence to reach the desired Vandermonde structure of the mode matrix. To handle the DV structure, RAP-MPM makes use of the equivalent medium approximation to unify the form of the mode vectors and then uses a projection scheme to strip them alternately. The algorithms were subjected to simulation scenarios of varying bandwidth, SNR, and $B\tau_2$. The results confirmed the wideband attribute of the algorithms, their unbiasedness over the considered SNR range, robustness to noise, rapid convergence, and high-resolution capability.

Acknowledgements

This work may contribute to the European Cooperation in Science and Technology (COST) Action TU1208 Civil Engineering Applications of Ground Penetrating Radar (http://www.gpradar.eu/ [Accessed: 2016-01-26]).

Author details

Cédric Le Bastard[1*], Khaled Chahine[2], Yide Wang[3], Vincent Baltazart[4], Nicolas Pinel[5], Christophe Bourlier[3] and Xavier Derobert[4]

*Address all correspondence to: cedric.lebastard@cerema.fr

1 Cerema, DTEROuest, DLRCA, Les Ponts de Cé, France and LUNAM Université de Nantes–IETR, UMR CNRS 6164– Polytech Nantes, Nantes, France

2 Signals and Systems Group, Department of Electrical and Electronics Engineering, Lebanese International University, Beirut, Lebanon

3 LUNAM, IETR, UMR CNRS 6164 – Polytech Nantes, Nantes, France

4 LUNAM, IFSTTAR, Nantes, France

5 Alyotech, France

References

[1] In: Proceedings of the 15th IEEE International Conference on Ground Penetrating Radar; 30 June–4 July 2014; Brussels. New York: IEEE; 2014.

[2] Daniels D.J. Ground-Penetrating Radar. 2nd ed. London, UK: Institute of Electrical Engineers; 2004. ISBN: 086341360.

[3] Jol H. Ground Penetrating Radar Theory and Applications. Amsterdam (The Netherlands) and Oxford (UK): Elsevier; 2008. ISBN: 978-0-444-53348-7.

[4] Persico R. Introduction to Ground Penetrating Radar: Inverse Scattering and Data Processing. New Jersey (USA): Wiley-IEEE Press; 2014. 392 p. ISBN: 978-1-118-30500-3.

[5] Benedetto A., Pajewski L. Civil Engineering Applications of Ground Penetrating Radar. New Jersey (USA): Springer; 2015. 371 p. ISBN: 978-3-319-04812-3.

[6] Van Der Kruk J., Streich R., Green A. Properties of surface waveguides derived from separate and joint inversion of dispersive TE and TM GPR data. Geophysics. 2006; 71: K19–K29. DOI: 10.1190/1.2168011.

[7] Lambot S., Slob E.C., Minet J., Jadoon K.Z., Vanclooster M., Vereecken H. Full-waveform modeling and inversion of ground penetrating radar data for non-invasive characterization of the soil hydrogeophysical properties. In: Viscara Rossel R.A., McBratney A.B., and Minasny B., editors. Proximal Soil Sensing, Developments in Soil Science Series. New Jersey (USA): Springer; 2010. p. 299–313.

[8] Radzevicius S. Practical 3-D migration and visualization for accurate imaging of complex geometries with GPR. Journal of Environmental & Engineering Geophysics. 2008; 13(2): 99–112. DOI: 10.2113/JEEG13.2.99.

[9] AL-Qadi I.L., Lahouar S. Measuring layer thicknesses with GPR theory to practice. Construction and Building Materials. 2005; 19(10): 763–772. DOI: 10.1016/j.conbuildmat.2005.06.005.

[10] Benedetto A., Benedetto F., De Blasiis M.R., Giunta G. Reliability of signal processing technique for pavement damages detection and classification using ground penetrating radar. IEEE Sensors Journal. 2005; 5(3): 471–480. DOI: 10.1109/JSEN.2005.846176.

[11] Lahouar S., Al-Qadi I.L. Automatic detection of multiple pavement layers from GPR data. NDT&E International. 2008; 41(2): 69–81. DOI: 10.1016/j.ndteint.2007.09.001.

[12] Spagnolini U., Rampa V. Multitarget detection/tracking for monostatic ground penetrating radar: Application to pavement profiling. IEEE Transactions on Geoscience and Remote Sensing. 1999; 37(1): 383–394. DOI: 10.1109/36.739074.

[13] Spagnolini U. Permittivity measurements of multilayered media with monostatic pulse radar. IEEE Transactions on Geoscience and Remote Sensing. 1997; 35(2): 454–463. DOI: 10.1109/36.563284.

[14] Saarenketo T., Scullion T. Road evaluation with ground penetrating radar. Journal of Applied Geophysics. 2000; 43(2–4): 119–138. DOI: 10.1016/S0926-9851(99)00052-X.

[15] Le Bastard C., Baltazart V., Wang Y., Saillard J. Thin pavement thickness estimation using GPR with high-resolution and super resolution methods. IEEE Transactions on

Geoscience and Remote Sensing. 2007; 45(8): 2511–2519. DOI: 10.1109/TGRS. 2007.900982.

[16] Le Bastard C., Baltazart V., Wang Y. Modified ESPRIT (M-ESPRIT) algorithm for time delay estimation in both any noise and any radar pulse context by a GPR radar. Signal Processing. 2010; 90: 173–179. DOI: 10.1016/j.sigpro.2009.06.007.

[17] Koudogbo F., Mametsa H., Combes P. Surface and volume scattering from natural and manmade rough surfaces in the process of setting up data base coefficients. In: IEEE, editor. International Geoscience and Remote Sensing Symposium; July 2003; Toulouse, France; 2003. p. 4211–4213. DOI: 10.1109/IGARSS.2003.1295466.

[18] Pinel N., Le Bastard C., Baltazart V., Bourlier C., Wang Y. Influence of layer roughness for road survey by ground penetrating radar at nadir: Theoretical study. IET Radar, Sonar and Navigation. 2011; 5(6): 650–656. DOI: 10.1049/iet-rsn.2010.0197.

[19] Bourlier C., Le Bastard C., Baltazart V. Generalization of PILE method to the EM scattering from stratified subsurface with rough interlayers – Application to the detection of debondings within pavement structure. IEEE Transactions on Geoscience and Remote Sensing. 2015; 53(7): 4104–4115. DOI: 10.1109/TGRS.2015.2390677.

[20] Adous M. Caractérisation electromagnetique des materiaux traites du genie civil dans la bande de frequence 50MHz-13GHz. Thesis. University of Nantes; 2006.

[21] Fauchard C. Utilisation de radars tres hautes frequences: application a l'auscultation non destructive des chaussées. Thesis. University of Nantes; 2001.

[22] Schmidt R.O. A signal subspace approach to multiple emitter location and spectral estimation. Thesis. Stanford University; 1981.

[23] Roy R., Paulraj A., Kaikath T. ESPRIT—A subspace rotation approach to estimation of parameters of cisoids in noise. IEEE Transactions on Acoustics, Speech and Signal Processing. 1986; 34(5): 1340–1342. DOI: 10.1109/TASSP.1986.1164935.

[24] Saarnisaari H. TLS-ESPRIT in a time delay estimation. In: Proceedings of the IEEE 47th Vehicular Technology Conference; Phoenix, USA; May 1997. p. 1619–1623. DOI: 10.1109/VETEC.1997.605832.

[25] Swindlehurst A.L. Time delay and spatial signature estimation using known asynchronous signals. IEEE Transactions on Signal Processing. 1998; 46(2): 449–462. DOI: 10.1109/78.655429.

[26] Roy R., Kailath T. Total least squares ESPRIT. In: Proceedings of the Asilomar Conference on Circuits, Systems and Computers; Pacific Grove, CA, USA; 1987. p. 297–301.

[27] Sun M., Le Bastard C., Wang Y., Pinel N. Time-delay estimation using ESPRIT with extended improved spatial smoothing techniques for radar signals. IEEE Geoscience and Remote Sensing Letters. 2016; 13(1): 73–77. DOI: 10.1109/LGRS.2015.2497378.

[28] Zha H. Fast algorithms for direction-of-arrival finding using large ESPRIT arrays. Signal Processing. 1996; 48: 111–121. DOI: 10.1016/0165-1684(95)00128-X.

[29] Le Bastard C., Wang Y., Baltazart V. Algorithmes haute résolution linéaires pour l'estimation des temps de retard dans un contexte de bruit quelconque. Traitement du Signal. 2008; 25(4): 253–263. Available from: http://hdl.handle.net/2042/28451 [2016-01-19]

[30] Marcos S., Marsal A., Benidir M. The propagator method for source bearing estimation. Signal Processing. 1995; 42: 121–138. DOI: 10.1016/0165-1684(94)00122-G.

[31] Schneider J.B. Plane waves in FDTD simulations and a nearly perfect. IEEE Transaction Antenna and Propagation. 2004; 52(12): 3280–3287. DOI: 10.1109/TAP.2004.836403.

[32] Guillanton E., Dauvignac J.-Y., Pichot C., Cashman J. A new design tapered slot antenna for ultra-wide band application. Microwave and Optical Technology Letters. 1998; 19(4): 286–289. DOI: 10.1002/(SICI)1098-2760(199811)19:4<286::AID-MOP12>3.0.CO;2-0.

[33] Bano M. Constant dielectric losses of ground-penetrating radar waves. Geophysical Journal International. 1996; 124(1): 279–288. DOI: 10.1111/j.1365-246X.1996.tb06370.x.

[34] Kjartansson E. Constant Q-wave propagation and attenuation. Journal of Geophysical Research. 1979; 84(B9): 4737–4748. DOI: 10.1029/JB084iB09p04737.

[35] Wang Y. Inverse Q-filter for seismic resolution enhancement. Geophysics 2006; 71(3): 51–60. DOI: 10.1190/1.2192912.

[36] Chahine K., Baltazart V., Wang Y. Subspace leakage suppression for joint parameter estimation of quality factors and time delays in dispersive media. Circuits, Systems, and Signal Processing. 2015; 1: 15. DOI: 10.1007/s00034-015-0180-8.

[37] Chahine K., Baltazart V., Wang Y. Interpolation-based matrix pencil method for parameter estimation of dispersive media in civil engineering. Signal Processing. 2010; 90(10): 2567–2580. DOI: 10.1016/j.sigpro.2010.03.003.

[38] Chahine K., Baltazart V., Wang Y. Parameter estimation of damped power-law phase signals via a recursive and alternately projected matrix pencil method. IEEE Transactions on Antennas and Propagation. 2011; 59(4): 1207–1216. DOI: 10.1109/TAP.2011.2109359.

[39] Chen H., Van Huffel S., Vandewalle J. Improved methods for exponential parameter estimation in the presence of known poles and noise. IEEE Transactions on Signal Processing. 1997; 45(5): 1390–1393. DOI: 10.1109/78.575717.

[40] Chen H., Van Huffel S., Van Den Boom A., Van Den Bosch P. Subspace-based parameter estimation of exponentially damped sinusoids using prior knowledge of frequency and phase. Signal Processing. 1997; 59: 129–136. DOI: 10.1016/S0165-1684(97)00085-6.

Chapter 10

Nondestructive Tests for Induction Machine Faults Diagnosis

Paulo Cezar Monteiro Lamim Filho,
Lane Maria Rabelo Baccarini and Robson Pederiva

Additional information is available at the end of the chapter

http://dx.doi.org/10.5772/63166

Abstract

A maintenance program must include several techniques of monitoring of the electric motor's conditions. Among these techniques, probably the two classic ones are related to megger and impulse test. Unfortunately, in both cases, inherent drawbacks can expose the electrical motor at a high voltage that could deteriorate insulation condition making difficult its use on industrial environment. As the electrical machines have several different components (e.g., bearings, rotor bars, shaft, and stator windings), the fault frequencies can be excited by mechanical and/or electrical faults making the identification of the real condition difficult. This chapter describes several methods of the nondestructive tests for induction motors based on the motor current signature analysis (MCSA), magnetic flux, and vibration analysis. The method of analysis is a good alternative tool for destructive tests and fault detection in induction motors. Numerical and experimental results demonstrate the effectiveness of the proposed technique. This chapter also presents a model suitable for computer simulation of induction motor in a healthy state and with general asymmetries that can be analyzed simultaneously. The model makes it possible to conduct research on different characteristics of engines and outstanding effects produced by the faults.

Keywords: Diagnosis technique, induction motor fault diagnosis, motor current signature analysis, magnetic flux, vibration, non-destructive tests

1. Introduction

Three-phase squirrel-cage induction motors are commonly used as electrical drives in industry because they are rugged, mechanically simple, and adaptable to a wide variety of operation conditions and simple to control.

Their design is commonly intended to have electrical and mechanical symmetries in the stator and the rotor for better coupling and higher efficiency [1]. However, the motors are exposed to different loading and environmental conditions. These conditions acting together with the natural aging of the motor may lead to many faults. The fault condition in a motor will damage the symmetrical property and induces an abnormal symptom during motor operation, such as mechanical vibration, temperature increase, irregular air-gap torque, output power variation, acoustic noise, line voltage changes, line current changes, and speed variations [1–5].

In induction electrical motors, surveys [6–9] report that the failures include internal faults (e.g., low insulation, stator interturn short circuits, worn-out/broken bearings, eccentricity, broken rotor bars, and mechanical faults) along with external faults (e.g., unbalanced supply voltage, over loading, blocked rotor, under voltage, and over voltage). These faults modify the modulation amplitudes, which are inherent in the vibration, current, and flux signals.

In order to obtain good results during the fault diagnosis of electrical machinery, a growing amount of research about new measurement instrumentation [10–13] and new signal processing techniques have been presented [14–17]. Their major goal is to improve or propose new alternative techniques to those that already exist. Among these techniques, probably the two classic ones are related to megger and impulse test. Unfortunately, in both cases, inherent drawbacks can expose the electrical motor at a high voltage that could deteriorate insulation condition making difficult their use in loco on industry plants.

An efficient and modern diagnosis technique should be nondestructive and requires only the acquisition of signals that are readily available in the motor or in the motor control center.

In the case of three-phase induction motors, which in particular represent around 85% of worldwide power consumption [18], it can be observed when using modern technologies such as the motor current signature analysis (MCSA), magnetic flux, and vibration analysis that the main faults are directly related to modulations around the power supply frequencies, rotor slot frequencies, and resonance frequencies.

This chapter describes several methods of nondestructive testing to diagnose faults of induction machine based on recent research conducted during this study. In addition, it presents a dynamic model suitable for computer simulation of induction machines in a healthy state with general asymmetries, such as power systems unbalance, stator interturn short circuit, rotor broken bars, and mechanical faults. The model is an important tool of new diagnostic techniques project because it provides analysis of the impact of failures on the characteristics of the electrical machinery.

This chapter is organized as follows. The dynamic model and the characteristic frequencies of the faults are presented in Section 2. A sound knowledge of dynamic model and characteristic frequencies will facilitate understanding the health condition of the electrical machine. The simulation and experimental results are described in Section 3. Finally, conclusions are presented in Section 4.

2. Fault models

The symmetrical machine model is well known in the literature. The dynamic model proposed by Baccarini et al. [19] accounts for both mechanical and electrical faults in induction machines. The model allowed several simulations in different conditions and outstanding effects produced by the faults. The asymmetries can be power systems unbalance, stator interturn short circuit, rotor broken bars, and mechanical faults (unbalance, misalignment, and mechanical looseness). A convenient selection of the state-space variable set enables description of the machine with a very simple set of equations. The model is described in Section 2.1. To facilitate understanding of the relationship failure/sign, the characteristic frequencies of the faults are described in Section 2.2.

2.1. Dynamic model

2.1.1. Unbalance supply voltage

Unbalance supply voltage is characterized by the nonequality of voltage magnitudes and/or when voltage angles among the three phases are different of $2\pi/3$. In this chapter, the voltage unbalance is represented by Eqs. (1)–(3), where w is the angular velocity and the subscript "s" denotes variables and parameters associated with the stator circuits. When the voltage is balanced the values K_{as}, K_{bs}, and K_{angle} are unities:

$$v_{as} = K_{as} V_{\max} \cos(wt), \tag{1}$$

$$v_{bs} = K_{bs} V_{\max} \cos\left(wt - K_{angle}\frac{2\pi}{3}\right), \tag{2}$$

$$v_{cs} = -(v_{as} + v_{bs}). \tag{3}$$

A change of variables that formulates a transformation of the three-phase variables of the machine circuits to two-phase transformation is given by Eqs. (4)–(6):

$$[f_{dq}] = K_s [f_{abs}], \tag{4}$$

$$\begin{bmatrix} f_q \\ f_d \\ f_0 \end{bmatrix} = \frac{2}{3} \begin{bmatrix} \cos\theta & \cos\left(\theta - \frac{2\pi}{3}\right) & \cos\left(\theta + \frac{2\pi}{3}\right) \\ \sin\theta & \sin\left(\theta - \frac{2\pi}{3}\right) & \sin\left(\theta + \frac{2\pi}{3}\right) \\ \frac{1}{2} & \frac{1}{2} & \frac{1}{2} \end{bmatrix} \begin{bmatrix} f_a \\ f_b \\ f_c \end{bmatrix}, \tag{5}$$

$$\theta = \int_0^t w(\varepsilon)\, d\varepsilon + \theta_0, \tag{6}$$

where f represents either voltage, current, flux linkage, or electric charge. The variable ε is a dummy variable of integration.

2.1.2. Symmetric model

The voltage v, the flux λ, and the current i can be expressed in arbitrary reference frame. The equations for the stator and rotor flux are expressed as

$$\frac{d\lambda_{qs}}{dt} = v_{qs} - r_s i_{qs} - w\lambda_{ds}, \tag{7}$$

$$\frac{d\lambda_{ds}}{dt} = v_{ds} - r_s i_{ds} + w\lambda_{qs}, \tag{8}$$

$$\frac{d\lambda_{qr}}{dt} = -r_r i_{qr} - (w - w_r)\lambda_{dr}, \tag{9}$$

$$\frac{d\lambda_{dr}}{dt} = -r_r i_{dr} + (w - w_r)\lambda_{qr}. \tag{10}$$

The subscript "r" indicates the variables, the parameters, and the transformation associated with rotor circuit. The machine electromagnetic torque T_e, the load torque T_{load}, and the rotor velocity w_r are related as

$$\frac{dw_r}{dt} = \frac{T_e - B_m w_r - T_{load}}{J}, \tag{11}$$

where J is the combined moment of inertia of the motor and the mechanical load. The constant B_m represents the viscous friction associated with the mechanical load and motor.

The Runge–Kutta fourth-order method is used to obtain the solution to the differential equations expressed in (7)–(11).The currents of the stator and the rotor are represented by Eqs. (12)–(15):

$$i_{qs} = \lambda_{qs} a_1 - \lambda_{qr} a_2, \tag{12}$$

$$i_{ds} = \lambda_{ds} a_1 - \lambda_{dr} a_2, \tag{13}$$

$$i_{qr} = \lambda_{qr} a_3 - \lambda_{qs} a_2, \tag{14}$$

$$i_{dr} = \lambda_{dr} a_3 - \lambda_{ds} a_2, \tag{15}$$

where $a_0 = L_s L_r - L_m^2$; $a_1 = \frac{L_r}{a_0}$; $a_2 = \frac{L_m}{a_0}$; $a_3 = \frac{L_s}{a_0}$.

The parameters L_s and L_r are the stator and rotor self-inductances, and L_m is the mutual inductance. The expression for the electromagnetic torque in terms of arbitrary reference frame for a p-pole machine may be expressed as

$$T = \frac{3}{2}\frac{p}{2}\left(\lambda_{ds} i_{qs} - \lambda_{qs} i_{ds}\right). \tag{16}$$

2.1.3. *Mechanical fault model*

The occurrence of motor mechanical faults (unbalance, misalignment, and mechanical looseness) results changes in the air-gap space harmonics distribution, which leads to a sideband currents in the current spectrum that can be written as

$$f_{mec} = f\left(1 \pm K_{mec}\frac{1-s}{p}\right), \tag{17}$$

where f is the stator supply frequency, $K_{mec} = 1, 2, 3, \ldots$ is the order number, and s represents the motor slip. The slip is defined as

$$s = \frac{w - w_r}{w}. \tag{18}$$

The interaction of those harmonics with the mainly sinusoidal supply voltage causes specific harmonics in the power and torque spectrum:

$$f_{torque} = K_{torque} \frac{1-s}{p}. \tag{19}$$

Considering the harmonics in torque spectrum, the mechanical faults are analyzed, in this model, by introducing a load torque modulation T_{mec} associated with the rotation frequency in the load torque. Thus, Eq. (11) becomes

$$\frac{dw_r}{dt} = \frac{T_e - B_m w_r - T_{load} - T_{mec}\sin\left(2\pi f_s t\right)}{J}. \tag{20}$$

2.1.4. Stator fault model

The shorted turns leakage inductance is assumed to be μL_{ls} where μ denotes the shorted turns fraction. The voltage v andd flux linkage λ equations of the stator and rotor windings are transformed to dq axes, rotating at an arbitrary speed $w = d\theta / dt$. The machine stator equations can be expressed in complex two-phase dq variables as follows:

$$\frac{d\lambda_{qs}}{dt} = v_{qs} - r_s i_{qs} - w\lambda_{ds} + \frac{2}{3}\mu r_s i_f \cos\theta, \tag{21}$$

$$\frac{d\lambda_{ds}}{dt} = v_{ds} - r_s i_{ds} + w\lambda_{qs} + \frac{2}{3}\mu r_s i_f \sin\theta. \tag{22}$$

The equations for rotor circuits are equal to those corresponding to a healthy motor as in Eqs. (9)–(10). For the shorted turns, the flux linkage equation is

$$\frac{d\lambda_{as2}}{dt} = r_f i_f - \mu r_s \left(i_{ds}\cos\theta + i_{qs}\sin\theta - i_f\right), \tag{23}$$

where r_f and i_f represent the fault resistance and the short-circuit current, respectively.

http://dx.doi.org/10.5772/63166

The stator and rotor currents are obtained from the following equations:

$$i_{qs} = \lambda_{qs} a_7 - \lambda_{qr} a_8 + \left(L_r a_{10} - L_m a_{11} \right) i_f \cos\theta, \tag{24}$$

$$i_{ds} = \lambda_{ds} a_7 - \lambda_{dr} a_8 + \left(L_r a_{10} - L_m a_{11} \right) i_f \sin\theta, \tag{25}$$

$$i_{qr} = \lambda_{qr} a_9 - \lambda_{qs} a_8 + \left(L_s a_{11} - L_m a_{10} \right) i_f \cos\theta, \tag{26}$$

$$i_{dr} = \lambda_{dr} a_9 - \lambda_{ds} a_8 + \left(L_s a_{11} - L_m a_{10} \right) i_f \sin\theta, \tag{27}$$

$$i_f = \left[-\lambda_{as2} + \left(a_{12} i_{qs} + a_{13} i_{qr} \right) \cos\theta + \left(a_{12} i_{ds} + a_{13} i_{dr} \right) \sin\theta \right] / a_{14}, \tag{28}$$

where $a_6 = 1 - \frac{L_m^2}{L_s L_r};\ a_7 = \frac{1}{a_6 L_s};\ a_8 = \frac{L_m a_7}{L_r};\ a_9 = \frac{1}{a_6 L_r};\ a_{10} = \frac{2}{3}\frac{\mu L_s}{a_0};\ a_{11} = \frac{2}{3}\frac{\mu L_m}{a_0};$
$a_{12} = \mu L_s;\ a_{13} = \mu L_m;\ a_{14} = \mu\left(L_s + \frac{2}{3}\mu L_m\right).$

The electromagnetic torque is expressed in dq variables as shown in Eq. (29): the first member of this equation is equal to those corresponding to a healthy motor and the second member is the additional component introduced by the fault:

$$T = \frac{3}{2}\frac{p}{2} L_m \left(i_{qs} i_{dr} - i_{ds} i_{qr} \right) + \frac{p}{2} \mu L_m i_f \left(i_{qr} \sin\theta - i_{dr} \cos\theta \right). \tag{29}$$

2.1.5. *Rotor fault model*

The rotor fault model is proposed by Cunha et al. [20]. The complex vector rotor current, Eq. (15) or Eqs. (26) and (27), is computed from the symmetric model in rotor-fixed reference frame as in Eq. (30). The T_{dq} transformation matrix is generated by a simple algorithm, Eq. (31):

$$\begin{bmatrix} i_{r1} \\ i_{r2} \\ \vdots \\ i_{rn} \end{bmatrix} = T_{dq}^{-1} \begin{bmatrix} i_{dr} \\ i_{qr} \\ \vdots \\ i_0 \end{bmatrix}, \tag{30}$$

$$T_{dq} = K_b \begin{bmatrix} \cos\theta & \cos\left(\theta - \frac{p}{2}\frac{2\pi}{n}\right) & \cos\left(\theta - \frac{p}{2}\frac{4\pi}{n}\right) & \cos\left(\theta - \frac{p}{2}\frac{n-1}{n}2\pi\right) \\ \sin\theta & \sin\left(\theta - \frac{p}{2}\frac{2\pi}{n}\right) & \cos\left(\theta - \frac{p}{2}\frac{4\pi}{n}\right) & \sin\left(\theta - \frac{p}{2}\frac{n-1}{n}2\pi\right) \\ f_{31} & f_{32} & 1 & 0 \\ f_{41} & f_{42} & 0 & 0 \\ f_{n1} & f_{n2} & 0 & 1 \end{bmatrix}, \tag{31}$$

where $K_b = \frac{n-1}{n}$.

It is known that the fault current is mainly divided in a few bars adjacent to the broken bar. Considering that the bar kth is broken, the zero current through this bar is achieved by changing the n-loop rotor current as in Eq. (32). Finally, the new n-dimensional rotor current vector is determined by Eq. (33) and fed back to the induction motor integration algorithm;

$$\begin{bmatrix} i_{r1}^{new} \\ i_{r2}^{new} \\ \cdots \\ i_{rk}^{new} \\ i_{r(k+1)}^{new} \\ \cdots \\ i_{rn} \end{bmatrix} = \begin{bmatrix} i_{r1} \\ i_{r2} \\ \cdots \\ \frac{i_{rk} + i_{r(k+1)}}{2} \\ \frac{i_{rk} + i_{r(k+1)}}{2} \\ \cdots \\ i_{rn} \end{bmatrix}, \tag{32}$$

$$\begin{bmatrix} i_{rd}^{new} \\ i_{rq}^{new} \\ i_{01} \\ i_{0k} \\ i_{0(k+1)} \\ \cdots \\ i_{0r} \end{bmatrix} = \begin{bmatrix} i_{r1}^{new} \\ i_{r2}^{new} \\ \cdots \\ i_{rk}^{new} \\ i_{r(k+1)}^{new} \\ \cdots \\ i_{rn}^{new} \end{bmatrix}. \tag{33}$$

2.1.6. Complete model

Figure 1 shows the flowchart of the computer program that is suitable for computer simulation of induction machines in a healthy state with general asymmetries: power systems unbalance, stator interturn short circuit, rotor broken bars, and mechanical faults. The parameters t and h are the total simulation and integration time values in seconds.

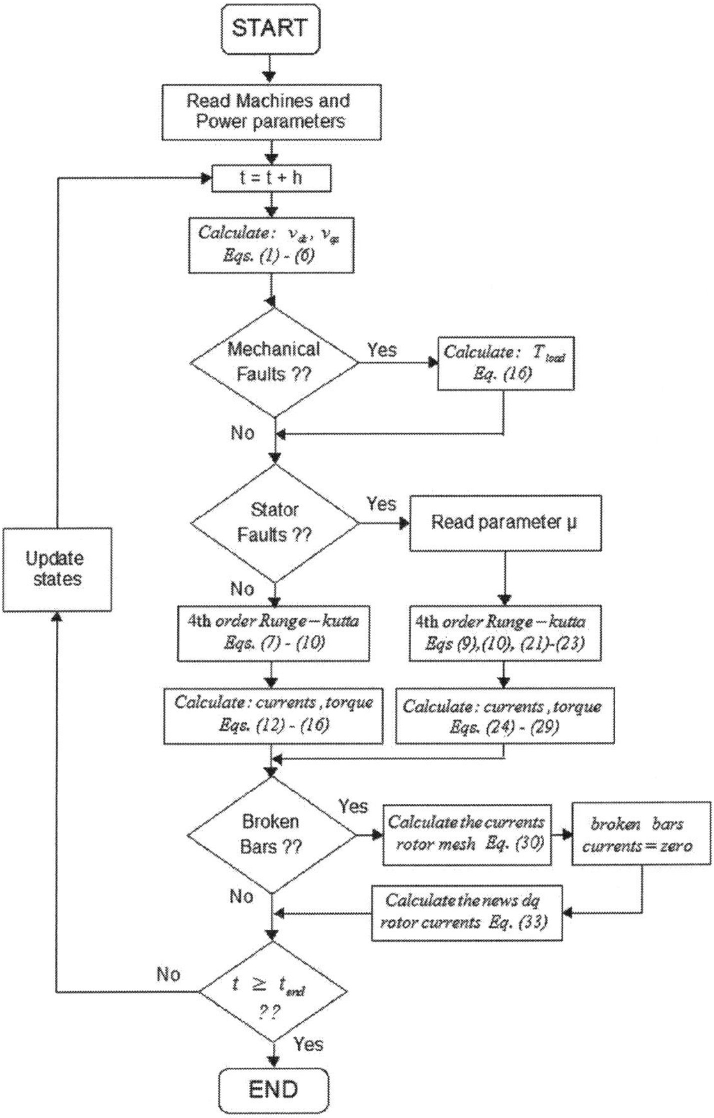

Figure 1. Flowchart of the computer program.

2.2. Characteristic frequencies of faults

2.2.1. Unbalance supply voltage

The study of the rotating magnetic field nature presented here is important to give a better understanding of the deterministic frequencies identification related to the unbalanced voltage supplies, using vibration and magnetic flux analysis.

The three-phase induction motor, composed by stator and rotor, usually works at the constant speed, but small changes can occur due to changes of the mechanical load linked to the shaft.

The stator is composed of laminations of high-grade steel sheet. A three-phase winding is put in slots cut on the inner surface of the stator frame. The rotor also consists of laminated ferromagnetic material, with slots cut on the outer surface.

The squirrel-cage windings consists of aluminum or copper bars embedded in the rotor slots and shorted at both ends by aluminum or copper end rings. The three-phase winding on the stator is a distributed winding. The winding of each phase is distributed over several slots. When current flows through a distributed winding it produces a sinusoidal distributed magnetomotive force (MMF) centered on the axis of the coil representing the phase winding. If an alternating current flows through the coil, it produces a pulsating MMF wave, whose amplitude and direction depend on the instantaneous value of the current flowing through the windings.

The definition of unbalanced voltage used by the power community is the ratio of the negative sequence voltage to the positive sequence voltage [21]. For a set of unbalanced voltages V_{ab}, V_{bc}, and V_{ca}, the positive and negative sequence voltages are given by

$$V_{ab1} = \frac{V_{ab} + aV_{bc} + a^2V_{ca}}{3}, \tag{34}$$

$$V_{ab2} = \frac{V_{ab} + a^2V_{bc} + aV_{ca}}{3}, \tag{35}$$

where $a = -0.5 + j0.855$, $a^2 = -0.5 + j0.855$, and V_{ab1} and V_{ab2} are positive and negative phase sequences, respectively.

Each set produces corresponding balanced currents, and two vectors represent the three-phase currents of the stator.

The positive sequence voltage is the same as for machine in a regular operation condition. However, the negative sequence will produce a reverse rotating field and the slip rotor will be $2 - s$. The motor behaves as the addition of two separate motors, one running at slip s with a terminal voltage of V_{ab1} per phase and the other running with a slip of $2 - s$ and a terminal voltage of V_{ab2}.

The frequency of the induced rotor voltage for this negative motor is approximately 120 Hz. This high frequency in the rotor causes the rotor current to concentrate in the top of the rotor conductors, thus increasing the effective rotor resistance and decreasing the effective rotor leakage reactance. The increase in the effective rotor resistance reflects a significant increase in rotor losses I^2R for a given value of rotor current [22].

Unbalance supply voltage is a usual industrial problem that generates electrical machine overheating, reducing lifetime, provoking vibration that thereafter generates mechanical wear, and noise.

The influence of voltage unbalance upon three-phase induction motors has long been a concern of electrical engineers. One of the main scopes of this chapter is about nondestructive test on the induction motor under voltage unbalance. The paper [21] reviews three definitions of voltage unbalance developed by NEMA, IEEE, and the power community, respectively. The differing definitions of voltage unbalance are analyzed in order to understand the implications of their use.

2.2.2. *Short circuit*

The presence of interturn short circuit and/or unbalanced voltage supply will give rise to a magnetic asymmetry that crosses the machine's air gap. These asymmetry can result in magnetizing current proportional to the rotor slot harmonic frequencies and their respective modulations in twice the line frequency, i.e., $(1 \pm \lambda n(1 - s)/p)fl \pm 2fl$, $\pm 4fl$, etc., where $\lambda = 1, 2, 3, \ldots$, n is the number of rotor bars or slot, p is the number of pole pairs, fl is the line frequency, and s is the rotor slip. These frequencies of stator current will generate a torque pulsation and consequently vibration that is transmitted through the motor frame. These considerations are based on studies conducted by Gojko and Penman [23] and Gupta and Culbert [24], for three-phase motor.

3. Nondestructive diagnostic method

Using the model described in Section 2.1, several analyses were performed and fault diagnosis algorithms were developed by Baccarini et al. [19,26]. The proposed methods allow the full engine diagnostics, i.e., verify the absence or the presence of the following failure conditions: initial short-circuit, broken bars, and mechanical failures.

The proposed methods were validated on a test bench. The results and the test bench are described in Section 3.1. The methods are nondestructive and need only the information of the current sensors, typically present in an industrial plant. Therefore, these techniques will be referred to herein, "the signal current methods."

The results and experimental analysis of Sections 2.2.1 and 2.2.2 are described in Section 3.2.

3.1. The signals current methods

3.1.1. Description of the experimental setup 1

The experimental system is set with special induction machine to simulate the failure, a direct current (dc) machine, a measuring system, encoder, computer, a three-phase varivolt, resistances, and the board acquisition.

A separate dc generator feeding a variable resistor provides a mechanical load. In order to allow tests to be performed at different load levels, the dc excitation current and load resistor are both controllable.

The motor was rewound to allow short-circuit simulation between different numbers of coil turns. The stator windings are connected in delta in all tests. In order to prevent damage of the stator windings, a resistor was used to limit the short-circuit current.

Either the mechanical structure where the motors are settled offers the possibility to move the two machines, in a way the system can be aligned or different degrees of misalignment can be tested. Shaft alignment in the setup was guaranteed by using a laser alignment tool.

The motor was initially set with the cage intact and several tests were realized with the symmetric rotor. The rotor bar fault has been caused by drilling holes into the aluminum bars.

3.1.2. Winding short-circuit fault and unbalance supply voltage

To validate the proposed method, simulations of different induction machine were carried out. The simulations presented here refer to a motor with the following nominal parameters: 3 CV, 220 V, 60 Hz, 4 poles, and 1710 rpm. It is the motor of the test bench. The Monte Carlo computational algorithm was used to simulate random motor operation conditions: mechanical load, different broken bar, degree of mechanical fault, different percentages of turns shorted, level of voltage unbalance, and noise in the measurement system.

Several experimental results are presented below to demonstrate the robustness and accuracy of the stator short-circuit model. The negative sequence impedance of a healthy induction motor is practically constant. However in the presence of short-circuit failure, the symmetry of the windings is lost and the value of negative impedance changes.

Multiple experimental tests were performed on different days and times. The negative sequence impedances for each test were calculated. The average value of impedance is the situation of lack of failures. Other series of tests were performed in the following conditions: the absence of short circuit with 3, 6, and 15 shorted turns. **Figure 2** shows the mean and square mean error of the negative components impedance for a healthy motor with 3, 6, and 15 shorted turns.

Results from the experiments are very encouraging. The performance was not affected by the voltage supply unbalances or inherent machine and/or monitoring system asymmetry. It can be seen that experimental results exhibit the same trend as predicted by the model.

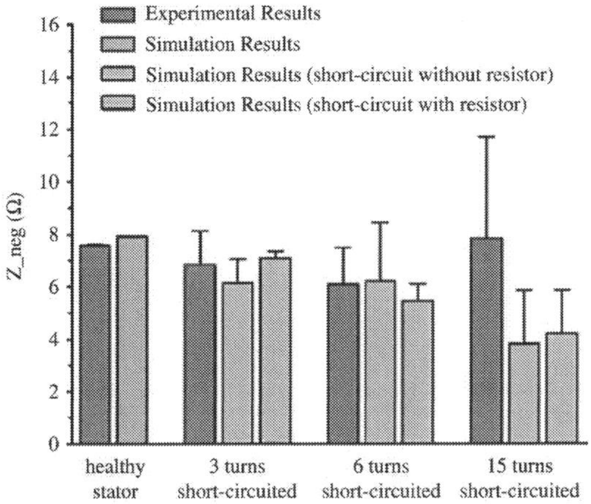

Figure 2. Mean and square mean error of the negative component impedance for healthy motor with 3, 6 and 15 shorted turns.

3.1.3. *Mechanical faults*

The presence of mechanical fault is analyzed in simulation and experimental tests. **Figure 3a** and **b** shows the simulation stator current spectrum of the motor with different levels of mechanical failures and operation at rated load. The nominal load frequency f_r is 28.5 Hz. In that way, the spectrum contains components near 31.5 and 88.5 Hz. These $f \pm f_r$ components point out the presence of mechanical fault and their amplitudes increase as the mechanical fault's level rises.

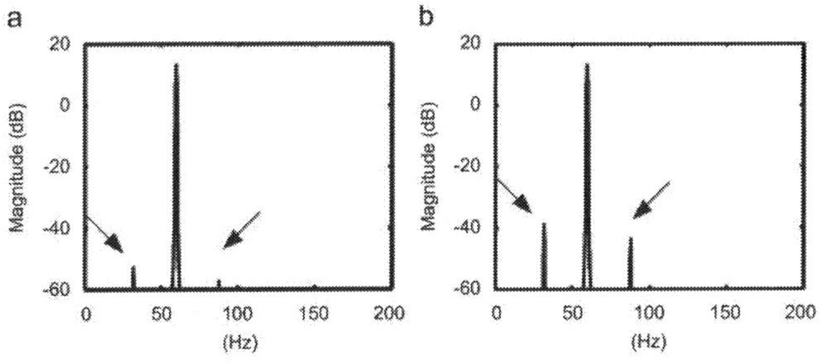

Figure 3. Simulation current frequency spectrum.

Figure 4 shows the phase current spectra for a motor with mechanical fault. The zoom spectra are centered on the fundamental frequency (60 Hz). The component 32 Hz ($f - f_r$) and the component 88 Hz ($f + f_r$) indicates the presence of mechanical fault.

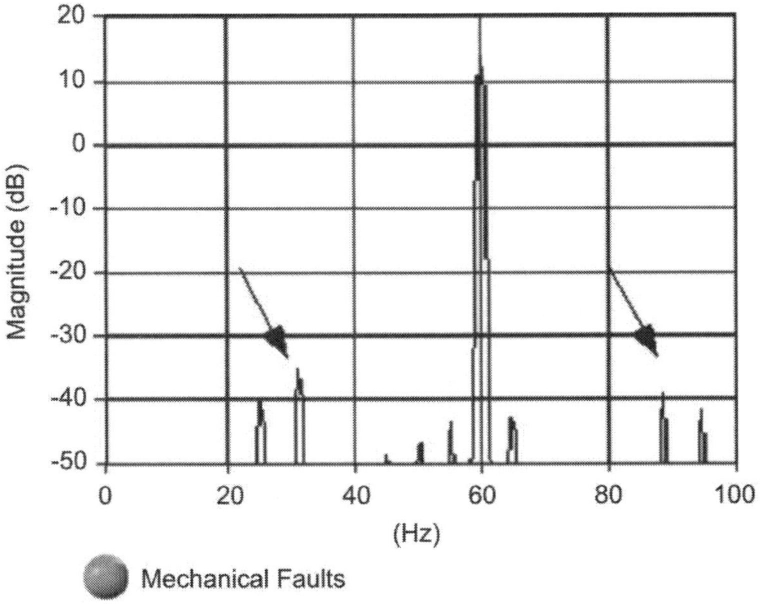

Figure 4. Experimental current frequency spectrum and mechanical faults.

3.1.4. Broken rotor bars

The best-known technique for the detection of broken bars in induction motors is related to monitoring the side bands around the fundamental frequency of the stator current. However, the stator current is a nonstationary signal as it changes over time. Therefore, the fault rotor bar detection using the frequency spectrum is a difficult task that needs a complex signal processing.

The Vienna Monitoring Method (VMM) was proposed for diagnosing faults in the rotor cage induction motors. The technique estimates the torque of induction machine using two different models: voltage model (T_v) and current model (T_c) [25]. In symmetric rotor condition, no faults, the difference torque between the two models ($\Delta T = T_v - T_c$) is almost zero. In the case of a rotor failure, the resultant torque ΔT oscillates at twice the slip frequency.

Baccarini et al. [26] also proposed a method for online monitoring of the induction motor in order to detect and locate a single broken rotor bar. Similar to the VMM, the technique does not use the frequency spectrum. The machine states are calculated with the help of two models. One model is designed to reject asymmetries in the rotor resistance. The other estimates the rotor flux using the discrete model proposed by Bottura et al. [27]. The presence of asymmetry in the rotor causes a different response in the form of a modulated torque deviation. The frequency of the modulation is determined by two times the slip frequency. A minimum of torque difference indicates the faulty rotor bar location.

3.1.4.1. Simulations results

Figure 5 shows the torque modulation, simulation results, for three load operating conditions (112, 78, and 45% of nominal load). The presence of broken rotor bars results in the presence of a noticeable deviation in the torque's magnitude. The torque residue has a constant term and an oscillating term that is related to the presence of broken bars.

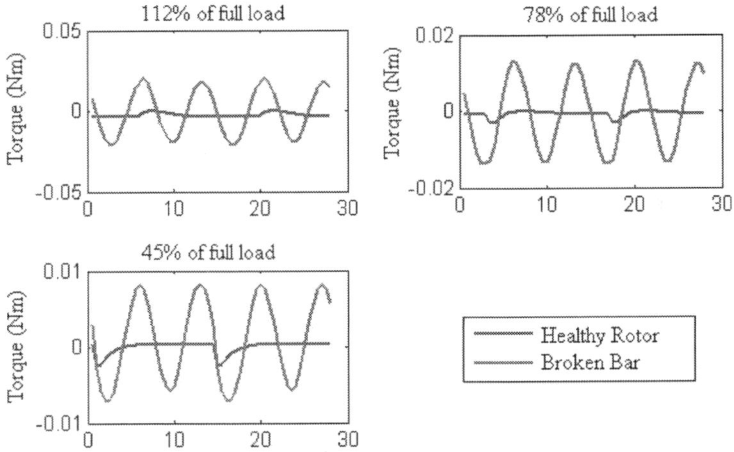

Figure 5. Induction motor torque deviation for healthy rotor and one broken rotor bar for three different load operating conditions and simulations results.

3.1.4.2. Experimental results

Several tests were performed to obtain the torque residue for a healthy rotor for different load operating conditions. **Figure 6** shows the torque residue for a healthy rotor. The value of the maximum residue gives the fault pattern. The presence of these residues is probably due to inherent rotor asymmetries.

The fault of the rotor bar has been caused by drilling a hole in an aluminum bar of a genuine motor and several tests were performed by different load conditions. The torque residues are shown in **Figure 7a** and **b** for nominal and 60% of nominal load condition. Due to the presence of broken bars, these values are higher than the reference standard (**Figure 6**). The modulation frequency is twice the slip frequency. For operation with low load (**Figure 7b**), the modulation frequency is lower compared to the nominal operating conditions (**Figure 7a**), as a result of the low value of the slip. The fault diagnosis is not compromised because of the low value of slip. The technique allows the diagnosis of the presence of cracks or broken bars for all operating conditions.

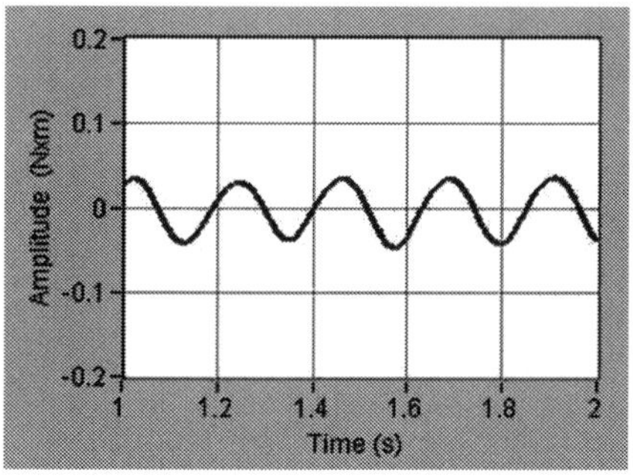

Figure 6. Full load motor operating condition torque residue for healthy rotor, fixed as pattern fault. Experimental results.

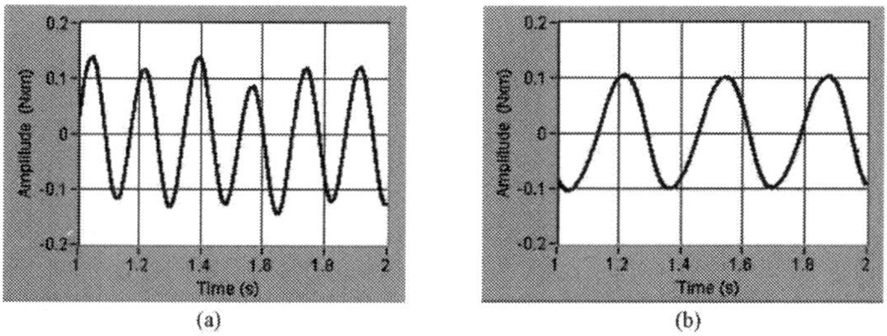

Figure 7. Torque residue for one broken rotor bar; experimental results: (a) full load motor operating condition; (b) 60% nominal load motor operating condition.

3.2. Characteristic frequencies analysis

3.2.1. Description of the experimental setup 2

The experimental systems is set with special three-phase induction motor to simulate the faults, 5 HP, 1730 rpm, 220 V, 13.8 A, 60 Hz, 4 poles, 44 bars, and 36 slots. A dc generator feeding a resistance bank is used as a load system. By varying the excitation current of the dc generator field and therefore its output voltage, a variation of the motor load can be obtained [11, 28].

A torque meter, 0–7500 rpm, bidirectional, and maximum toque of 1000 LB-IN was used to ensure the same operating conditions in all tests. To simulate a low isolation, among turns from the same phase four tappings in a coil were extracted and this makes possible the control of the turns in short circuit.

The flux signals were acquired by a magnetic flux sensor [11]. The vibration signals were acquired by an accelerometer, sensitivity 10.13 mV/g, and frequency range from 1 Hz to 20 kHz. It was observed from the spectra of vibration that all tests had a good repeatability and that there were no variations of mechanical origin that would interfere in the spectra of vibration, ensuring a perfect analysis of the results.

Magnetic flux and vibration signals were collected (total of 300) from a series of 10 tests at each excitement (without fault, two, four, and eight turns short circuits, and voltage unbalance) and randomly repeated under the same load conditions (100, 90, and 80% of load).

The signs of magnetic flux and vibration were submitted to an antialiasing filter with 2.5 kHz of cut frequency and 5 kHz of sampling frequency.

3.2.2. *Magnetic flux and vibration*

The main rotor slot frequencies and their side bands described in Section 2.2.1 and 2.2.2 are computed at $\lambda = 1$, $n = 44$, $s = 0.036$, $p = 2$, and $fl = 60$ (1212.48 and 1332.48 Hz). The spectra computed by FFT for the magnetic flux and vibration in linear scale with the motor working without fault and full load are shown in **Figure 8**. Unlike the vibration spectrum, the magnetic flux spectrum clearly shows the frequencies 1212.48 and 1332.48 Hz. This is because the magnetic flux signals have a strong presence of the rotor slot frequencies and their side bands [28].

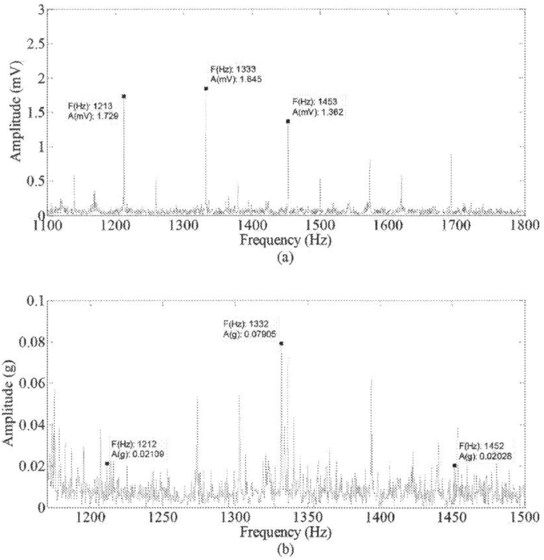

Figure 8. Spectra without fault, with machine operating at nominal speed, and under full load: (a) magnetic flux; (b) vibration.

The visualization of the components of rotor slot harmonic frequencies, **Figure 8**, that had been most excited by the short circuit and voltage unbalance imperfections turns out to be extremely difficult.

Given this difficulty, Lamim Filho et al. [11] have proposed the application of the envelope analysis by Hilbert transform (HT) [29], which is very used in the detection of mechanical faults, for the visualization of the rotor slot frequency components that have been most excited by the electrical faults.

Thus, by applying Hilbert transform in the frequency range from 1 to 2 kHz for magnetic flux and vibration analysis, low-frequency flux and vibration spectra were obtained in which the rotor slot frequency components became extremely easy to visualize and compare with each other.

The spectra of the magnetic flux and vibration after the application of envelope analysis for the motor working at a full-load condition without fault, short circuit of eight turns, and voltage unbalanced (V_{ab} = 200 V, V_{bc} = 200 V, and Vca = 220 V) are shown in **Figures 9** and **10**, respectively.

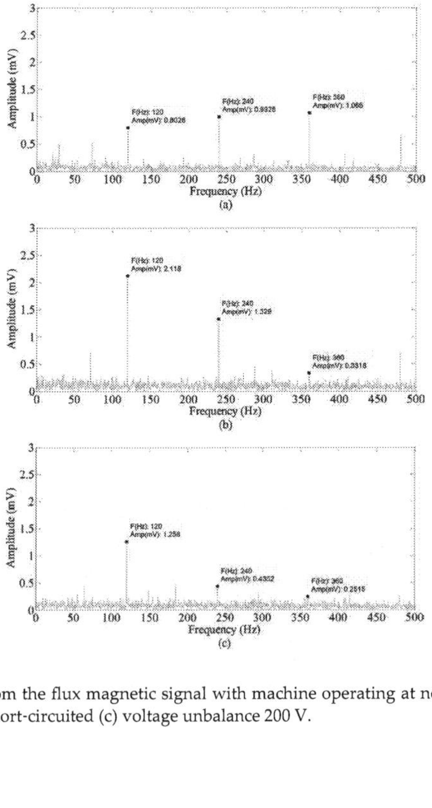

Figure 9. Envelope analysis from the flux magnetic signal with machine operating at nominal speed and full load: (a) without fault; (b) eight turns short-circuited (c) voltage unbalance 200 V.

http://dx.doi.org/10.5772/63166

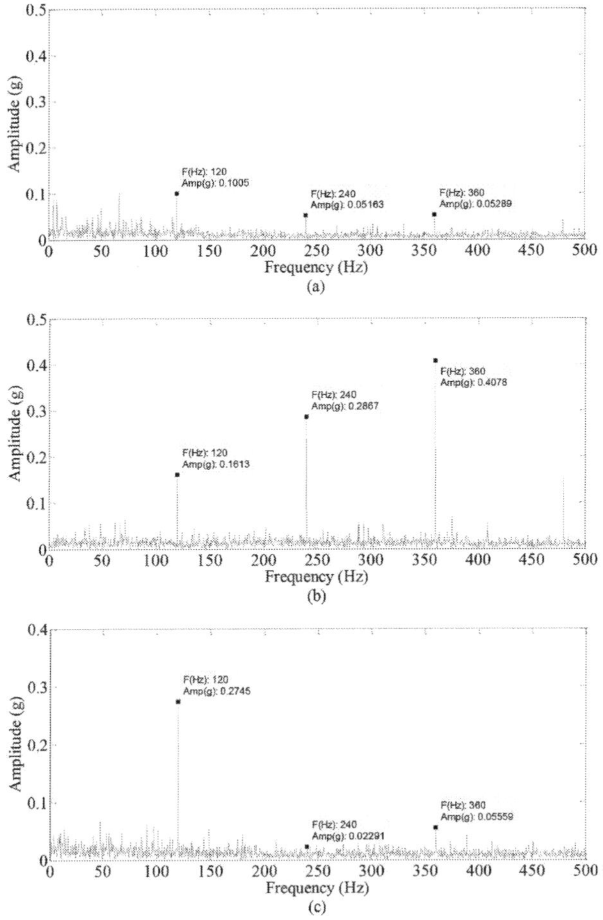

Figure 10. Envelope analysis from the vibration signal with machine operating at nominal speed and full load: (a) without fault; (b) eight turns short-circuited (c) voltage unbalance 200 V.

After comparing the spectra of the magnetic flux signals and the spectra of the vibration signals, it could be verified that the components of frequency demodulated in 2*fl* (120 Hz), 4*fl* (240 Hz), and 6*fl* (360 Hz) were excited the most by the insertion of the short circuit and voltage unbalanced (Figures 10 and 11). Therefore, these harmonics will be considered as the characteristic frequencies for the identification of short circuit and voltage unbalanced, and will be referred to as second (2dh), fourth (4dh), and sixth demodulated harmonic (6dh). For the magnetic flux and vibration analysis, the graphs of tendency for the motor working with 100, 90, and 80% of load are shown in **Figures 11** to **13**, respectively.

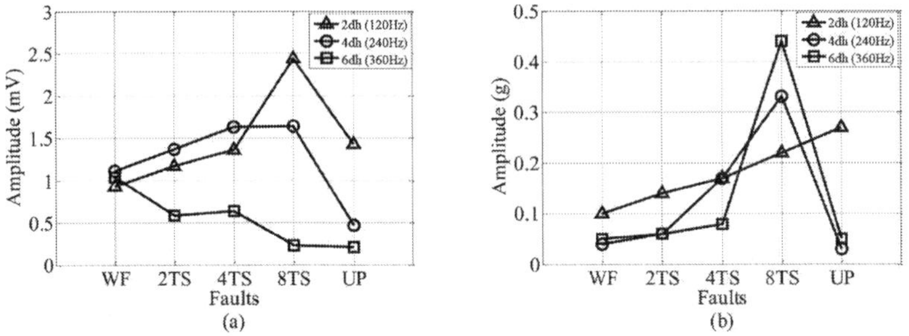

Figure 11. Tendency of the faults introduced into the motor with 100% of load, without fault (WF), two turns short (2TS), four turns short (4TS), eight turns short (8TS), and voltage unbalanced (UP): (a) magnetic flux; (b) vibration.

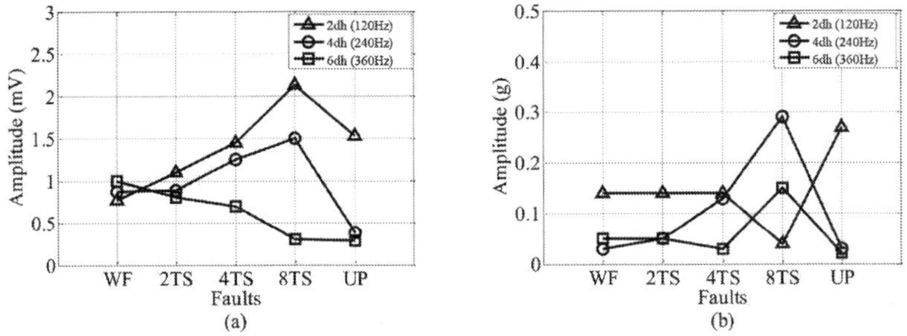

Figure 12. Tendency of the faults introduced into the motor with 90% of load, without fault (WF), two turns short (2TS), four turns short (4TS), eight turns short (8TS) and voltage unbalanced (UP): (a) magnetic flux; (b) vibration.

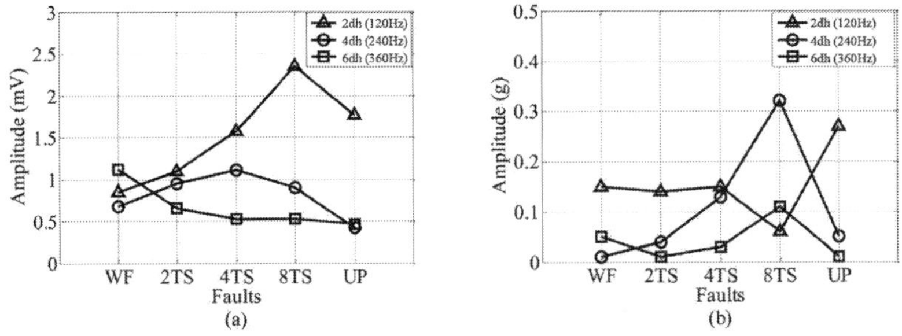

Figure 13. Tendency of the faults introduced into the motor with 80% of load, without fault (WF), two turns short (2TS), four turns short (4TS), eight turns short (8TS) and voltage unbalanced (UP): (a) magnetic flux; (b) vibration.

The averages of the amplitudes of the fault characteristic frequencies of the ten tests conducted under conditions without fault, two turns short circuit, four turns short circuit, eight turns short circuit, and voltage unbalanced using Hilbert transform.

4. Conclusion

A good start of any reliable diagnosis method is an understanding of the electric, magnetic, and mechanical behavior of the machine in healthy and under fault conditions. Therefore, this chapter describes a dynamic model for asymmetric motor operating conditions. The model permits the simultaneous introduction of electrical faults (stator interturn short circuit, broken rotor bars, and/or voltage unbalance) and/or mechanical faults (unbalance, mixed eccentricity, and/or shaft misalignments) for three-phase induction motors, and further detection of a specific fault in a mixture of patterned fault signs.

Further simulations are carried out to check the validation of these computational models. Then, a strategy to detect and diagnose faults is presented and tested on the rig. The approach is easy to apply and all that is required is measurement of the three line currents and voltages. Therefore, it can be used in a continuous basis without interfering with normal motor operation.

In addition, the characteristic frequency of faults is analytically studied and shown in the magnetic flux and vibration signals as a nondestructive test for induction motors.

The spectral analysis after the application of the Hilbert transform makes it possible to highlight a clear way for all modulations related to considered characteristic frequency of faults, unlike the analysis of the spectra of the vibration and magnetic flux in its raw form applying only the fast Fourier transform.

The experimental results show that this methodology can be adapted for different motors and used as a nondestructive test in real predictive maintenance programs in several industries segments.

Author details

Paulo Cezar Monteiro Lamim Filho[1*], Lane Maria Rabelo Baccarini[1] and Robson Pederiva[2]

*Address all correspondence to: plamim@yahoo.com

1 Department of Electrical Engineering, Federal University of São João del Rei, São João del Rei-MG, Brazil

2 Faculty of Mechanical Engineering, Department of Integrated Systems, University of Campinas, Campinas-SP, Brazil

References

[1] Choi S. Introduction. In: Electric Machines: Modeling, Condition Monitoring, and Fault Diagnosis. 1st ed. Taylor & Francis Group; 2013, Florida, USA. pp. 1–8.

[2] Kliman GB, Koegl RA, Stein J, Endicott RD, Madden MW. Noninvasive detection of broken rotor bars in operating induction motors. IEEE Transactions on Energy Conversions. 1988;3:873–879.

[3] Nandi S, Toliyat HA, Li X. Condition monitoring and fault diagnosis of electrical machines—a review. IEEE Transactions on Energy Conversion. 2005;20(4):719–729.

[4] Siddique A, Yadava GS, Singh B. A review of stator fault monitoring techniques of induction motors. IEEE Transactions on Energy Conversion. 2005;20:106–114.

[5] Benbouzid MEH. A review of induction motors signature analysis as a medium for faults detection. IEEE Transactions on Industrial Electronics. 2000;47:984–993.

[6] Zhongming Y. A review of induction motor fault diagnosis.. In: The Third International Power Electronics and Motion Control Conference; 2000. pp. 1353–1358.

[7] Nandi S, Toliyat HA. Novel frequency-domain based technique to detect stator inter turn faults in induction machines using stator-induced voltages after switch-off. IEEE Transaction on Industry Application. 2002;38(1):101–109.

[8] Zhang P, Du Y, Habetler TG, Lu B. A survey of condition monitoring and protection methods for medium-voltage induction motors. IEEE Transaction on Industry Application. 2011;47(1):34–46.

[9] Benbouzid MEH, Kliman GB. What stator current processing-based technique to use for induction motor rotor faults diagnosis? IEEE Transaction on Energy Conversion. 2003;18:238–244.

[10] Hou L, Bergmann NW. Novel industrial wireless sensor networks for machine condition monitoring and fault diagnosis. IEEE Transaction on Instrumentation and Measurement. 2012;61(10):2787–2798.

[11] Lamim Filho PCM, Pederiva R, Brito JN. Detection of stator winding faults in induction machines using flux and vibration analysis. Mechanical Systems and Signal Processing. 2014;42:377–387.

[12] Dias CG, Chabu I. Spectral analysis using a Hall effect sensor for diagnosing broken bars in large induction motors. IEEE Transaction on Instrumentation and Measurement. 2014;63(12):2890-2902.

[13] Cabanas MF, Pedrayes F, Melero MG, García CHR, Cano JM, Orcajo GA, Norniella JG. Unambiguous detection of broken bars in asynchronous motors by means of a flux measurement-based procedure. IEEE Transaction on Instrumentation and Measurement. 2011;60(03):891–899.

[14] Climente-Alarcon V, Antonino-Daviu JA, Haavisto A, Arkkio A. Particle filter-based estimation of instantaneous frequency for the diagnosis of electrical asymmetries in induction machines. IEEE Transaction on Instrumentation and Measurement. 2014;63(10):2454–2463.

[15] Valles-Novo R, Rangel-Magdaleno JJ, Ramirez-Cortes JM, Peregrina-Barreto H, Morales-Caporal R. Empirical mode decomposition analysis for broken-bar detection on squirrel cage induction motors. IEEE Transaction on Instrumentation and Measurement. 2015;64(05):1118–1128.

[16] Ponci F, Monti A, Cristaldi L, Lazzaroni M. Diagnostic of a faulty induction motor drive via wavelet decomposition. IEEE Transaction on Instrumentation and Measurement. 2007;56(06):2606–2615.

[17] Cristaldi L, Faifer M, Lazzaroni M, Monti A, Toscani S. An inverter-fed induction motor diagnostic tool based on time-domain current analysis. IEEE Transaction on Instrumentation and Measurement. 2009;58(05):1454–1461.

[18] Rangel-Magdaleno JJ, Peregrina-Barreto H, Ramirez-Cortes JM, Gomez-Gil P, Morales-Caporal R. FPGA-based broken bars detection on induction motors under different load using motor current signature analysis and mathematical morphology. IEEE Transaction on Instrumentation and Measurement. 2014;63(05):1032–1040.

[19] Baccarini LMR, Menezes BR, Caminhas WM. Fault induction dynamic model suitable for computer simulation: simulation results and experimental validation. Mechanical Systems and Signal Processing. 2010;24(1):300–311.

[20] Cunha CCM, Lyra ROC, Filho BC. Simulation and analysis of induction machines with rotor asymmetries. IEEE Transactions on Industry Applications. 2005;41(1):18–24.

[21] Pillay P, Hofmann P, Manyage M. Derating of induction motors operating with a combination of unbalanced voltages and over or undervoltages. IEEE Transactions on Energy Conversion. 2002;17(4):485–490.

[22] Wang YJ. An analytical study on steady state performance of an induction motor connected to unbalanced three-phase voltage. In: Power Engineering Society Winter Meeting; 2000, Singapore. pp. 159–164.

[23] Gojko MJ, Penman J. The detection of inter-turn short circuits in the windings of operating motors. IEEE Transaction on Industrial Electronics. 2000;43(5):1078–1084.

[24] Gupta BY, Culbert IM. Assessment of insulation condition in rotating machine stators. IEEE Transaction on Energy Conversion. 1993;7(3):500–505.

[25] Wiese RS, Schagginger M, Kral C, Pirker F. The integration of machine fault detection into an indirect field oriented induction machine drive control scheme—The Vienna monitoring method. In: Industry Application Conference. Thirty-Third IAS; 1998, Saint Louis. pp. 278–285.

[26] Baccarini LMR, Tavares PB, Menezes BR, Caminhas WM. Sliding mode observer for on-line broken rotor bar detection. Electric Power Systems Research. 2010;80(9):1089–1095.

[27] Bottura CP, Silvino JL, Resende P. A flux observer for induction machines based on a time-variant discrete model. IEEE Transaction on Industry Application. 1993;29(5):343–353.

[28] Batista FB, Lamim Filho PCM, Pederiva R, Silva VAD. An empirical demodulation for electrical fault detection in induction motors. IEEE Transactions on Instrumentation and Measurement. 2016;PP(99):1–11.

[29] Feldman M. Hilbert transform in vibration analysis. Mechanical Systems and Signal Processing. 2011;25:735–802.